Plant cell culture

Plant molecular biology

Plasmids

Post-implantation mammalian development

Prostaglandins and related substances

Protein function

Protein purification applications

Protein purification methods

Protein sequencing

Protein structure

Proteolytic enzymes

Radioisotopes in Biology

Receptor biochemistry

Receptor—effector coupling

Ribosomes and protein synthesis

Solid phase peptide synthesis

Spectrophotometry and spectrofluorimetry

Steroid hormones

Teratocarcinomas and embryonic stem cells

Transcription and translation

Virology

Yeast

Receptor Biochemistry

A Practical Approach

Edited by

E. C. HULME

Division of Physical Biochemistry, National Institute for Medical Research, Mill Hill, London NW7 1AA, UK

◯ **IRL PRESS**
——at——
OXFORD UNIVERSITY PRESS
Oxford New York Tokyo

Oxford University Press, Walton Street, Oxford OX2 6DP

Oxford is a trade mark of Oxford University Press

*Published in the United States
by Oxford University Press, New York*

© *Oxford University Press 1990*

British Library Cataloguing in Publication Data
Receptor biochemistry.
1. Organisms. Receptors. Biochemical aspects
I. Hulme, E. C.   II. Series 574.875
ISBN 0−19−963092−5
ISBN 0−19−963093−3 (pbk)

Library of Congress Cataloging in Publication Data
Receptor biochemistry : a practical approach / edited by E.C. Hulme.
p. cm.—(Practical approach series)
Includes bibliographical references.
1. Neurotransmitter receptors.   2. Neural receptors.   3. Cell receptors.
I. Hulme, E. C.
[DNLM: 1. Molecular Biology.   2. Receptors, Adrenergic.
3. Receptors, Endogenous Substances.
WL 102.8 R294]
QP364.4.R4275 1990
599'.0188—dc20
89−72206   CIP
ISBN 0−19−963092−5
ISBN 0−19−963093−3 (pbk.)

Typeset and printed by Information Press Ltd, Eynsham, Oxford

Preface

STRUCTURAL studies on cell-surface receptors of pharmacological importance have made great strides in recent years. To a considerable extent, this has resulted from the application of the powerful and universal techniques of DNA cloning to the study of these scarce and difficult molecules. As a result, the amino-acid sequences of a large and increasing number of cell-surface receptors have been, and are being determined.

An important outcome has been to emphasize the evolutionary and structural similarities that exist at the sequence level within the various classes of cell-surface hormone, neurotransmitter, and growth factor receptors, to the extent that certain broad categories of structural generalization are now probably acceptable on the basis of sequence data alone. A related advance has been the application of techniques of site-directed mutagenesis, and the construction of chimeric receptors. These have been at their most powerful when they have, in effect, involved the swapping of whole domains between related receptors with different functional properties, and have enabled an outline to be inked in of those parts of specific receptor sequences that are probably involved in effector functions, or interactions of various kinds.

In interpretational terms, however, these approaches suffer from a lack of detailed information about the three-dimensional structures of cell surface receptor proteins. They are also at a more subtle and fundamental disadvantage in the study of receptors which, by definition, are proteins that are capable of a conformational transition between a ground, or inactive state, and an excited or activated state. The disadvantage resides in the fact that it is impossible, a posteriori, to disentangle the postulated effects of a given mutation on the ground-state three-dimensional structure of the protein from alterations of the thermodynamics and kinetics of receptor activation. Both classes of perturbation may alter both ligand and effector interactions. Additionally, mutagenesis studies have a limited part to play in establishing the nature and role of secondary structure, and post-translational modifications.

The conclusion to be drawn is the obvious one that, as in enzymology, receptor structures will only be properly understood as a result of the application of physical methods, most particularly crystallographic and other spatially resolving techniques, at the protein level. The high-level expression of cloned receptors will have an important role here in enabling large-enough amounts of these rare proteins to be produced to make such studies possible. At present, many people believe that this is likely to entail over-expression of receptors in mammalian cell lines. However, this is unlikely to produce receptors at levels much in excess of tens of pmoles per mg of membrane protein. While this may lighten the task of receptor purification, it will not abolish the problem. In the longer term, it may be possible to use transgenic techniques to exploit the few situations in which cell-surface receptors are produced naturally at a very high density, as in the case of rhodopsin in the retinal rod outer segments, or nicotinic acetylcholine receptors in electric organs. Alternatively, novel expression systems may come to the rescue.

For these reasons, the techniques of receptor protein chemistry, and particularly of receptor purification, are overdue for a return to centre stage. It is fair to say that in the furore created by molecular neurobiology, the fundamental enabling role of receptor biochemistry at the protein level has been overshadowed. However, without this essential work future progress will be limited by the lack of a structural context.

The purification, handling, fragmentation, and sequencing of hydrophobic, membrane-embedded proteins such as receptors remains a difficult craft. In this book, an attempt has been made to bring together the practical experience of some of the world's most accomplished practitioners in this problematic area. The book is structured in such a way that Chapter 1 provides an introduction to the basic concepts of receptor structure and biochemistry. In this chapter, a number of the basic protocols are introduced. However, an equally important function is to provide cross-references to other sections of the book in which these protocols are illustrated with respect to a variety of different receptors (in Chapters 2 – 9). Chapter 10 discusses approaches that have been applied successfully to obtaining structural information from receptor proteins. Chapter 11 provides a brief introduction to the expression of recombinant receptors in cell lines in culture. Chapter 12 reviews the thorny question of how receptor sequences, available in increasing abundance from the sequencing of cloned genes, may be interpreted in structural terms. In addition, a short appendix is provided which outlines the techniques of receptor-ligand binding.

Most of the techniques that have proved to be of everyday value in the receptor biochemistry laboratory have been covered in this book. However, it is inevitable that there are both some omissions, and repetitions, where essentially the same protocol occurs in different contexts. Despite these deficiencies, it is to be hoped that there are sufficient useful, specific practical guidelines, or even generalizations, in this book to encourage anyone embarking on this difficult and thorny path. An attempt has been made to provide a reasonable amount of cross-referencing within the book, and with relevant sections in other books in the Practical Approach series. Hopefully, this will increase the book's utility. Other books in the Practical Approach series that are relevant to the subject matter in this book are those on *Biological membranes, Neurochemistry, Protein structure, Protein function, DNA cloning, Animal cell culture,* and *Affinity chromatography.*

Finally, I would like to thank the authors for their excellent contributions to this book, and for their tolerance of editorial interference.

E. C. HULME

Contents

List of contributors   xvii

Abbreviations   xix

1 Solubilization, purification, and molecular characterization of receptors: principles and strategy   1

T. Haga, K. Haga and E. C. Hulme

1. Introduction   1
Receptor groups   3
Illustrations   8

2. Solubilization   8
Choice of detergents   8
Group 1 receptors   10
Group 2 receptors   11
Group 3 receptors   12
Testing a range of detergents   15
Estimation of the molecular size of solubilized receptors   19
Identification of solubilized sites by binding studies   20

3. Purification of receptors   21
Receptor sources and membrane preparations   21
Protease inhibitors   22
General approaches to purification   23
Lectin chromatography   23
Gel-permeation chromatography   24
Ion-exchange, hydroxyapatite, heparin adsorption, and hydrophobic chromatography   24
SDS-PAGE and isoelectric focusing   25
Concentration of dilute solutions of receptors   26

4. Affinity chromatography   29
Synthesis of ligands with functional groups   29
Preparation of an affinity gel   32
Elution of receptors from affinity gels   35
Testing new affinity gels   36
Affinity chromatography with polypeptide ligands   38
Elution of receptors from affinity gels made from immobilized polypeptide ligands   39

5. Molecular characterization of receptors   40
The ligand-binding activity of purified receptors   40

Protein estimation and the polypeptide composition of purified
preparations 40
Determination of amino-acid and oligonucleotide sequences of receptors 42
Expression of cDNA 43
Reconstitution of receptors and identification of their functions 44
New strategies for the molecular characterization of receptors 45

Acknowledgements 46

References 46

2 Purification and molecular characterization of muscarinic acetylcholine receptors 51

T. Haga, K. Haga and E. C. Hulme

1. Introduction 51

2. Preparation of the affinity gel 51
Identification of ABT 53

3. The preparation and solubilization of membranes 55
Synaptic membranes from porcine cerebra 55
Membranes from porcine atria 57
Yield of membrane mAChRs 57
Solubilization 58
Assay for ligand-binding activity of solubilized mAChRs 58
Yield of solubilized mAChRs 59

4. Purification of mAChRs from solubilized extracts 59
Affinity chromatography 59
Further purification 61
Determination of protein concentration with fluorescamine 61
Yield and specific binding activity of purified mAChR 61

5. Electrophoretic analysis 62
[^{125}I]Iodination of mAChRs 62
Affinity labelling of purified mAChRs with [^{3}H]propylbenzilyl-
choline mustard 63
SDS-PAGE by the Laemmli method 64
Detection of the mAChR band 65

6. Estimation of the molecular size of mAChRs 66
Estimation from amino-acid sequences 66
Estimation by SDS-PAGE 66
Estimation from hydrodynamic properties 67

7. Cloning, sequencing, and expression of cDNAs for
mAChRs 75
mAChR I and mAChR II subtypes 75
Other subtypes 76
Expression of cDNA clones for mAChRs 76

8. Ligand-binding activity of purified mAChRs 77

References 77

3 Dopamine receptors: isolation and molecular characterization 79

P. G. Strange and R. A. Williamson

1. Introduction 79
2. Solubilization of dopamine receptors 79
   D_1 dopamine receptor 79
   D_2 dopamine receptor 80
3. Fractionation of soluble dopamine receptors 86
   Lectin affinity chromatography 86
   Ligand affinity chromatography 88
4. Molecular properties of D_1 and D_2 dopamine receptors 95
   Acknowledgements 96
   References 96

4 Opioid receptors 99

C. D. Demoliou-Mason and E. A. Barnard

1. Introduction 99
2. Solubilization 99
   Buffer systems for membrane preparation and solubilization 100
   Solubilization with ionic detergents 101
   Solubilization with the zwitterionic detergent, CHAPS 102
   Solubilization with non-ionic detergents 103
   Evaluation of the detergents used 105
   [^{3}H]ligand-binding assay for solubilized opioid receptors 105
3. Purification studies 108
   Molecular exclusion chromatography and sucrose gradient centrifugation 108
   Lectin-affinity chromatography 110
   Hydrophobic chromatography 110
   Affinity cross-linking studies 111
   Affinity chromatography 116
4. The status of the opioid receptors at present 118
   Molecular studies: outstanding issues 118
   Opioid-receptor heterogeneity 119
   Approaches to the molecular biology of the opioid receptors 120
   Conclusions 121
   References 122

5 Photoaffinity labelling and purification of the β-adrenergic receptor 125

J. L. Benovic

1. Introduction 125

2. Tissue source 125
3. Radioligand binding 126
Radioligand binding to particulate β-adrenergic receptors 127
Radioligand binding to soluble β-adrenergic receptors 128
4. Membrane preparation 128
5. Photoaffinity labelling of the β-adrenergic receptor 129
Photoaffinity labelling of particulate β-adrenergic receptors 132
Photoaffinity labelling of soluble β-adrenergic receptors 133
6. Solubilization procedures 133
7. Purification of the β-adrenergic receptor 135
Synthesis of the sepharose−alprenolol affinity resin 136
Chromatography of solubilized β-adrenergic receptor on
Sepharose-alprenolol 138
Chromatography of affinity-purified β-adrenergic receptor on
gel-permeation HPLC 138
Acknowledgements 139
References 139

6 α_2-Adrenergic receptor purification 141
J. W. Regan and H. Matsui

1. Introduction 141
2. Synthesis of yohimbinic acid – Sepharose 145
Ethylenediamine-activated Sepharose 145
Coupling of yohimbinic acid to the activated Sepharose 147
3. Membrane preparation and solubilization 148
Membranes 148
Solubilization 148
4. Receptor purification 149
Affinity chromatography 149
Heparin−agarose chromatography 151
Inverse affinity chromatography 151
WGA−agarose chromatography 152
5. Comments 152
6. Assay procedures 156
Receptor binding 156
Protein 157
Radioiodination 158
Acknowledgements 160
References 160

7 Purification of nicotinic acetylcholine receptors   163

S. Hertling-Jaweed, G. Bandini, and F. Hucho

1. Introduction—the nicotinic acetylcholine receptor as a model receptor   163
   - The nAChR molecule   163
   - Open questions   164
   - Methods available and methods described here   165
2. Assays and quantification of nAChR   165
   - Iodination of α-bungarotoxin (BgTX)   165
   - BgTX-binding activity   166
   - Acetylcholine-binding assay   168
3. Receptor purification   169
   - Rapid preparation of crude receptor extract   169
   - Pure AChR-rich membranes   169
   - Extraction of peripheral proteins   171
   - Detergent extraction   171
   - Chromatography on a Cibacron Blue Sepharose column   171
   - Affinity chromatography   172
   - Conclusion   175
   - References   175

8 Purification and molecular characterization of the γ-aminobutyric acid$_A$ receptor   177

F. A. Stephenson

1. Introduction   177
2. GABA$_A$-receptor solubilization and purification   178
   - Solubilization   178
   - Isolation of the GABA$_A$-receptor   181
3. Molecular characterization of the purified GABA$_A$ receptor   185
   - Assessment of purity by isoelectric focusing   185
   - Molecular weight determination of the GABA$_A$ receptor   188
   - Subunit composition of the GABA$_A$ receptor   191
   - The carbohydrate properties of the GABA$_A$ receptor   196
4. Conclusions   199
   - Acknowledgements   200
   - References   200

9 Purification of the epidermal growth factor receptor from A431 cells   203

G. N. Panayotou and M. Gregoriou

1. Introduction   203
2. Preparation of the Affinity Matrix   203
   Purification of EGF   203
   Preparation of EGF−Affi-Gel   204
3. Solubilization and Purification of the EGF Receptor   205
   Culture of A431 cells   205
   Extraction of the EGF receptor   206
   Affinity chromatography using EGF−Affi-Gel-10   207
   References   211

10 Peptide mapping and the generation and isolation of sequenceable peptides from receptors   213

M. Wheatley

1. Introduction   213
2. Sample preparation for peptide mapping   213
   Radioiodination   213
   Oxidizing methods of radioiodination   214
   Conjugation of a prelabelled structure   217
   Removal of free radioiodine   218
   Reduction and S-carboxymethylation   221
3. One-dimensional peptide mapping   222
   Enzymatic cleavage methods   223
   Chemical cleavage methods   228
   Separation of cleavage products by slab SDS-PAGE   234
   Stock solutions   235
   Composition of SDS-polyacrylamide gels   236
   Pouring and running gels   236
   Detection, visualization, and quantitation of peptides after SDS-PAGE   241
4. Two-dimensional peptide maps   247
   Fluoresceination of standard proteins   247
   Re-digesting peptides generated by 1D peptide maps   248
5. Glycosylated peptides   250
   Using immobolized lectin to construct peptide maps   250
   Deglycosylation of peptides   251
6. Sequencing peptides after SDS-PAGE   251
   Electroelution of peptides from gel slices   251
   Preparing peptides for solid-phase sequencing   254
   Preparing peptides for gas-phase sequencing   255

7. Purification of sequenceable peptides without using
SDS-PAGE   256
  Proteolytic cleavage of receptors in the absence of SDS   256
  Isolating peptides by HPLC and FPLC   257
  Isolating peptides by Sephadex LH   258

Acknowledgements   259

References   259

11 Expression of receptor genes in
cultured cells   263

C. M. Fraser

1. Introduction   263
2. Choosing a recipient cell line for use in transfection   263
3. Vectors used in gene expression studies   264
  General features   264
  Choice of promoter sequences   264
  Inducible promoter sequences   265
4. Transient expression of receptor genes in cultured cells   266
5. Stable expression of receptor genes in cultured cells   266
  gpt selection   267
  Neomycin-resistance selection   267
  Thymidine kinase selection   268
  Dihydrofolate reductase selection   268
6. Baculovirus expression systems   268
7. General methodology for expression of receptor genes in
cultured cells   269
  Preparation of plasmid DNA   269
  Transfection of monolayer cells by the calcium phosphate method   269
  Solutions for calcium phosphate precipitation of DNA   271
  Preparation of DNA$-$CaPO$_4$ precipitate   271
  Other methods for transfection of cultured cells with plasmid DNA 271
8. Summary and conclusions   272
9. Appendix   272
  Medium for thymidine kinase selection   272
  Medium for *gpt* selection   272
  Partial list of plasmid vectors suitable for transfection of cultured
  mammalian cells   273
  Cell lines available from the American Type Culture   273
  Collection for use in receptor expression studies   273

References   274

12 Structural deductions from receptor sequences 277
N. M. Green

1. Introduction 277
2. Subdivision of the sequence 278
   Hydrophobicity 278
   Unusual features of transmembrane helices 280
3. Homology searches and pattern matching 281
4. Functional sites in receptors 282
   Potential sites for covalent modification 283
   Proteolytic cleavage sites 283
   Calcium-binding sites 285
   Nucleotide-binding sites 285
   Cysteine-rich domains 286
   Unusual repetitive elements 286
5. Higher-order structure 286
   Secondary structure prediction 286
   Hydrophobicity patterns 288
   Domain types and boundaries 290
   Domains of receptors 292
6. Software for sequence analysis 293
   References 297

Appendix. Receptor-binding studies, a brief outline 303
E. C. Hulme

1. Basic principles 303
2. The ligand off-rate may determine the assay method 305
3. Criteria for receptor-specific binding 306
4. Tracer ligand saturation curve 306
   Receptor preparation 307
   Setting up the binding curve 308
   Replicate incubations 308
5. Preliminary analysis and interpretation 310
6. Separation methods: particulate preparations 312
   Filtration assays 312
   Centrifugation assays 313

7. Separation methods: soluble preparations   313
    Gel-filtration   313
    Filtration on polyethyleneimine-treated filters   314
    Precipitation by polyethylene glycol   314
    Charcoal adsorption   315
  Reference   315

Index   317

Contributors

G. BANDINI
Institute of Biochemistry,
Department of Neurochemistry, Free University of Berlin, Thielallee 63,
D-1000 Berlin 33, Germany.

E. A. BARNARD
MRC Molecular Neurobiology Unit,
Medical Research Council Centre, Hills Road, Cambridge CB2 2QH, UK.

J. L. BENOVIC
Fels Institute for Cancer Research and Molecular Biology,
Temple University School of Medicine,
Philadelphia, PA 19140, USA.

C. D. DEMOLIOU-MASON
MRC Molecular Neurobiology Unit,
Medical Research Council Centre, Hills Road, Cambridge CB2 2QH, UK.

C. M. FRASER
Section on Molecular Neurobiology,
Laboratory of Physiologic and Pharmacologic Studies,
National Institute on Alcohol Abuse and Alcoholism,
12501 Washington Ave.,
Rockville, MD 20852, USA.

N. M. GREEN
Laboratory of Protein Structure,
National Institute for Medical Research, Mill Hill, London NW7 1AA, UK.

M. GREGORIOU
Laboratory of Molecular Biophysics,
Department of Zoology, The Rex Richards Building, University of Oxford,
South Parks Road, Oxford OX1 3QU, UK.

K. HAGA
Department of Biochemistry,
Institute of Brain Research, Faculty of Medicine, University of Tokyo,
Hongo 7-3-1 Bunkyo-ku, Tokyo 113, Japan.

T. HAGA
Department of Biochemistry,
Institute of Brain Research, Faculty of Medicine, University of Tokyo,
Hongo 7-3-1 Bunkyo-ku, Tokyo 113, Japan.

S. HERTLING-JAWEED

Institute of Biochemistry,
Department of Neurochemistry, Free University of Berlin, Thielallee 63,
D-1000 Berlin 33, Germany.

F. HUCHO

Institute of Biochemistry,
Department of Neurochemistry, Free University of Berlin, Thielallee 63,
D-1000 Berlin 33, Germany.

E. C. HULME

Division of Physical Biochemistry,
National Institute for Medical Research, Mill Hill, London NW7 1AA, UK.

H. MATSUI

Department of Neuropsychiatry,
St. Marianna University School of Medicine, Kawasaki, Japan.

G. N. PANAYOTOU

Ludwig Institute for Cancer Research,
Courtauld Building, 91 Riding House Street, London W1P 8BT, UK.

J. W. REGAN

Department of Pharmacology and Toxicology,
College of Pharmacy, University of Arizona,
Tucson, AZ 85721, USA.

F. A. STEPHENSON

Department of Pharmaceutical Chemistry,
The School of Pharmacy, 29/39 Brunswick Square, London WC1N 1AX, UK.

P. G. STRANGE

Biological Laboratory,
The University, Canterbury, Kent CT2 7NJ, UK.

M. WHEATLEY

School of Biochemistry,
University of Birmingham, PO Box 363, Birmingham B15 2TT, UK.

R. A. WILLIAMSON

Biological Laboratory,
The University, Canterbury, Kent CT2 7NJ, UK.

Abbreviations

ABT	aminobenzhydryloxytropane
α_2R	α_2-adrenergic receptor
βAR	beta-adrenergic receptor
BSA	bovine serum albumin
BZ	benzodiazepine
cDNA	complementary DNA
CHAPS	3-(3-cholamidopropyl)dimethylammonio-1-propanesulphonate
CHAPSO	3-(3-cholamidopropyl)dimethylammonio-2-hydroxy-1-propanesulphonate
CMC	critical micellar concentration
CNBr	cyanogen bromide
ConA	concanavalin A
DA	dopamine
D_2R	dopamine D_2 receptor
DEAE	diethylaminoethyl
DHA	dihydroalprenolol
DHFR	dihydrofolate reductase
DMS	dimethyl suberimidate
DOC	deoxycholate
dpm	radioactive decays/min
DTT	dithiothreitol
EDTA	ethylene diamine tetra-acetic acid
EGF	epidermal growth factor
EGFR	epidermal growth factor receptor
EGTA	ethylene glycol-bis (β-aminoethyl ether)N,N,N',N'-tetra-acetic acid
FPLC	fast protein liquid chromatography
GABA	gamma aminobutyric acid
$GABA_AR$	$gaba_A$ receptor
GlyDOC	glycodeoxycholate
HA	hydroxyapatite
HD	hepes − digitonin
HAT	hypoxanthine, aminopterin, thymidine
Hepes	N-2-hydroxyethylpiperazine-N'-2-ethanesulphonic acid
HPLC	high-performance liquid chromatography
HPRT	hypoxanthine phosphoribosyl transferase
IEF	iso-electric focusing
IgE	immunoglobulin E
IMP	inosine monophosphate
K	affinity (association) constant
K_d	dissociation constant
KPB	potassium phosphate buffer
mAChR	muscarinic acetylcholine receptor

NaBH$_4$	sodium borohydride
nAChR	Nicotinic acetylcholine receptor
NGF	nerve growth factor
nm	nanometres
NMS	(-)-*N*-methylscopolamine
PAGE	polyacrylamide gel electrophoresis
PEG	polyethylene glycol
pI	isoelectric point
PMSF	phenylmethylsulphonyl fluoride
PrBCM	propylbenzilylcholine mustard
PSV	partial specific volume
PTX	pertussis toxin
PVDF	polyvinylidene difluoride
QNB	3-quinuclidinyl benzilate
Rf	elution position relative to front
SDS	sodium dodecyl sulphate
SEM	standard error of mean
S$_{20,w}$	sedimentation coefficient at 20°C in water
TEMED	tetramethyl ethylene diamine
TES	*N*-tris(hydroxymethyl)methyl-2-aminoethanesulphonic acid
TFA	trifluoroacetic acid
TK	thymidine kinase
TLC	thin-layer chromatography
WGA	wheat-germ agglutinin

1

Solubilization, purification, and molecular characterization of receptors: principles and strategy

T. HAGA, K. HAGA, and E. C. HULME

1. Introduction

Most hormone and neurotransmitter receptors are transmembrane glycoproteins. The exceptions are receptors for steroids and thyroid hormones, which are soluble DNA-binding proteins and will not be dealt with in this book. The function of transmembrane receptors is to communicate the information that hormones or neurotransmitter molecules are present on the outside of cells to the inside, so that there is a sense in which such receptors can be regarded as allosteric proteins, whose functions are elicited when they bind allosteric ligands, namely the cognate hormones and neurotransmitters.

The ligand-binding activities of receptors in intact tissues, or on cells or membranes, can be characterized directly by binding of labelled agonists or antagonists, or indirectly by competition experiments. The principles are detailed in a companion book in this series, *Receptor—ligand interactions, a practical approach* (1). However, the readily measurable cellular or membrane responses elicited by hormones or neurotransmitters are usually secondary consequences of the immediate ligand-induced conformational change, and, clearly, neither the structures, nor even the primary functions of the receptor molecules involved, can be deduced from them. As a result, until recently we were in almost complete ignorance of receptor structure.

Very recently, this situation has begun to change with the characterization of some important receptors in molecular terms. This chapter outlines the methods that are required for the molecular characterization of receptors. Illustrative examples of the principles so defined are provided elsewhere in this book. These examples include:

(a) Receptors coupled to GTP-binding proteins, namely the muscarinic acetylcholine receptors (mAChRs; Chapter 2), the dopamine D_2 receptor (Chapter 3), the opioid receptors (Chapter 4), the β_2-adrenergic receptors (Chapter 5), and the α_2-adrenergic receptor (Chapter 6).

(b) Oligomeric receptors that incorporate an ion channel within their structure, namely the nicotinic acetylcholine receptor (Chapter 7) and the $GABA_A$ receptor (Chapter 8).

(c) A protein representative of the class of receptors with tyrosine kinase activity, and a single transmembrane segment, namely the epidermal growth factor (EGF) receptor (Chapter 9).

Examples of the reconstitution of purified G-protein-coupled receptors with the appropriate GTP-binding proteins are given in another volume in this series, *Receptor-effector coupling, a practical approach* (2).

Usually, the responses elicited by agonist binding to receptors disappear after solubilization of the membranes and may not be recovered until the receptor can be reconstituted in a suitable lipid bilayer. Further to complicate matters, even the immediate functional consequences of receptor activation may be unknown. Because of this, the strategy for the molecular characterization of receptor molecules is compelled to progress through a number of steps:

(a) Solubilization of the receptor polypeptide in a form in which the ligand binding activity is preserved.

(b) Isolation and purification of the polypeptides carrying the binding sites.

(c) Determination of the structure of the receptor polypeptide(s).

(d) Reconstitution, and identification and study of the function of the receptor polypeptide(s).

A major difficulty is that even the ligand-binding activity is vulnerable to denaturation during the solubilization procedure and may only be partially recovered in solubilized preparations (see Section 2 of this chapter).

The amounts of receptors present in tissues are generally low, except in a few special cases, such as those of nicotinic acetylcholine receptors (nAChRs) in electric organs (Chaper 7) or rhodopsin in the retina. The specific activity of ligand binding to receptors in membrane preparations is rarely in excess of 1 pmol/mg protein, which means that 10^4-fold purification is usually the minimum needed to obtain homogeneity, assuming a receptor molecular weight of *c*. 10^5. Thus, the development of a purification procedure of sufficient specificity is essential. Usually, only affinity chromatography is adequate. The application of affinity chromatographic techniques is described in Section 4 of this chapter. The application of these principles to specific receptor types is the major aim of this book.

The development of new methods for high-level expression of receptor genes in transformed clonal cell lines may lighten the task of receptor purification in the future. This rapidly developing approach is outlined in Chapter 11. To date, it has been applied successfully to the isolation of a mAChR subtype, using a construct in which the receptor domain was linked to an inducible gene, that for dihydrofolate reductase (3), and to the isolation of the EGF receptor, by its insertion in one of the new class of Baculovirus vectors, followed by expression in an insect cell line (4). By these means, expression of up to 7×10^6 receptors/cell has been obtained.

To date, the primary structures of receptors, i.e. their amino-acid sequences, have been deduced mainly from the nucleotide sequences of cDNA clones, isolated from cDNA libraries by hybridization with probes based on partial amino-acid sequences determined from purified receptors (see Chapter 10). The functions of receptors have been examined by reconstitution of the purified proteins in artificial lipid bilayers, or by expressing mRNA transcribed from receptor cDNA in oocytes, or other suitable cells. These subjects are discussed in Sections 5.2−5.4 of this chapter. New strategies, developed very recently, include the isolation of cDNA clones by expression screening of cDNA libraries in suitable cells, usually oocytes, using the receptor function, typically the opening of a specific type of ion channel, to detect the message. This topic is introduced briefly in Section 5.5. The study of receptor function by microinjection of mRNA into *Xenopus* oocytes has been covered in detail in a companion book in this series (5), and will not be dealt with further here.

1.1 Receptor groups

The primary pharmacological classification of receptors is made on the basis of the hormones or neurotransmitters with which they interact. However, we know that a given hormone or transmitter may interact with more than one class of receptor. The acetylcholine receptors provide a classical example: nicotinic AChRs and muscarinic AChRs are structurally completely different from one another.

The strategies for the solubilization, purification, and molecular characterization of different receptor types are determined more by their structural and functional classification than by their detailed ligand specificities. If it is possible to determine the class to which a given receptor belongs, the information can be used to help design a strategy for its solubilization, purification, and molecular characterization.

The structures and functions of a number of transmembrane receptors have now been elucidated. Structural and functional analysis enables most of them to be classified into three groups, or families, illustrated in *Tables 1−3*. The three groups are:

(a) Receptors with a single transmembrane segment (group 1).

(b) Oligomeric receptors incorporating both ligand-binding sites and an ion channel (group 2).

(c) Receptors which exert their effects by coupling to GTP-binding proteins (G-proteins) (group 3).

A number of receptors which have yet to be fully characterized are believed to fall into one or other of these three groups, although additional groups may be discovered as knowledge increases, cf. the IgE receptor (6). There are several characteristics common to receptors in each group.

Table 1 summarizes information relevant to group 1 receptors. The endogenous ligands for group 1 receptors are polypeptide hormones, whose action is related to metabolic changes, or cell proliferation and survival (22). Group 1 receptors have an extracellular ligand-binding domain associated with a single transmembrane

---

Table 1. Group 1 receptors (receptors with a single transmembrane segment).

(1) Structural characteristics

    (a) Monomer (EGFR, PDGFR, IGF_2R, ANPR, LDLR) or hetero-oligomer (IR, IGF_1R, IL_2R, NGFR)

    (b) A single transmembrane segment per subunit

(2) Function

    Tyrosine kinase (EGFR, PDGFR, IR, IGF_1R, IGF_2R); ligand uptake (LDLR), guanylyl cyclase (ANPR) or unknown (IL_2R, NGFR, IGF_2R)

(3) Detergent used for solubilization

    Triton X-100 or Nonidet P-40

(4) Sequenced receptors and principal purification methods

Insulin R (11,12):	AC^a with mAb^b (13) and AC with insulin (14)
IGF_1R (15):	AC with mAb (16)
EGFR (17):	AC with mAb (17) and AC with EGF (Chapter 9)
PDGFR (18):	AC with mAb against phosphotyrosine (19)
IGF_2R (18):	AC with IGF_2 (8)
ANPR (20):	AC with ANP (21)

Abbreviations: [a] AC, affinity chromatography; [b] mAb, anti-receptor monoclonal antibodies; I, insulin; IGF, insulin-like growth factor; EGF, epidermal growth factor; PDGF, platelet-derived growth factor; ANP, atrial natriuretic peptide, LDL, low-density lipoprotein, NGF, nerve growth factor; IL_2, interleukin-2.

---

domain. Most of the receptors of this class characterized to date have an intracellular domain which has tyrosine kinase catalytic activity. However, there are many important exceptions, for instance the interleukin-2, IGF_2, nerve growth factor, and low-density lipoprotein receptors whose intracellular domains are both smaller and unrelated to tyrosine kinase sequences (7,8,9,10). The mode of operation of the IL_2 and NGF receptors is unknown. The IGF_2 receptor is identical to the mannose-6-phosphate receptor which is implicated in the targeting of glycoproteins to the lysosomes (8). In this case, as in the case of the LDL receptor, internalization of the receptor-ligand complex is an important part of the function. This is also probably true for the NGF and IL_2 receptors. Very recently the structure of the receptor for atrial natriuretic peptide (ANP) has been deduced and found to contain a guanylate cyclase catalytic site in the intracellular domain, as well as an extracellular ligand-binding domain and a single transmembrane segment (21).

There is significant structural variation amongst the tyrosine kinases. Thus, the EGF and platelet-derived growth factor (PDGF) receptors both exemplify structures in which the binding site and tyrosine kinase domain are encoded by a single polypeptide chain. In contrast, the ligand-binding and tyrosine kinase domains of the insulin receptor are separate, the ligand-binding domain being formed by the α-chain, which is disulphide-bonded to the membrane-spanning β-chain, two α/β units in turn being disulphide-bonded together to form an $\alpha_2\beta_2$ heterotetramer. The EGF-receptor binding domain contains two long cysteine-rich internal repeats (17),

Table 2. Group 2 receptors (receptor – ion channel complex).

(1) Structural characteristics
    (a) Hetero-oligomers
    (b) At least four transmembrane segments per subunit; at least 20 transmembrane segments per functional unit

(2) Function
    Receptor – ion channel complex

(3) Detergent used for solubilization
    Sodium cholate, sodium deoxycholate, or Triton X-100

(4) Sequenced receptors and principal purification methods

nAChR (23):	membrane purification (24; Chapter 7)
GABA$_A$R (25):	AC[a] with low molecular weight ligand (26; Chapter 8)
GlyR (27):	AC with low molecular weight ligand (28)

[a] AC, affinity chromatography.

whilst the PDGF-receptor binding domain is based on a larger number of cysteine-rich regions (18), yielding a fivefold repeated immunoglobulin-like domain.

Activation of those group 1 receptors with tyrosine kinase activity results in enhancement of phosphorylation of tyrosine residues in cellular proteins, including the receptors themselves. Although the tyrosine kinase activity of this group of receptors is known to be essential to their function, the key substrates that mediate the response, and thus the exact mechanisms of action, are unknown. Activation of group 3 receptors also leads to the activation of protein kinases, through increases in intracellular second messengers such as cAMP, cGMP, Ca^{2+}, and diacylglycerol. However, the substrates for these kinases are serine or threonine residues rather than tyrosine. Phosphoaminoacid analysis therefore enables these receptors to be distinguished from the group 1 receptors.

As shown in *Table 2*, the primary structures of several group 2 receptors have now been determined. These include the nAChRs, which activate a cation (Na^+/K^+) channel and the GABA$_A$ and glyRs, both of which operate chloride conductances. In each case, there are numerous variants of the receptor subunits, which can be combined in a range of stoichiometries to generate receptors with different channel properties (e.g. conductance, or open channel lifetime) and pharmacological characteristics in different tissues. The nAChRs from electric organs and skeletal muscle are the best-characterized members of this group, and have been shown to consist of five subunits with an $\alpha_2\beta\gamma\delta$ stoichiometry. The oligomer forms a rosette structure with approximate pentagonal symmetry, which incorporates the ion channel, selectivity filter, and gating mechanism, as well as at least two acetylcholine binding sites (29). It is clear that there are embryonic and adult (ϵ) forms of the γ subunit. Up to five α and four β chain variants have been found in the central nervous system. It remains to be seen how the various subunits of the nAChRs, GABA$_A$Rs and glyRs

Table 3. Group 3 receptors (receptors linked to G-proteins).

(1) Structural characteristics
    (a) Probably monomeric
    (b) Seven transmembrane segments

(2) Function
    Activation of GTP-binding proteins (G proteins)

(3) Detergent used for solubilization
    Digitonin, or digitonin/cholate mixtures, CHAPS, CHAPSO, dodecyl maltoside

(4) Sequenced receptors and principal purification methods

Rhodopsin (30):	sequenced by chemical methods
βAR (31,32):	AC[a] with low mol. wt ligand (33, Chapter 5)
mAChR (34,35):	AC with low mol. wt ligand (36,37, Chapter 2)
α_2R (38):	AC with low mol. wt ligand (39, Chapter 6)
α_1R (40):	AC with low mol. wt ligand (41)
SKR (42):	isolation of cDNA by positive screening
$5HT_{1c}$R (43):	isolation of cDNA by positive screening
Angiotensin$_2$R:	expression of *mas* oncogene sequence (44)

Several cDNAs or genes encoding receptors have been isolated by homology screening using DNA probes encoding related receptors. These include nAChRs (45), mAChRs (46), β_1Rs (47), α_2Rs (48), D_2Rs (49), and $5HT_{1A}$Rs (50).
[a] AC, Affinity chromatography.

are combined *in vivo* to form functional channels. However, it is suspected that all will have the archetypal pentameric structure.

The endogenous ligands for group 2 receptors are typical neurotransmitters, exemplified by amino acids such as GABA, glycine, and glutamate, which act by changing membrane potential. Changes in membrane potential may also be induced by the activation of group 3 receptors. However, a functional distinction is that the responses of group 2 receptors are fast, whereas the responses of group 3 receptors, which are mediated by G-proteins, are much slower.

The endogenous ligands for the group 3 receptors (*Table 3*) include both hormones and neurotransmitters, and also external stimulants such as light and odourants. A very large number of receptors are now thought to belong to this group. They include receptors for adrenaline and noradrenaline (α_1, α_2, β_1, β_2), acetylcholine (muscarinic receptors m1−m5), dopamine (D_1, D_2), 5-hydroxytryptamine ($5HT_{1A}$, $5HT_{1C}$, $5HT_2$), adenosine (A_1, A_2), GABA ($GABA_B$), peptides (substance P, substance K, enkephalins, endorphins, and others). The amino-acid sequence of the retinal rod photoreceptor, rhodopsin, was initially determined by protein-sequencing methods (cf. Chapter 10), while those for the retinal cone opsins, the β receptors, the mAChRs, the α_2 receptor, the 5HT receptors, the substance K receptor, the D_2 receptor have been determined from the sequences of their cDNA clones. In one case, that of the MAS oncogene, the clone was isolated on the basis of its mitogenic potential, and subsequently identified as an angiotensin$_2$ receptor by expression studies (44). The results of these studies suggest that all of these receptors are variants on a single archetypal structure with seven transmembrane segments (51).

Both direct and indirect evidence shows that the group 3 receptors interact directly with members of a large class of heterotrimeric GTP-binding proteins, the G-proteins (52), which are often found in association with the internal surface of the plasma membrane. It is not known, in most cases, precisely which of the seven or more species of G-proteins interacts with any given receptor. The purification of G-proteins, and the reconstitution of receptor−G-protein interactions is described elsewhere (2).

The formation of a specific complex between a group 3 receptor and its cognate G-protein is accompanied by an increase in agonist binding affinity. In contrast, the equilibrium binding of antagonists is usually unaffected. The high-affinity agonist-binding state is switched to low affinity by the addition of GTP or GTP analogues. Thus, GTP-sensitive high-affinity binding of agonists to membrane preparations is a good indication that the relevant receptor belongs to group 3. However, the converse is not necessarily true, because in some cases the effect of GTP on agonist binding is not large. For instance, the GTP-induced shift in the binding curves for agonists to mAChRs in cerebral cortex is small, and difficult to detect using the usual protocol for measurement of agonist-binding curves, namely competitive inhibition of the binding of a radiolabelled antagonist by an agonist (1). In contrast, the effect of GTP analogues can be detected clearly in the case of m2 mAChRs in cardiac membranes, or after solubilization and reconstitution.

Small effects of GTP analogues are more easily detected by labelling the high-affinity receptor−G-protein complex directly with a radiolabelled agonist than by using a labelled antagonist, which fails to distinguish between the receptor−G-protein complex, and the free receptor. Other activities that may be measured are the stimulation by agonists of the GTP-ase activity of the G-protein, or the acceleration of binding of [^{35}S]GTPγS (a labelled, non-hydrolyzable GTP analogue), both of which are measures of the cycle of G-protein activation and deactivation, which is catalysed by the receptor−agonist complex (see reference 2).

G-proteins are known to mediate the activation and inhibition of adenylate cyclase, and the hydrolysis of polyphosphoinositides, whose breakdown produces inositol polyphosphates, which liberate calcium ions from internal stores (53), and diacylglycerol, which stimulates protein kinase C. Thus, most if not all of the receptors which alter cAMP levels, or mobilize intracellular calcium can be taken to belong to group 3. In addition, activated G-proteins are known to open certain ion channels in the plasma membrane, particularly K^+ and Ca^{2+} channels, either by a direct interaction (54,55) or indirectly, through second messengers such as cAMP, Ca^{2+}, diacylglycerol, or arachidonic acid metabolites liberated by phospholipase A2 activation (56). Many of these responses are mimicked by GTPγS, which produces irreversible activation of G-proteins. In some cases, channels can be opened by the direct application of activated G-proteins to membrane patches. The study of ion channels by patch clamping is described in reference 2.

A number of these G-protein effects are inhibited by pretreatment with pertussis toxin (PTX, also known as islet-activating protein), which catalyses the ADP-ribosylation of a cysteine residue near the C-terminus of some, but not all G-protein α-chains. These characteristic effects of PTX, if present, indicate that the receptor response is mediated through a G-protein, and that the receptor belongs to group 3. However, the absence of a PTX effect does not rule out G-protein involvement in the response. Other toxins, particularly cholera toxin (which activates some G-proteins that are not susceptible to PTX) and, very recently, botulinum neurotoxin [which ADP-ribosylates a new class of low molecular weight GTP-binding proteins (57)] may also be valuable. Much of this territory is unexplored. ADP-ribosylation of G-proteins is described elsewhere (2).

1.2 Illustrations

In the following sections, we will introduce the procedures used for solubilization, purification, and characterization of the receptors listed in *Tables 1-3*, taking a fairly empirical approach. A more theoretical and systematic treatment can be found in references 58 (detergents), 59 (solubilization), 60 (affinity chromatography), and 61 (DNA cloning). Companion volumes in *The practical approach series* are well worth consulting on some of these issues (5,62,63). In most cases, the practicalities are illustrated with respect to mAChRs. This reflects the authors' experience, and is justified by the observation that many of the receptors awaiting characterization belong, like mAChRs, to group 3. However, examples of all the techniques are to be found throughout the book. Cross-references are given to point these out.

2. Solubilization

2.1 Choice of detergents

The receptors that concern us are transmembrane glycoproteins. Even in enriched membrane fractions, they are minor species, co-inserted in the lipid bilayer with a c. 10^4-fold excess, or greater, of other proteins. A critical step in the purification of these molecules is the dispersal of the bilayer, thus separating the various components so that they can be fractionated. This inevitably involves the use of detergents. The sonication of membranes, or their treatment with chelating agents (e.g. 1 – 10 mM EDTA, or EGTA), or high concentrations of salts (e.g. 1 – 2 M NaCl) or chaotropic agents (e.g. 10 mM lithium diiodosalicylate, see Chapter 7, Section 3.3), while valuable in extracting peripheral membrane proteins (i.e. proteins that are not inserted in the lipid bilayer and whose binding may be mediated by Ca^{2+} or other ionic interactions) is insufficient to solubilize integral membrane proteins, such as receptors. However, prolonged incubations of membrane fractions may, though endogenous proteolysis, lead to the liberation of water-soluble fragments of receptors, which may retain ligand-binding activity. This is particularly the case with the group 1 receptors, whose ligand-binding domains are discrete, water-soluble entities. Methods for the control of proteolysis are given below. Incubation with chelating agents can conveniently form part of the protocol for membrane preparation.

Detergents are amphiphilic molecules. Like phospholipids, they have a hydrophilic 'head' and a hydrophobic 'tail', or in some important cases, a hydrophilic face and a hydrophobic face (A-face/B-face structure). The effect of detergent addition on membranes depends on the properties of the detergent, the properties of the membrane, and the absolute and relative concentrations of detergent, phospholipid, and protein in the system. As the concentration of detergent is raised, the following steps may be envisaged to occur.

(a) At low detergent concentrations, below those at which the detergent molecules aggregate to form micelles (the critical micellar concentration, CMC), the detergent molecules will be in solution in the aqueous phase, and will start to partition into the membrane. Here they will compete with the endogenous phospholipids and cholesterol molecules for binding sites on integral membrane proteins. Peripheral membrane proteins may start to become solubilized.

(b) As the detergent concentration in the aqueous phase increases, so its concentration in the lipid phase will increase. Fragmentation of the membrane, and solubilization of some integral proteins will commence.

(c) As the detergent concentration increases above the CMC, micelles will form in the aqueous phase. Large-scale solubilization of integral membrane proteins and lipids will occur, with formation of mixed protein−detergent−phospholipid micelles in addition to detergent−phospholipid micelles. These micelles will also contain cholesterol. Typically, different proteins will become soluble at different detergent concentrations. Some proteins, e.g. proteins attached to cytoskeletal elements, may not be solubilized at all.

(d) Further increases in detergent concentration will be taken up largely by an increase in the concentration of detergent micelles. However, there will also be some increase in free detergent concentration. As the detergent to phospholipid ratio increases, there may be a tendency for endogenous lipid molecules to be stripped away from the solubilized membrane proteins, to be replaced by detergent molecules. This may affect their properties or function, and may inactivate them. Extensive lipid stripping may lead to the formation of large hydrophobic protein aggregates. To avoid this, it may be necessary to control the detergent : phospholipid ratio carefully, possibly through the addition of exogenous phospholipids. Ideally, the aim would be to replace all the receptor-bound lipid molecules by detergent molecules without altering the functional properties. In practice, this aim is not achieved. Even the most robust properties of receptors, such as their ligand-binding properties, are altered on solubilization.

(e) As a generalization, the most important parameter in determining percentage solubilization by a given detergent is the molar ratio of detergent present in micellar form to the total phospholipid concentration. It should be emphasized that solubilization may not mean solubilization in an active form. The latter question must be approached empirically. Guidance in this matter is given below.

Some important properties of the detergents that have been used in receptor purification are given in *Tables 4−6*.

Table 4. Detergents used for receptor solubilization and purification: *t*-octyl-phenyl-polyoxyethylenes (Tritons).

Detergent	Receptor type	CMC mM	Micellar wt	Mol. wt	PSV (ml/g)	For use see
Triton X-100	1,2	0.3	90 000	625	0.908	Chapters 4,7,
Nonidet P-40	1,2	0.3	90 000	603	0.908	8,9;
Triton X-405	3	0.3	90 000	2142	0.908	reference 64

Comments:

(a) Detergents of the Triton series break weak protein – protein interactions and tend to strip away endogenous lipids. They destroy the activity of group 3 receptors unless a receptor – ligand complex is first formed.

(b) Their high micellar weight and low CMC impedes removal by dialysis. Gel-filtration on a gel that resolves the receptor – detergent complex from free detergent may be effective. Sepharose 4B or 6B, Sephacryl S300, Superose 6 or 12, Ultrogel ACA34 (all from Pharmacia/LKB) can be tried. Bio-Beads SM2 or SM4 (Bio Rad) or Extractigel (Pierce) absorb Triton-type detergents. Extractigel excludes peptides of mol. wt greater than 10 000 but will remove free detergent. Addition of these gels to dialysis media accelerates dialysis of detergents.

(c) Tritons have high A_{280} nm, because of the aromatic ring. Aldrich and Calbiochem market a cyclohexyl derivative that is free of this defect.

(d) Tritons interfere with the Lowry protein assay, unless SDS is added to maintain solubility (see Chapter 3).

(e) The CMC is lowered by raising salt concentrations or temperature. Micelles may aggregate at the cloud point. With Triton X-114, this phenomenon is useful in phase partitioning (reference 62, Chapter 6).

(f) Polyoxyethylene-type detergents accumulate oxidizing impurities. These may oxidize thiol groups in proteins. They can be removed as follows (65):
1. Make a 10% solution of detergent in water.
2. Stir for 2 h at room temperature with 1% $NaHSO_3$.
3. Make the solution to 10% w/v NaCl.
4. Extract the solution with an equal volume of dichloroethane. Collect the organic layer, using a separating funnel.
5. Wash the organic layer three times with an equal volume of 1% NaOH in 10% NaCl, then wash to neutrality with 10% NaCl. Check for removal of $NaHSO_3$ by dilution of 1 ml of the final aqueous phase into 2 ml of 0.1 M phosphate buffer pH 7.0 containing 0.3 mM dithionitrobenzoic acid, and following its reduction at 412 nm. There should be no reduction for at least 3 min at 30°C.
6. Dry the organic layer over anhydrous Na_2SO_4 for 24 h, and rotary evaporate (20 mm Hg, 40°C). Re-oxidation of the detergent is prevented by addition of 0.2 mol % butylated hydroxytoluene.

(g) Triton series detergents are cheap and supplied by most of the usual laboratory suppliers. Calbiochem and Pierce supply purified non-ionic detergents which are advertised as low in peroxides.

2.2 Group 1 receptors

Receptors belonging to group 1 are best solubilized with non-ionic detergents such as Triton X-100, Nonidet P40, and *n*-octyl glucoside. The ligand-binding activity and tyrosine kinase activities remain intact. Since these receptors have only a single transmembrane segment, with the ligand-binding site and catalytic sites situated on discrete extracellular or intracellular domains, it is reasonable to assume that non-ionic detergents bind to the two functional sites so little that their activities are not affected significantly. Instead, the detergents bind preferentially to the hydrophobic transmembrane domain. The group 1 receptors are therefore relatively robust to solubilization.

Table 5. Detergents used for receptor solubilization and purification: n-alkylglycosides.

Detergent	Receptor type	CMC mM	Micellar wt	Mol. wt	PSV ml/g	For use see
Octyl glucoside	1,2,4	23	8000	292	0.859	Chapters 7,9; reference 66
Dodecyl maltoside	3	0.1 – 0.6	50 000	511	0.809	reference 67

Comments:

(a) Octyl glucoside is readily removed by dialysis, and is useful in reconstitution experiments (2). Octyl thioglucoside is also available (Calbiochem) and is said to be more stable.

(b) Alkyl glucosides are expensive for large-scale work. They are available from most specialist Biochemical suppliers (Calbiochem markets highly purified 'Ultrol' grades).

(c) Octyl glucoside is only marginally effective on group 3 receptors. Dodecyl maltoside may be better for this purpose (67).

Table 6. Detergents used for receptor solubilization and purification: detergents with a steroid nucleus.

Detergent	Receptor group	CMC mM	Micellar wt	Mol. wt	PSV ml/g	For use see
Cholate(Na)	1,2(3)	14	900 – 1800	431	0.77	Chapters 2,3; references 93 – 95
Deoxycholate(Na)	1,2(3)	5	1 700 – 4200	415	0.78	
CHAPS	1,2,3	8	6 200	615	0.81	Chapter 4,8; references 93 – 95
CHAPSO	1,2,3	8	9 960	631	0.79	reference 76
Digitonin	3	–	70 000	1229	0.74	Chapters 2,4,5,6

Comments:

(a) All of the above detergents have an A-face/B-face structure. They have a complex micellar structure, and may coat hydrophobic interfaces instead of forming spherical micelles. The uncharged glycosidic detergent digitonin complexes specifically with cholesterol and can solubilize cholesterol-containing membranes. See text for more information.

(b) Cholate and deoxycholate are anionic detergents with pKs of c. 6.5. They may complex (precipitate) with metal cations, and are insoluble when protonated. Deoxycholate precipitates in the cold. Raising ionic strength markedly lowers the CMC of cholate, and increases its solubilizing capacity for group 3 receptors.

(c) CHAPS and CHAPSO are zwitterionic sulphobetaine derivatives of cholate. As with digitonin, their lack of net charge permits separations such as ion exchange and isoelectric focusing to be performed. CHAPS and CHAPSO are capable of preserving easily disrupted protein – protein interactions.

(d) Cholate, deoxylcholate, CHAPS, and CHAPSO are readily dialysable. Digitonin is not.

(e) Cholate, deoxycholate, CHAPS, CHAPSO, and variants of these are available from the usual lab suppliers. Calbiochem markets highly purified 'Ultrol' grades. Impurities can be removed from cholic acid by passing a 2.5% w/v solution of the sodium salt in deionized water through a DEAE-Sepharose column (0.8 ml bed volume per g of cholate). The eluate is acidified to pH 4 to precipitate the cholic acid. The precipitate is filtered and washed with water, and then with anhydrous ether before drying, firstly in air and then under vacuum.

2.3 Group 2 receptors

The three receptors in group 2 can be solubilized in their active forms with sodium cholate or deoxycholate. Triton X-100 has also been used for the purification of nAChRs, GABA$_A$Rs, and glyRs. The addition of exogenous phospholipids is necessary to preserve full functionality. Such lipids may either be derived from the tissue from which the receptor is solubilized, or from other sources, such as soya bean azolectin (see Chapter 8 for the use of CHAPS/azolectin mixtures in purifying the GABA$_A$ receptor) or egg phosphatidyl choline [e.g. purification of the nicotinic AChR in 2% cholate plus 5 mg/ml phosphatidyl choline (68)]. Cholesterol plays a major role in maintaining the activity of these receptors (69) (see also Chapter 8, Section 2). Detailed methods for the extraction and handling of membrane phospholipids, and the reconstitution of nicotinic receptors and rhodopsin are given in *Biological membranes, a practical approach*, Chapters 4 and 5 (62), while Criado *et al*. (69) give an excellent description of the methodological considerations relevant to the maintenance of the agonist-induced conformational changes in the nicotinic AChR after reconstitution, and are particularly informative about the handling of cholesterol.

The GABA$_A$ receptor, when solubilized either in cholate, or in Triton/X-100, preserves the ability to bind both the agonist [^{3}H]muscimol, and the allosteric ligand [^{3}H]flunitrazepam, which binds to a separate benzodiazepine-binding site. In contrast, the binding of the chloride channel-blocking ligand [^{35}S]t-butyl bicyclophosphorothionate is lost in these detergents. However, all three activities are retained when the receptor is solubilized in CHAPS (see Chapter 8).

Both the agonist and the benzodiazepine-binding sites are thought to reside on extracellular domains of the receptor, while the channel blocker presumably binds in the transmembrane domain, which consists of four transmembrane segments per subunit. The transmembrane domain, being more hydrophobic, would be expected to manifest a higher affinity for detergents than is shown by the extracellular sequences, and may be more susceptible to interference by some species of detergents. CHAPS may interfere less drastically with the structure of the transmembrane domain than some other detergents, and may favour the retention of inter-subunit interactions. CHAPS, however, produces less complete solubilization than Triton X-100, for instance the yield of muscimol-binding activity is only 25–35% as opposed to 55% in cholate. Further details are given in Chapter 8.

2.4 Group 3 receptors

The ligand-binding activity of group 3 receptors is more labile to detergents than that of receptors in groups 1 and 2, and is usually completely lost in the presence of such common detergents as Triton X-100 and sodium cholate, although Triton X-100 has been reported to solubilize opioid receptors (Chapter 4) and Triton X-405 has been used successfully in the solubilization of muscarinic receptors for hydrodynamic studies (64). The origin of this lability remains enigmatic, as well as presenting a practical problem.

Determination of the structures of several of the group 3 receptors has revealed that their extracellular domains are smaller than those of the group 1 or 2 receptors. In contrast, the relative proportions of transmembrane sequences are larger. This suggests that the extracellular domains of the group 3 receptors are not sufficient to provide a ligand-binding site, and that the transmembrane domains also participate. Evidence supporting this assumption has been accumulating. This would explain, at least partly, the lability to detergents of the ligand-binding activity of group 3 receptors.

2.4.1 Digitonin

All of the receptors listed in *Table 3* can be solubilized with digitonin or digitonin supplemented with sodium cholate, and purified in solutions containing digitonin. Digitonin is the best established and most reliable of the detergents for keeping the ligand-binding activity of this group of receptors intact. However, there are several problems with digitonin, in addition to the fact that it is very expensive. A practical guide to the use of digitonin is given below.

(a) Prepare an aqueous suspension of 2% digitonin in water. Boil for 5−10 min to obtain a slightly turbid solution. Cool for 1 h and filter off any precipitate.

(b) The solution so obtained can be used directly for the solubilization of membranes. The solubilized supernatant remains clear through subsequent purification procedures. A typical concentration of digitonin is 1−1.5% w/v, with membranes at 5 mg protein/ml.

(c) Such a solution is not suitable for column or affinity chromatography, because precipitation continues for up to 1 week. If the solution is stored at 4°C for a week, and then carefully decanted through a fluted Whatman no. 1 filter paper, or similar, the resultant preparation can be used for most purposes.

(d) To determine the concentration of digitonin in solution, measure the weight of an aliquot of filtrate (e.g. 1 ml) after lyophilization. 50−90% of the digitonin should be recovered, depending on the batch. Before buying a large batch of digitonin, obtain a small sample from the manufacturer, and check its solubility, and its ability to solubilize the receptor of interest.

(e) A digitonin preparation which is soluble in water at room temperature at 4−10% is available from WAKO pure chemical company, but is twice as expensive as most other brands. However, in our experience, the investment of the extra money seems worthwhile.

(f) Digitonin can be used in combination with sodium cholate (e.g. at a 10 : 1 ratio, say 1% digitonin/0.1% cholate) and remains in solution better in this form. However, the addition of cholate should be checked for adverse effects on receptor stability. The use of digitonin/CHAPS or digitonin/CHAPSO combinations remains to be investigated.

(g) A method for the preparation of water-soluble digitonin from less soluble commercial samples has been reported (70). However, we have found it only marginally useful. Digitonin is not a pure chemical, consisting of a mixture of digitonin and gitonin (71). It is important to have optimal ratios of these two components if good water solubility and solubilizing activity are to be obtained (72). Variations in this ratio are no doubt the main reason for the variability of commercial preparations of this detergent.

2.4.2 Other detergents

Fatty acid esters of sugars have been used for the purification of rhodopsin (73) and β-adrenergic receptors (74). In preliminary experiments, mAChRs were solubilized and purified in myristoyl sucrose, but the binding activity was less stable in this detergent than in digitonin. Of the alkyl glycosides, octyl-β-D-glucopyranoside is only marginally effective in the solubilization of mAChRs, but dodecyl-β-D-maltoside has been reported to be as effective as digitonin in the solubilization of cardiac mAChRs (67). Octyl glucoside is particularly useful in reconstitution studies involving group 3 receptors (see reference 2, Chapter 4). Calbiochem markets a range of these detergents with different chain lengths, and also supplies a useful monograph on detergents.

Several other detergents have been used for the solubilization of group 3 receptors. These include lysophosphatidyl choline and CHAPS (for a review, see reference 75). In the case of mAChRs, 25% of the receptors were solubilized in an active form by lysophosphatidyl choline, while the remaining 75% were solubilized, but inactivated. When CHAPS was used, the ligand-binding activity was recovered in the supernatant after centrifugation at 100 000 g for 2 h, but in the void volume after gel-filtration chromatography, indicating aggregation. However, mAChRs have been successfully solubilized in CHAPSO (0.5%) (76). Both the free, monomeric receptor-binding subunit, and a receptor—G-protein complex have been identified (see reference 2, Chapter 2). Once again there was molecular weight heterogeneity, indicating the occurrence of aggregation. CHAPSO solubilization leads to reversible denaturation of a proportion of the binding sites, which can be recovered by detergent exchange, replacing CHAPSO by digitonin by gel-filtration on a column of Sephadex G50 fine, as described below (see Section 2.5). The stability of group 3 receptors in CHAPS or CHAPSO is improved by the addition of exogenous phospholipids. Both detergents are useful in reconstitution procedures.

The ligand-binding activity of β-adrenergic receptors and mAChRs is labile in non-ionic detergents such as Lubrol PX, and Triton X-405, although preformed complexes between receptors and high-affinity ligands are stable enough to be used for molecular size estimation (64,77). Thus, the receptor—ligand complex is more stable than the free receptor in detergents. For example, the ligand-binding activity of mAChRs could be recovered when membranes were solubilized with cholate or deoxycholate in the presence of muscarinic ligands, and the detergents subsequently removed, a step which is readily performed by gel-filtration or dialysis in the case of cholate and deoxycholate, which have relatively high CMCs (*Table 6*). Another tactic is to reduce the effective concentration of cholate necessary for solubilization by the addition of a high salt concentration, which reduces the CMC of the detergent. For example, dopamine D_2 receptors and muscarinic receptors can be solubilized with 0.3% instead of 1% cholate in the presence of 1 M NaCl. The addition of exogenous phospholipids helps to prevent aggregation. This methodology is described in Chapter 3, and Chapter 4, Section 2.5.

2.5 Testing a range of detergents

As the forgoing discussion illustrates, knowledge of the group to which the receptor belongs provides some guidance in the selection of detergents for attempted solubilization. However, if the group classification is unknown, a range of detergents, including non-ionic types (e.g. Triton X-100, alkyl glycosides), anionic (e.g. cholate/deoxycholate), zwitterionic (CHAPS/CHAPSO) and digitonin should be tested at several different concentrations. If no single detergent appears to be suitable, it may be beneficial to test mixtures of different detergents at low concentrations. The protocol used for the screening of detergents for the solubilization of mAChRs is as follows. It is readily adaptable to other receptor types.

Protocol 1. Screening detergents

1. Suspend the membranes at $5-10$ mg/ml protein in buffer at $0-4°C$. The choice of buffer system is determined by the receptor-binding properties, and by the demands of subsequent purification procedures. In most cases, suitable buffers are such as 20 mM sodium Hepes, pH 7.5, 20 mM Tris-Cl, pH 7.5, or 20 mM sodium phosphate, pH $7.0-7.5$. It has been stated that the use of phosphate buffers favours solubilization, but, in our experience, this does not seem to be critical. For screening purposes, it is probably unnecessary to add complex proteolysis inhibitor cocktails. The presence of EDTA (1 mM) to inhibit calcium-activated proteases and of PMSF or benzamidine (0.1 mM) to inhibit serine proteases it is often beneficial. Mg^{2+} ions (e.g. 2 mM) can be added when they are required (e.g. for the interaction between group 3 receptors and G-proteins); EDTA has a 100-fold higher affinity for Ca^{2+} than for Mg^{2+}, and hence contaminating Ca^{2+} ions can be removed by EDTA (e.g. 1 mM) even in the presence of Mg^{2+} ions (e.g. 2 mM).

2. Prelabel the membranes with radiolabelled ligand in the presence and absence of a suitable unlabelled ligand to determine the total and non-specific binding. Subtract non-specific from total binding to obtain a measure of receptor-specific binding. If possible, use a high enough concentration of labelled ligand to saturate the binding site (c. $10 \times K_d$), so that the binding capacity is measured. The unlabelled ligand should be used at a concentration of $1000 \times K_d$. The normal criteria for receptor-specific binding must be fulfilled. Make sure that the incubation is long enough for equilibrium to be reached. Full details are given in reference 1, and a summary of assay methods in the appendix to this volume.

3. Add 0.5 ml of the prelabelled membrane suspension to a series of centrifuge tubes at $4°C$.

4. Add an equal volume of detergent solution in the same buffer, to obtain a series of different final detergent concentrations, e.g. 0, 0.1, 0.2, 0.5, 1.0, 2.0% w/v.

5. Mix and incubate for 60 min at 4°C. Solubilization is assisted by stirring or mixing, e.g. with a rotating-wheel end-over-end mixer.

6. Centrifuge the tubes for 30−60 min at 100 000 g. The normally accepted criterion of solubilization is that the binding activity should remain in solution after centrifugation at 100 000 g for 60 min. For screening purposes a lower speed spin, e.g. 10 000 g, 5 min in a microfuge, may be acceptable.

7. Apply 0.2 ml of supernatant to a 2 ml column of Sephadex G50 F equilibrated with the detergent of interest at 0.1%. When the sample has run in, apply 2×0.2 ml of detergent-buffer, and then elute the void volume fraction with 0.5 ml of detergent buffer. This procedure is carried out conveniently with the column standing in a scintillation vial. Remove the column, add 10 ml of aqueous scintillation cocktail and count. Sephadex G50 M may be substituted for G50 F when working with hydrophilic ligands, which do not partition into detergent micelles. This gives a more rapid separation. Alternative methods of assaying solubilized binding sites are cross-referenced in the appendix.

   Several manufacturers (e.g. BioRad, Kontes) supply appropriate plastic disposable columns. We have found the use of Kontes Disposaflex columns with porous polythene sinters to be convenient. The gel bed is packed by pipetting in 4 ml of a 1 : 1 slurry of Sephadex swollen in water containing 0.02% sodium azide as a preservative, and allowing the bed to settle before washing it through with water, and equilibrating it with 2 bed volumes of detergent buffer. After use, the columns can be washed through with c. 5 bed volumes of water, and re-equilibrated. The end fittings of these columns can be removed to allow the gel to be washed out before repacking, and they are resistant enough to permit cleaning by immersion in 50% nitric acid for 24 h, which is useful for unblocking clogged sinters.

8. Subtract non-specific from total binding for each detergent concentration to assess the recovery of the receptor−ligand complex. Work out the recovery of specifically bound ligand in absolute terms, thus:

   bound ligand = (dpm, total−dpm,non-spec.) $\times$ 5/(2220 $\times$ spec.act.) pmol/ml

9. Resuspend and count an aliquot of the pellets, to work out the recovery of unsolubilized receptors.

10. Measure the concentration of protein in the solubilized supernatant. A convenient, if crude, measure is obtained by measurement of UV absorbance at 280 nm against a detergent−buffer blank (if necessary, dilute the supernatant to get the absorbance on scale). Ignoring any contributions from non-protein impurities or detergent, the concentration of protein in solution is approximately equal to the A_{280} nm. Alternatively, use the Lowry method as given in Chapter 3, Section 2.

11. Repeat *steps 1* and *3−6* above without first prelabelling the receptors in the membrane. Instead, incubate the solubilized supernatant in the absence and presence of unlabelled ligand. Again, use a concentration of the labelled ligand

that you believe will saturate the binding sites. Incubate under appropriate conditions, e.g. 2 h at 4°C, and assay the binding by gel-filtration as described above, or see Appendix. The pellet can be resuspended and assayed for residual binding to check overall recovery.

Steps *1−7* are performed to determine the concentrations of detergent required for solubilization of the binding sites. If specifically bound ligand is not detected in *7*, either the receptor−ligand complex has failed to solubilize, in which case the binding activity will be recovered in *9*, or else it is unstable in the detergent, in which case it is pointless to perform step *11*. If binding activity is detected after prelabelling the receptors in the membranes, but not on postlabelling the solubilized supernatant, attempts to stabilize the receptors can be made. These should include the addition of sucrose or glycerol (10−30%, see also Chapter 4, Section 2.5) or the inclusion of exogenous phospholipids in the detergent buffer (see Chapters 3 and 8 for examples). It is important to test as many different conditions as possible, so it may be advisable to dispense with the high-speed spin for screening purposes. Whether or not the receptors have been genuinely solubilized can be checked later. To compare different procedures, it is always worth drawing up detailed recovery tables.

It is important to establish the mildest conditions compatible with true solubilization in order that receptor stability be maintained. This is a prerequisite for prolonged purification procedures. Genuine solubilization may be defined as the achievement of a monodisperse solution of receptor−detergent−phospholipid complexes, in micellar form, which can be separated from other components of the solution by the application of suitable separation techniques. In some cases, the entity solubilized may be a one-to-one complex of the receptor with its cognate effector macromolecule. Thus in the case of group 3 receptors, there are numerous examples of solubilization of receptor−G-protein complexes (2).

An effort to define a criterion for the minimum concentration of detergent necessary to effectively solubilize membrane macromolecules was made by Rivnay and Metzger (66), who introduced the rho parameter, defined as

$$\varrho = (\text{detergent} - \text{CMC}_{\text{eff}})/(\text{phospholipid})$$

CMC_{eff} is the effective CMC of the detergent under the solubilization conditions, which is usually slightly smaller than the CMC in aqueous solution. For practical purposes, the CMC in water can be taken as a guide (see *Tables 4−6*). The phospholipid phosphate content of the membranes is determined by the method of Bartlett (see reference 5, Chapter 7). To obtain maximum solubilization, ϱ values of *c*. 1.6−10 appear to be required (see reference 62, Chapter 5 for a fuller discussion of this parameter and its importance).

For individual membrane components, solubilization curves tend to be highly co-operative in appearance. An example is shown in *Figure 1*, which shows solubilization

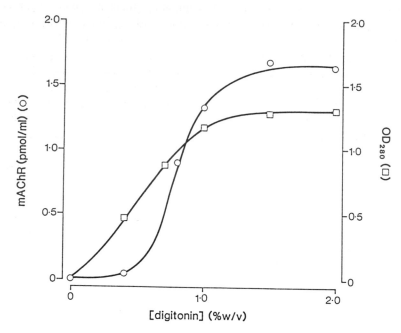

Figure 1. Rat forebrain membranes (5 mg/ml) in 20 mM Na-Hepes buffer, containing 1 mM EDTA, pH 7.5, were solubilized by stirring with increasing concentrations of digitonin for 30 min at 4°C. After centrifugation at 100 000 g for 60 min, the binding activity (20 nM [^{3}H]-NMS, 120 min, 4°C) and protein content (A_{280} nm) of the supernatant were measured. The maximum recovery of solubilized mAChRs was 35%.

of muscarinic receptors from rat forebrain membranes by digitonin. It is worth noting that the overall dependence of protein solubilization on detergent concentration is the summation of these individual components, and since each protein will partition into the detergent micelles at a slightly different free detergent concentration, the total protein extraction curve tends to show a less steep dependence on detergent concentration. In this case, a double extraction procedure may be possible, in which the membranes are first exposed to a concentration of detergent insufficient to solubilize the receptor, but which will remove some easily extracted contaminants. The receptor is then solubilized by re-extraction with a higher concentration of detergent. This can yield an effective purification step, and may remove important contaminants. Examples are given in Chapters 5 and 6. However, a possible drawback is that even low concentrations of detergents can activate membrane-bound proteases which will then work on the membrane-bound receptor. The action of these proteases will only be attenuated after solubilization. Thus speed at this stage is essential, as is the addition of protease inhibitors (see Section 3.2).

The value of ϱ is obviously dictated by the concentration of membrane phospholipid in the solubilization mixture. This is proportional to the concentration of membrane protein, so that the detergent : protein ratio can also be used as a measure of ϱ.

18

One implication is that an increase in the concentration of membranes will entail a roughly proportional increase in the detergent concentration if solubilization efficiency is to be maintained.

In several important instances, it is necessary to add exogenous phospholipids at the solubilization stage to maintain receptor stability or function (see Chapters 3 (Section 3) and 8). This reflects the ability of detergents to strip away essential endogenous phospholipid molecules. The hazard may be avoided to some degree by operating at a minimum value of ϱ. To do this, it may be desirable to add relatively pure phospholipid preparations, e.g. egg or soya bean phosphatidyl choline, azolectin, or cholesterol. Alternatively, phospholipid extracts can be prepared from the tissue of interest. The methodology for this, and for the separation and assay of phospholipids is described in detail by Higgins (reference 62, Chapter 4). A more extensive discussion of the role of detergents and phospholipids in solubilization and reconstitution is given in reference 62, Chapter 5. It should be noted that the unique ability of digitonin to substitute for endogenous lipids at group 3 receptors may be a reflection of its ability to mimic the interaction of these receptors with cholesterol, which appears in general to be very important for the function of both group 2 and group 3 receptors.

2.6 Estimation of the molecular size of solubilized receptors

After membranes have been treated with detergents, the receptors may be found in several different states:

(a) membrane fragments;

(b) non-specific aggregates with other membrane components;

(c) well-defined complexes with specific membrane components;

(d) independent molecules segregated from other membrane components.

In addition, phospholipid molecules may remain bound to the receptors, or be partly or completely replaced by detergent molecules in each of these states.

If the receptors are to be isolated, they need to be segregated from other protein components, especially from non-specifically interacting proteins. The recovery of receptors in the supernatant after centrifugation at 100 000 g for 1 h is a conventional criterion for solubilization. However, its fulfilment does not necessarily indicate effective segregation of the receptors from other components. Analysis by gel-filtration (e.g. Ultrogel AcA34), or sucrose density gradient centrifugation (5–20% sucrose, 100 000 g, 20 h) provides a more critical criterion. A substantial fraction of the ligand-binding activity should be recovered within the gel or in the gradient, although sometimes activity is recovered in the void volume, or at the bottom of the gradient, indicating aggregation.

The molecular size of receptors in solubilized preparations can be estimated by a combination of gel-filtration chromatography and sucrose density gradient centrifugation in H_2O and D_2O. A detailed protocol is given in Chapter 2, Section

6. The use of one or two of these procedures alone is not enough to estimate the true receptor molecular weight because a correction has to be made for the asymmetric shape of the receptor—detergent complex and for the contribution of bound detergent to the molecular size. Corrections can only be made if the partial specific volume of the detergent is significantly different from that of the protein. This is not the case for digitonin, for instance (see *Table 6*). Molecular size estimation not only provides a strict criterion for solubilization, but also yields information as to whether the receptors are monomers or oligomers, provided that the monomer molecular weight has been determined, e.g. by SDS-PAGE or target size analysis. The use of SDS-PAGE to determine receptor molecular weights is discussed in Chapter 2, Section 6 and Chapter 10, Section 3.3. The technique of radiation inactivation can be applied to estimation of the molecular weight of receptors in membrane preparations as well as in the purified state. This technique is not described in the present volume. A brief outline is given by Findlay (62). The technique is described in detail in reference 78.

2.7 Identification of solubilized sites by binding studies

The binding of labelled ligands to solubilized preparations can be assayed by the following techniques:

(a) Gel-filtration on Sephadex G50 F or M, as described in Section 2.5.

(b) Precipitation with polyethylene glycol, ammonium sulphate or protamine sulphate followed by filtration on glass-fibre filters. Examples of this technique are given in Chapters 4 and 8. A brief description is given in the appendix and fuller details are found in reference 1.

(c) Adsorption to polyethyleneimine-treated filter discs, or to DEAE filter discs. See Appendix and Chapter 7 for a brief description, and reference 1 for a fuller account.

(d) By adsorption of the free ligand but not the receptor—ligand complex, using activated charcoal. This technique is described in Chapter 3.

The reader will find further details under the heading of individual receptors. It should be noted that the normal criteria for the definition of receptor-specific binding must be rigorously applied (see Appendix).

The affinities of receptors for their ligands often decrease on solubilization. Therefore the solubilized sites may need to be identified as receptor-binding sites by correlation of the rank order of potency of several ligands before and after solubilization, rather than by their absolute affinities. To be statistically significant, such a correlation should be based on at least six ligands. For example, the affinities of ligands for solubilized mAChRs decreased in parallel, but no different degrees depending on the species of detergent used; the average ratios of the apparent dissociation constants of ligands for membranes and solubilized preparations were 4, 20, and 35 for preparations solubilized with sodium cholate, digitonin, and lysophosphatidyl choline respectively. The ligand-binding activities of receptors solubilized with sodium cholate were assayed after removing the detergent (72).

Solubilized sites are unequivocally identified as receptors or receptor subunits if an effector function mediated by the sites can be demonstrated. For example, the tyrosine kinase activity of some group 1 receptors, and the interaction of group 3 receptors with G-proteins has been shown to be preserved in solubilized preparations (79). However, confirmation of function is often obtainable only after the sites have been purified.

3. Purification of receptors

3.1 Receptor sources and membrane preparations

The purification of receptors is facilitated by the use of tissues containing a high receptor density. nAChRs and rhodopsin were purified much earlier than other receptors simply because of the availability of electric organs and retinas. As another example, the identification and purification of EGFRs was facilitated by the use of A431 epidermal carcinoma cells, which possess $10-50$ times more EGFR than other cell lines (Chapter 9, and reference 17).

The isolation of membrane fractions enriched in receptors is helpful. In the case of nAChRs, a membrane preparation containing the receptors but essentially free from other membrane components can be isolated from electric organs (see Chapter 7). This makes further purification after solubilization much easier. The preparation of purified membranes is especially important for tissues such as muscle which contain a large proportion of other insoluble components. The preparation of mAChR-enriched membranes from atria is a good example (36; see Chapter 2). Examples of membrane preparations are given throughout this book (Chapter 2, Section 3; Chapter 3, Section 2; Chapter 5, section 4; Chapter 6, Section 3; Chapter 7, Section 3; Chapter 8, Section 2). However, it should be noted that the use of highly sophisticated membrane fractionation techniques is difficult on the large scale, and rarely justifies a great deal of effort.

The available amounts of receptor sources, as compared with the amounts of purified receptor necessary for a chosen purpose, is an important factor when choosing starting materials. If purified receptors are to be used for the determination of partial peptide sequences, it is desirable to prepare *c*. 1 nmol of purified protein, although as little as 100 pmol have been used for this purpose, and even smaller amounts might be possible if cultured cells labelled with radioactive amino acids can be used as sources of radioactive receptors (for example, see the case of IL$_2$Rs (7)). In our experience, the higher figure of approximately 1 nmol is realistic. Since the overall yield of purified receptors is typically around 10% of the total population present in the membrane before purification (e.g. in the case of the receptors shown in *Tables 1−3*), a membrane preparation containing at least 10 nmol of the receptor of interest is required to provide enough starting material for purification for sequencing.

Table 7. Protease inhibitors used in receptor purification.

Inhibitor	Working concentration	Protease group
Amastatin	1 – 10 μg/ml	aminoexopeptidases
Antipain	1 – 10 μg/ml	trypsin,papain,cathepsin B
Aprotinin	1 – 10 μg/ml	kallikrein
Bacitracin	100 μg/ml	effective non-specific inhibitor
Benzamidine	up to 10 mM	serine proteases
Benzethonium chloride	0.1 mM	bacteriostatic
Benzylmalic acid	1 – 10 μg/ml	carboxypeptidases
Bestatin	1 μg/ml	aminopeptidase B, leucine aminopeptidase
Chymostatin	1 – 10 μg/ml	chymotrypsin,papain
Di-isopropyl fluorophosphate	0.1 mM	serine proteases
Diprotin A + B	10 – 50 μg/ml	dipeptidyl aminopeptidase
Elastatinal	10 μg/ml	elastases
EDTA	1 – 10 mM	metalloproteases, Ca^{2+}-dependent SH proteases
EGTA	1 – 10 mM	metalloproteases, Ca^{2+}-dependent SH proteases
Iodoacetate	1 – 10 mM	SH proteases
Iodoacetamide	1 – 10 mM	SH proteases
Leupeptin	1 – 10 μg/ml	serine and SH proteases
Pepstatin A	1 – 10 μg/ml	aspartic proteases
N-ethyl maleimide	1 mM	SH proteases
Phenylmethane sulphonyl fluoride	0.1 – 1 mM	serine proteases
Phosphoramidon	1 – 10 μg/ml	thermolysin (Zn proteases)
Sodium tetrathionate	5 mM	SH proteases
TPCK	10 mM	chymotrypsin
TLCK	10 mM	chymotrypsin
Trypsin inhibitor	500 μg/ml	trypsin
2-phenanthroline	1 – 10 mM	metalloproteases, Ca^{2+}-dependent SH proteases

3.2 Protease inhibitors

In the course of membrane preparation and solubilization, it is important to prevent the hydrolysis of receptors by endogenous proteases. To inhibit Ca^{2+}-activated and serine proteases, EDTA (1 – 10 mM) and phenylmethylsulphonyl fluoride (PMSF, 0.1 – 1 mM) or benzamidine (0.1 mM) are routinely included in the medium. Several antibiotics, such as pepstatin, leupeptin, chymostatin, and trasylol (aprotinin) have also been used (80). It is recommended to use these antibiotics in initial attempts at purification, although some of them will turn out to be dispensable, and it may be too expensive to use all of them for a large-scale purification. A comprehensive list of protease inhibitors is given in *Table 7*. Most workers develop their favourite cocktail of inhibitors, and numerous examples occur in other parts of this book. As a guide, it seems wise to aim to inhibit each of the various classes of protease, i.e. serine proteases (PMSF), metalloproteases (EDTA), Ca^{2+}-activated proteases (EDTA), aspartic proteases (pepstatin), SH proteases (leupeptin), and kallikreins (aprotinin, also inhibited by di-isopropylfluorophosphate). Note that SH reagents, while they will inhibit SH proteases, will also modify receptors.

3.3 General approaches to purification

Affinity chromatography has been used as the principal method for the purification of all of the receptors shown in *Tables 1−3*. In each case, the affinity chromatographic step contributed 100- to 1000-fold purification to the overall scheme. However, a single-step procedure employing affinity chromatography alone is usually insufficient to isolate receptors. Usually, it must be combined with other procedures, such as gel-permeation HPLC, and ion-exchange, hydroxyapatite, or lectin column chromatography.

Different categories of ligands have been used for the preparation of affinity gels. These include:

(a) The hormones themselves, e.g. polypeptides for group 1 receptors, cf. Chapter 9.

(b) Analogues of hormones or neurotransmitters, particularly various low molecular weight antagonists that bind to group 2 or 3 receptors (see Chapters 2−8).

(c) Other ligands that bind to subsidiary or allosteric sites on the receptor, e.g. a benzodiazepine analogue for $GABA_A$ receptors (Chapter 8) and snake venom for nAChRs (Chapter 7).

(d) Monoclonal or polyclonal antibodies to the receptor (e.g. monoclonal antibodies against the EGF receptor) or to a modification of the receptor (e.g. against phosphotyrosine for purification of the PDGF receptor; see *Table 1*). Protocols for raising and using antiphosphotyrosine antibodies are given in reference 2, Chapter 8.

3.4 Lectin chromatography

Lectin, in particular wheat-germ agglutinin (WGA), chromatography has frequently been used for the partial purification of receptors prior to affinity chromatography. In the case of the cardiac mAChR, approximately 16-fold purification was obtained by this technique (36). All of the receptors listed in *Tables 1−3* have been deduced to be glycoproteins, either because putative glycosylation sites (Asn−X−Thr/Ser) have been identified in appropriate (i.e. extracellular) parts of the deduced amino-acid sequences, or more directly because they have been shown to bind lectins. This suggests that most, if not all, receptors are glycoproteins, and that, in general, lectin column chromatography can be used for the partial purification of receptors.

In the case of mAChRs, a number of different lectins were examined, but only WGA was found to be useful. In fact, WGA-Sepharose has been used in the purification of most of the receptors listed in *Tables 1−3*. Several specific examples are to be found in this book (Chapter 3, Section 3; Chapter 4, Section 3; Chapter 6, Section 4). The specificity of WGA is for terminal *N*-acetyl-glucosamine and sialic acid. Specific elution can be achieved with the conjugate sugar, i.e. *N*-acetyl-glucosamine (20−400 mM), or *N*-acetylchitotriose (*c*. 20 mM; the high cost of this material rules out its large-scale use). For most purposes, WGA-agarose 6MB eluted with *N*-acetyl-glucosamine (Sigma) seems to work well.

Another lectin specific for *N*-acetyl-glucosamine, *Datura stramonium* lectin, has been used in the purification of dopamine D$_2$ receptors (81), while concanavalin A has been used in the purification of insulin receptors (*Table 1*). Lectins are also useful in peptide-mapping studies, and may be used to identify glycosylated peptides on Western blots. Specific examples are given in Chapter 8, Section 3 and in Chapter 10, Section 5. Lectin specificity may be altered by digestion with specific endoglycosidases, which expose a new set of sugar residues or remove them entirely. A comprehensive list of lectins and their properties is provided in the Sigma catalogue. A useful description of the methodology involved in the use of different lectins and specific glycosidases in the analysis of the carbohydrate composition of the β-adrenergic receptor is given by Cervantes-Olivier *et al.* (82).

3.5 Gel-permeation chromatography

Gel-permeation chromatography on silica high-performance gel-permeation columns (e.g. TSK 2000, 3000, and 4000 SW columns, Toyo Soda) is frequently used in receptor purification. A specific example is given in Chapter 5, Section 7. Toyo Soda now markets an improved range of TSK columns, the TSK SWXL range, which have higher resolution. It should be noted that the presence of detergents (e.g. digitonin) raises the effective molecular weight of receptors during gel-filtration, as it is the receptor−detergent complex, not the free receptor, whose effective molecular weight determines the separation. The Pharmacia Superose 6 and 12 FPLC columns are also extremely useful for gel-permeation separations of receptors, generally giving good recoveries in a biocompatible format. The excellent resolution of FPLC gel-filtration on Superose 12 in 0.1% SDS makes this step a very useful aid to final clean-up of receptor preparations before digestion for sequencing. The use of this protocol is described in Chapter 10.

3.6 Ion-exchange, hydroxyapatite, heparin adsorption and hydrophobic chromatography

Ion-exchange and hydroxyapatite chromatography have both been used extensively in the purification of receptors. Most receptor proteins tend to have a low isoelectric point, (pI less than 5). A major contribution to this is provided by the sialic acid residues. Most receptors will therefore adsorb to DEAE Sepharose at neutral pH and can be eluted by a salt gradient, e.g. 20−400 mM NaCl. A specific example is given in Chapter 8, Section 2. Pre-chromatography on DEAE Sephacel has been used successfully to remove protein impurities that bind strongly to key positively charged residues in an affinity gel by an ion-exchange mechanism, thus interfering with genuine affinity chromatography (83). Ion-exchange steps scale-up well, and can be used at an early stage of purification. However, the overall purification that results is usually rather low (approximately two-fold). Nevertheless, an ion-exchange step may remove key impurities, and by increasing the receptor concentration, enhance adsorption in a subsequent affinity chromatographic step. High-resolution HPLC ion-exchange columns have been little used in receptor purification.

Hydroxyapatite chromatography involves the adsorption of the protein to free calcium and phosphate sites in a crystalline calcium phosphate matrix. The method depends for its success on differences in the charge densities of different proteins. Proteins are eluted, after an initial high ionic strength (200 mM NaCl) wash to remove non-specifically adsorbed proteins, using a linear or step gradient (e.g. 20–200 mM) of sodium phosphate at neutral pH in detergent buffer. In general, it provides good purification of receptor proteins. Up to tenfold purification is common. A specific example of the use of hydroxyapatite chromatography is given in Chapter 2, Section 3. Small columns of hydroxyapatite or DEAE Sephacel can be used to harvest receptors from dilute solutions eluted from affinity columns. Again, this tactic is described in Chapter 2. Hydroxyapatite columns for HPLC are also available, e.g. BioRad HTP.

An interesting, and unusual (in the receptor field) chromatographic procedure is the use of heparin sepharose in the purification of the α_2 receptor (Chapter 6, Section 4). The mechanism of interaction of the receptor with the gel is not known for sure, but it is possibly of more general applicability.

Hydrophobic interaction chromatography has also been used with some success in the purification of receptors. In this technique, the proteins are applied to a column of, for example, phenyl Sepharose in a high ionic-strength medium (e.g. 1–2 M ammonium sulphate) which favours hydrophobic interactions. Elution is achieved by reducing the salt concentration, or by increasing the detergent concentration. An example is given in Chapter 4, Section 3.3.

Of the techniques mentioned above, ion-exchange, hydroxyapatite, and hydrophobic interaction chromatography lend themselves well to scaling-up. Lectin affinity chromatography can also be done on quite a large scale, though the gels are expensive. Any of these steps may be used in advance of affinity chromatography, to provide some initial purification, and to reduce volumes to more manageable dimensions. However, chromatography on expensive gels such as heparin agarose (Sigma), or intrinsically low-capacity steps such as HPLC gel-permeation chromatography are best left to the terminal, post-affinity chromatography steps of the purification procedure. A classical example of how different protein purification methods may be combined to give very substantial overall purification of a receptor even without affinity chromatography is provided by the purification of muscarinic receptors from porcine atria (36). Obviously, the disadvantage of the combination of sequential steps is overall low yield.

3.7 SDS-PAGE and isoelectric focusing

SDS-PAGE can be used as a purification step as well as to provide a criterion of purity. Although it is a very efficient purification method, the receptors will be denatured. The receptor content of the preparation will therefore have to be identified by the use of antibodies. Alternatively, the receptors may, in ideal cases, have been prelabelled with a specific active-site-directed affinity label (see Chapters 2, 4, and 5 for examples). Clearly, receptors purified by SDS-PAGE cannot be used for functional studies. A further restriction is that SDS-PAGE is a low-capacity technique, suitable for sub-milligram quantities of protein. Thus, SDS-PAGE is most often used as the final step in purification, in order to obtain a homogeneous preparation of receptors for digestion and sequencing. The techniques of SDS-PAGE and electro-elution are described in Chapter 10, Sections 3 and 6.

A further technique that may be considered is isoelectric focusing. Unlike SDS-PAGE, IEF is carried out in non-ionic or zwitterionic detergents containing ampholytes, and is therefore compatible with the recovery of native receptor protein. An example of the analytical use of IEF in a polyacrylamide gel support in the assessment of purity is given in Chapter 8, Section 3. IEF can, however, also be used preparatively, and a new commercial device, the Rotofor, is now marketed (Bio-Rad) which allows IEF to be carried out in free solution, without the use of a gel bed or sucrose gradient to stabilize the pH gradient, and may be very useful in this regard. IEF is, potentially, a high-resolution technique, capable of handling milligram quantities of protein. It has been rather little exploited in the purification of receptors. A drawback is the relatively high cost of the ampholytes. A little-exploited variant is chromatofocusing, in which the pH gradient is formed on a column. Pharmacia provide information on this technique.

3.8 Concentration of dilute solutions of receptors

Concentration of the receptor solution is often required between purification steps, or after the final purification step. Concentration using membrane ultrafilters under nitrogen pressure is often used. Suitable devices include stirred cells, marketed by Millipore and Amicon. These are used in conjunction with ultrafiltration membranes which vary both in their molecular-weight cut-off, and in their tendency to adsorb hydrophobic molecules, such as receptors. Examples of the use of Amicon YM-30 ultrafiltration membranes to concentrate receptor solutions are given in Chapters 5 and 6. We have found the use of Millipore immersible ultrafilters convenient, and their use is described below. Concentration on the larger scale is also possible, using cassette systems, in which the ultrafiltration membranes are stacked up, thus increasing the surface area. Such devices may be more useful than the stirred cells for concentration of large volumes (e.g. 200−500 ml), entailing less manipulation, avoiding stirring and denaturation of the sample. Small-scale concentration of samples, e.g. from 2 ml to 50 μl, can be accomplished conveniently using devices such as the Amicon Centricon, or Millipore Ultrafree centrifugal concentrators. Here, the solution to be concentrated is filtered by centrifugal force (*c.* 5000 *g*) in a bench centrifuge. In general, a well-equipped laboratory will need a range of these devices, including stirred concentrators of various sizes, immersible ultrafilters, and centrifugal concentrators.

Lyophilization following dialysis may also be used if it does not denature the receptors. However, detergents will also be concentrated, because the CMCs of most detergents are too low to permit effective dialysis (*Tables 4−6*). The presence of concentrated detergent may affect receptor function, or interfere with further purification, partial hydrolysis, and separation of peptides.

A small column of hydroxyapatite or DEAE Sephacel can be used to recover receptors from a large volume of medium, without concomitant concentration of detergent. An advantage of hydroxyapatite, compared with DEAE Sephadex, is that it can be used in the presence of the monovalent salts which are usually present in the media. The disadvantage of hydroxypatite gel is that the receptor may have to be eluted with a high concentration of phosphate buffer (e.g. 0.5 M in the case of the mAChR, see Chapter 2).

A typical problem is to recover receptors from, say, 150 ml of a solution containing 1−2 pmol/ml binding sites, concentrating them 100−200-fold for further procedures, such as affinity labelling, protein modification, SDS-PAGE, digestion, high-resolution gel-filtration, etc. To concentrate a very dilute solution of hydrophobic protein without incurring serious losses of material is not, in practice, a trivial matter.

In our experience, amongst the most useful tools for this purpose are the Millipore CX series ultrafiltration bulbs. These are immersed in the solution to be concentrated, and evacuated using a vacuum line or pump. Low molecular weight solutes arc drawn away through the ultrafiltration membrane, while high molecular weight solutes remain in the concentrate. The bulbs are attached to an agitator (Millipore) which endows them with a gentle vibrating motion, avoiding the formation of a polarized layer, without causing frothing and denaturation. Bulbs with different molecular weight cut-offs are available, e.g. 10 kDa (CX10), 30 kDa (CX30). For concentrating a molecule with a molecular weight of 70 000, e.g. the mAChR, we have found the CX10 to give the best overall recovery of activity. The use of CX30 ultrafilters, although 2−3 times quicker, gives considerable (*c.* 30%) losses.

(a) All components of the apparatus which come into contact with the receptor solution must be completely clean. This is best ensured by using fresh, pre-sterilized disposable polypropylene or polycarbonate tubes to contain the receptor solution. We find the use of ELKAY tubes convenient. The ultrafiltration bulbs themselves are disposable, and can be handled in a no-touch manner. The plastic connecting tubes, and silicone rubber tubing seals should be cleaned by immersing them in detergent buffer contained in a measuring cylinder and sonicating for several minutes in a cleaning bath before rinsing them. Wear plastic gloves to connect up components of the apparatus. Follow the manufacturers instructions in handling the bulbs and connecting them to the agitator (see Figure 2 for a diagrammatic illustration).

(b) Before use, immerse the bulbs in a large volume of clean distilled water (e.g. HPLC-grade water), start the agitator, evacuate the bulbs, and draw through water for about 15 min to remove preservatives from the ultrafiltration membrane.

(c) Place the receptor solution in *c.* 50 ml aliquots in clean disposable plastic tubes cooled in ice. Immerse two CX10 bulbs in each tube so that they are just covered. Evacuate via a large reservoir and trap using a vacuum pump, and turn on the

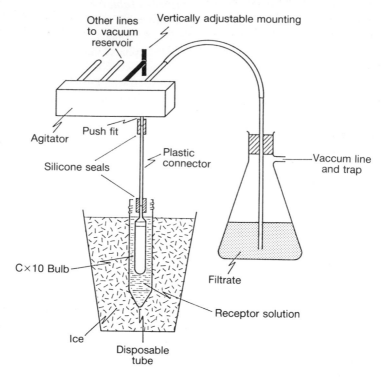

Figure 2. Concentration of dilute solutions of receptor proteins using immersible ultrafilters.

agitator. The stability of receptor solution is ensured by concentrating in the presence of a low molecular weight ligand to protect the binding site. This is natural after affinity chromatography, if a ligand has been used for specific elution of the receptor. The ligand will pass through the ultrafiltration membrane, so that its concentration will not rise, provided that it is not retained in detergent micelles. Non-micellar detergent will also pass through the membrane. Check the liquid level periodically, and adjust the level of the bulbs to keep them covered as far as possible.

(d) As concentration proceeds, it is possible to pool the contents of the tubes, ending up with all of the receptor solution in one tube. Before terminating the use of a particular set of bulbs, release the vacuum (NB do this before turning off the pump, or the pump oil will suck back), and agitate the bulbs in the concentrated solution for a few minutes to wash off any adherent receptor. To facilitate this, it is convenient to have each pair of bulbs on a separately controllable lead to the vacuum reservoir.

(e) By transferring the whole of the solution to a 20 ml polypropylene test tube, it is possible to concentrate the receptor solution to *c*. 5 ml before losses begin to mount. Typical recoveries are *c*. 90% for a ten- to twenty-fold concentration.

(f) Further concentration can be obtained by the use of 2 ml centrifugal ultrafilters (e.g. Centricon 10 or 30, Amicon). Again, these should be washed with clean distilled or deionized water before use, following the manufacturer's instructions. These devices are well designed, and give near-quantitative recoveries of receptor.

4. Affinity chromatography

4.1 Synthesis of ligands with functional groups

The ligand used for affinity chromatography should have at least two properties:

(a) a high affinity for the relevant receptor;

(b) a functional group for coupling to gel matrices.

A large number of drugs with high affinities are known for 'classical' receptors such as mAChRs and βARs, but most of these ligands lack functional groups such as $-SH$, $-NH_2$, COOH, or double bonds. Consequently, in these cases, it is usually necessary to design and synthesize compounds which include appropriate functional groups. In contrast, the synthesis of an affinity gel for the purification of peptide receptors, such as those in group 1 (*Table 1*) is, in theory, more straightforward. In many cases, the peptide itself can be immobilized to generate a useful gel. Examples are shown in *Tables 9* and *10*. If they are available, monoclonal antibodies can also be immobilized to generate affinity columns to purify receptors, and these may well become the reagent of choice in such cases.

In designing a ligand molecule suitable for immobilization to form an affinity gel, it is critical to determine the optimum location for the functional group. The introduction of the functional group, and the attachment of a linker to it will more often than not affect the binding of the ligand to the receptor. In the course of developing an affinity gel for mAChRs, we synthesized seven analogues of a model compound (BP, 2-benzhydryloxy-*N*-ethyl piperazine), each of which had an amino group at a different position. The affinities of these compounds for mAChRs decreased by five- to twentyfold, compared to the parent compounds, as a result of the introduction of the amino group. The effect of differences in position was relatively small when the modification was confined to the introduction of the amino group, but became much greater when a heptanoic acid moiety was coupled to the amino function. As shown in *Table 8*, the effect was much greater when the amino group was inserted near to the piperazine group than when it was placed near to the phenyl ring: the ratio of the apparent dissociation constants for ligands with and without the linker was 120-fold and ninefold, respectively. This suggested that the affinity ligand should be designed with the functional group on the phenyl ring rather than on the piperazine group.

Another important consideration in designing ligands is their absolute affinity for the receptor. The ligand should have a high-enough affinity for the receptor to allow it to be well retained on the gel, but not such a high affinity as to cause difficulty

Table 8. Affinities for sulubilized mAchRs of several muscarinic ligands with an amino group.

Chemical Structure		Kd (nM) $R = H$	Kd (nM) $R - CO(CH_2)_6CH_3$
CHOCH₂CH₂N	(BP)	84	
CHOCH₂CH₂N NR		420	50 000
NHR CHOCH₂CH₂N	(ABP)	280	2 500
NH₂ CHO N-CH₃	(ABT)	7	
NH₂ CHO N	(ABQ)	35	
H₂N OH C-CH₂CH₂N	(APCPP)	4	

when trying to elute the bound receptors from the gel. The latter problem is often encountered when polypeptides are immobilized to form affinity gels. However, when low molecular weight ligands are used, it is rare to find that the affinity is too high. In the case of mAChRs, three classes of compounds with an aminophenyl group, and a different base functionality, namely ABT, ABQ, and ABP were synthesized (see *Table 8*). Their apparent dissociation constants for solubilized receptors were estimated to be 7, 35, and 300 nM, respectively. In experiments using these compounds, the best affinity gel was obtained using the compound with the highest affinity for the receptors.

Table 9 summarizes the affinity chromatography used for the purification of the glyR, GABA$_A$R, βAR, α_2R, and mAChR. The apparent dissociation constants of the free ligands for the solubilized receptors ranged from 2.4 to 48 nM. In all cases, the receptors were specifically eluted using their ligands, indicating that the affinities of the gels for the receptors were not so high as to render them unusable. In fact, the affinity of the ligand usually decreases on linking to a spacer arm, and then decreases further on immobilization on the gel matrix. The extent of the decrease may differ from one ligand to another. Despite this, the above values of the dissociation constants of the free ligands are distributed in a relatively narrow range, and can be used as a practical guide in the evaluation of affinity ligands [see also *Affinity chromatography, a practical approach (63)*].

Although many successful applications of affinity chromatography have been reported, many times this number of unsuccessful attempts must have been made. As far as the design of affinity ligands is concerned, the factors considered above, namely the affinity of the ligand for the receptor and the positioning of the functional

Table 9. Affinity chromatography with low molecular weight ligands used for purification of receptors.

Receptors	GlyR (84)	GABA$_A$R (85)	βAR (86)	α_2R (87)	mAChR (88)
Ligands bound to gel	Strychnine	Benzodiazepine	Alprenolol	3-benzazepine	ABT
Density (μmol/ml gel) : K$_d^a$ (nM)	— : 10.4	0.4 : 40	1 – 3 : 2.4	1.6 : 48	3 : 7
Ratio of density^b to K$_d$ ($\times$ 10^3)	—	10	400 – 1300	30	400
Gel	Affi-Gel 10	Adipic dihydrazide agarose (Chapter 8)	Sepharose 4B coupled with butanediol diglycidyl ether (Chapter 5)	Sepharose 4B coupled with butanediol diglycidyl ether (Chapter 6)	Epoxy – Sepharose (Chapter 2)
Receptors applied to gels					
pmol/ml gel : pmol/μmol ligand	70 – 270 : —	25 : 50 – 83	7 – 8 : 2 – 8	14 : 9	15 : 5
Ligands used for elution	glycine	chlorazepate	isoproterenol	phentolamine	BP
Concentration (mM) : K$_d^a$ (nM)	200 : 6100	10 : ?	1 : 16	0.05 : 5	1 : 84
Ratio of conc. to K$_d$ ($\times$ 10^3)	30		60	10	10
Purification fold	1130	>300	106	200	920
Recovery of receptors (%)	19	14 – 20	34	40 – 50	26

^{a} K$_d$ for the binding of solubilized receptors and free ligands.
^{b} A density of 1 μmol/ml gel is taken to be equal to 1 mM.

Table 10. Spacer arms used in affinity gels for receptor purification[a].

Group 1 receptors

EGFR $M-O-CH_2-CONH-(CH_2)_2-NHCO-(CH_2)_2-CONH-EFG$
Affi-Gel-10-EGF (Chapter 9)

IGF_2R $M-O-CH_2-CONH-(CH_2)_3-N-(CH_2)_3-NHCO-(CH_2)_2-CONH-IGF2$
                                                $|$
                                            CH_3
Affi-Gel-15-IGF_2 (96)

Insulin R $M-NH-(CH_2)_3-NH-(CH_2)_3-NHCO-CH_2-CH_2-CONH-INSULIN$ (97)

Group 2 receptors

$GABA_AR$ $M-NH-NHCO-(CH_2)_4-CONH-NHCO-CH_2-RO7-1986/1$
Adipic acid dihydrazide agarose-benzodiazepine (Chapter 8)

Glycine R $M-O-CH_2-CONH-(CH_2)_2-NHCO-(CH_2)_2-CONH-STRYCHNINE$
Affi-Gel-10-aminostrychnine (84)

nAChR   CNBr-activated Sepharose + Naja Toxin (98)

Group 3 receptors

βAR $M-O-CH_2-CHOH-CH_2-O-(CH_2)_4-O-CH_2-CHOH-CH_2-S-ALPRENOLOL$
Thiol-substituted epoxy-Sepharose-alprenolol (Chapter 5)

α_2R $M-O-CH_2-CHOH-CH_2-O-(CH_2)_4-O-CH_2-CHOH-CH-NH-(CH_2)_2-NH-CO-$
YOHIMBINIC ACID
Ethylene-diamine-substituted epoxy-Sepharose-yohimbinic acid (Chapter 6)

mAChR $M-O-CH_2-CHOH-CH_2-O-(CH_2)_4-O-CH_2-CHOH-CH_2NH(A)BT$
Epoxy-Sepharose-ABT (Chapter 2)

D_2R $M-NH-(CH_2)_6-NHCO-(CH_2)_2-COO-HALOPERIDOL$
Aminohexyl Sepharose-succinylhaloperidol (Chapter 3)

[a] M indicates the gel matrix.

group used in immobilization, should always be considered in the first place. However, these two factors may not always be sufficient, as shown in the following example. A compound named APCPP (*Table 8*) was synthesized with an amino group in the appropriate position. This compound had the highest affinity for mAChRs of all the compounds which we synthesized, and had, consequently, been considered to be the most promising ligand for synthesis of an affinity gel. Despite this, it was not possible to prepare a useful affinity gel from it. Since we now know that ABT can be used as an affinity ligand, it is possible to guess that there is a dead-end pocket in muscarinic receptor molecules which fits the phenyl group, but not the cyclohexyl group. However, the problem is that this is impossible to know in advance. This indicates that attempts to screen a number of different compounds with a range of chemical structures is almost certain to be necessary.

4.2 Preparation of an affinity gel

Whether the affinity ligand is a small molecule or a peptide, the introduction of a spacer arm between the ligand and the gel matrix is almost always necessary to avoid steric hindrance between the matrix and the receptor. A number of different kinds of spacer arm are available. Methods for their use are provided in literature available from the manufacturers, including Pierce Chemical Co., BioRad, and Pharmacia/LKB. Pharmacia provide a useful introductory booklet. The appropriate spacer arm may differ from receptor to receptor. It is difficult to generalize in this area. A number of examples of successful affinity gels, many of them drawn from different parts of this book, are shown in *Table 10*.

In the case of the group 1 receptors, all known examples of which are peptide receptors, a successful approach is often to couple the peptide to an immobilized *N*-hydroxysuccinimide ester, such as Affi-Gel-10 or Affi-Gel-15 (BioRad). The Affi-Gels couple to amino functionalities in the peptide, under slightly basic conditions. Affi-Gel-10 is suitable for coupling neutral or basic peptides with isoelectric points in the range 6.5 − 11 whereas Affi-Gel-15 is suitable for peptides with isoelectric points less than 6.5. The preparation of an EGF-affinity gel exemplifies the procedure, and is described in Chapter 9.

In the case of group 2 receptors, Affi-Gel-10 has also been used to immobilize the channel blocker, strychnine, to generate an affinity gel for the purification of glycine receptors (28) while GABA$_A$ receptors can be purified by affinity chromatography on a gel created by the immobilization of a benzodiazepine ligand on an adipic acid dihydrazide resin (Chapter 8). The benzodiazepine binds to an allosteric site on the GABA$_A$ receptor rather than to the transmitter binding site. Nicotinic ACh receptors have been purified from low-abundance tissues such as skeletal muscle and brain using immobilized neurotoxins, such as cobra toxin, which block the action of acetylcholine at the neuromuscular junction (cf. Chapter 7). It should be noted that in few of the above cases did purification of a group 2 receptor employ a straightforward 'classical' competitive antagonist. This reflects the relatively low affinity of the classical antagonists compared to ligands such as the channel blockers, which may have more scope for multiple interactions with the receptor oligomer.

In contrast, group 3 receptors have been purified by the immobilization of 'classical' ligands, either a competitive antagonist (βAR, α_2R, m1 and m2 mAChRs) or a peptide agonist (opioid R). The most commonly used spacer arm is based on epoxy-activated Sepharose. Epoxy-activated Sepharose 6B can be purchased from Pharmacia. However, the use of activated Sepharose 4B has been preferred by most workers. This is readily prepared from Sepharose 4B by treatment with butane diol bis-glycidyl ether (see protocols in Chapters 2, 5, and 6). Several different kinds of chemistry have been devised to exploit this linkage.

(a) A diamine spacer can be introduced, and used to derivatize a carboxylate group in the ligand (cf. purification of the α_2R, Chapter 6).

(b) The epoxide group can be derivatized directly by nucleophilic attack by an amino-containing ligand under basic conditions (cf. purification of the mAChR, Chapter 2).

(c) The epoxide ring can be opened by nucleophilic attack using thiosulphate anion, and the thiosulphate anion replaced with a sulphydryl group by reduction with dithiothreitol. The sulphydryl group can then be induced to add to a double bond in the ligand molecule by a persulphate-catalysed free-radical mechanism (e.g. to an allyl group, cf. the purification of the β_2AR, Chapters 5, 6).

It is notable that in each of the above cases, the ligand is thought to bind within the central cavity of the seven transmembrane helix receptor structures, and is somewhat analogous to the binding of channel blockers to the group 2 receptors.

In all of the above cases, reference to *Table 10* shows that the spacer arm contains at least 12 bonds, and is, in nearly every case, hydrophilic. This is presumably dictated by the necessity for the ligand to approach the binding site via a hydrophilic pore or cavity.

In each case, the linkages used in the spacer and in the coupling of the spacer to the ligand were stable, e.g. ether, amide, $C-N$ or $C-S$ linkages, which are not readily hydrolysed. This is vital to avoid leaching of the ligand from the gel, and is one reason why the epoxy-activated Sepharose chemistry is successful.

Again these generalities may be illustrated with respect to the muscarinic receptors. In this case, two different spacer arms of different lengths $[-(CH_2)_6-$ and $(CH_2)_{12}]$ did not show any appreciable difference with respect to the amount of receptor bound to the gel. In contrast, there was a clear difference between the hydrophobic and hydrophilic spacer arms, in that when the hydrophobic spacer was used, the amount of receptor bound was greater, but the non-specific binding of bulk protein was also increased, resulting in a lower purification when compared to a gel in which the hydrophilic spacer arm was used. This indicates that bulk contaminants, as well as receptors, bind non-specifically to the hydrophobic arms. This is particularly important in receptor purification, in the sense that the bulk contaminants are also likely to be hydrophobic membrane proteins. Again, this underlines the desirability of using hydrophilic spacer arms (*Tables 9* and *10*).

Another important factor is electrostatic interaction. It is necessary to reduce the electric charge of the spacer arm as far as possible, to minimize non-specific ion-exchange interactions with bulk contaminants. This precaution is often overlooked, because electrostatic interactions can be reduced by using a high ionic strength medium for washing the gel. However, it becomes important when the specific interaction of the receptor with the gel also involves an electrostatic interaction, which is true in the majority of cases. The use of high ionic strength media will then tend to reduce both the specific and the non-specific interactions. In the case of the mAChRs, we initially blocked the unreacted epoxide groups with ethanolamine. However, this procedure created an amino group that is positively charged at neutral pH. In later experiments we blocked the unreacted epoxide groups with a prolonged incubation in alkaline solution and then ethanol, so avoiding this potential problem. The difference in the properties of the two gels was striking. The purification attainable with the ethanolamine-blocked gel was $20-200$-fold, but for the ethanol-blocked gel, the comparable figure was 1000-fold. It is probable, but not certain, that the difference in purification efficiency reflects the absence of the excess positive charges.

The density of ligand substitution of the matrix needs to be determined. A compromise needs to be struck between a high ligand density, which may promote non-specific binding, and a low density, which may give inadequate retention of the receptor. In practice it is not always easy to determine the optimal level of substitution. In the cases shown in *Table 9*, the amount of ligand bound to the gel matrix ranged between 0.4 and 3 μmol/ml gel. It is clear that only a tiny fraction of the immobilized ligand molecules is effective in binding receptors, indicating the operation of adverse steric factors. The reason for this is far from understood.

4.3 Elution of receptors from affinity gels

When receptors are eluted from affinity gels by the use of specific ligands, important parameters to consider are the ratio $C(A)/Kd(A)$ of the ligand concentration $C(A)$ to the $K_d(A)$ for the eluting ligand (A) and the ratio $C(B)/Kd(B)$ for the ligand (B) bound to the gel. The $K_d(B)$ value is difficult to determine for the bound ligand and usually only the value for free ligand $K_d(B')$ is known. In the cases listed in *Table 9*, the $C(A)/K_d(A)$ values were $3-40$ times smaller than the $C(B)/K_d(B')$ values, in spite of the expectation that the elution of the receptors will be retarded if $C(A)/K_d(A)$ is smaller than $C(B)/Kd(B)$. This indicates that the affinities of the ligands for receptors decrease substantially with their binding to the gel, i.e. the $K_d(B)$ values are greater than the $K_d(B')$ values. In addition, a substantial fraction of the immobilized ligand may not be available for receptor binding, i.e. the effective $C(B)$ is smaller than the chemically estimated value. Consistent with this, the capacities of these affinity gels are fairly low. Only $7-83$ pmol of receptors were taken up per μmol of bound ligand. The upshot is that the species and concentration of eluting ligand, as well as the capacity of the gel for taking up receptor, have to be determined experimentally. It is rarely wise to use a ligand of very high affinity to elute an affinity gel. Not only will the kinetics of ligand binding be slow, subsequent removal of the eluting ligand from the receptor binding site will be difficult. Eluting ligand K_ds in the range 10 nM to 1 mM are usable. The possibility of using a ligand which interacts allosterically with the receptor's ligand-binding site should not be overlooked.

For the receptors shown in *Table 9*, the uptake of the binding activity onto the gel was usually in excess of 75%. However, recovery of activity from the gel ranged from 14 to 50%. The reason for the low recovery is not known. One possible explanation is that a fraction of the receptors was bound too tightly to be eluted. If true, this would imply that gel-immobilized ligand molecules display a wide spectrum of affinities, ranging from values which are too high to permit elution, to values which are too low to give good uptake. Whatever the mechanism, the low capacities and low recovery values of affinity gels indicate that there is a real possibility of improving their performance in the future.

As far as the composition of the medium is concerned, the ionic strength is important in a case such as that of the mAChR, where ligand binding is affected by salt concentration owing to the ionic interaction between the positively charged headgroup and the negatively charged receptor-binding site. A medium of high ionic strength cannot be used in this case, in spite of the advantages that it offers in suppressing non-specific ionic interactions between the gel and contaminant proteins. In such cases, it is necessary to choose the salt concentration to give a compromise between the specific uptake of receptors and the non-specific binding of contaminants.

4.4 Testing new affinity gels

When a new affinity gel is to be developed, many parameters require to be tested. These include the nature of the immobilized ligand, the type of gel matrix, the type of spacer arm, the concentration of immobilized ligand, and the composition of the incubation medium. This being so, it is necessary to screen a large number of different gel preparations, and for this, a simple procedure is needed. Here we give the protocol that was used in the development of an affinity gel for purification of the mAChR.

Protocol 2. Binding of receptors to an affinity gel

1. Take 0, 50, or 100 μl of a 50% v/v gel suspension in buffer, and adjust the volume to 600 μl in microfuge tubes by the addition of appropriate detergent-containing media.

2. Add solubilized receptors (0.1 − 0.5 pmol), with the binding sites either free, or preblocked with a radiolabelled ligand. The ligand should have a high affinity, and a slow off-rate, to ensure that the receptor site remains blocked.

3. Shake the gel suspension with the receptor solution for 4 h at 4°C (or 60 min at 30°C if the receptor stability permits this). This is best done by gentle end-over-end mixing, or rotation.

4. Spin down the gel using a microcentrigure (1 min, 15 000 g). Assay the binding activity and protein concentration in the supernatant. See Section 5.2 for comments on protein assays.

5. It is useful to perform a control incubation using an equivalent volume of an unsubstituted gel of the appropriate type, e.g. Sepharose 4B. This provides a control for partitioning of the receptor into the gel, and other non-specific losses. It is also possible to measure the A_{280} of the supernatant that has been exposed to the affinity gel against the supernatant that has been exposed to the non-substituted gel. This gives a more sensitive measure of the uptake of protein onto the gel.

6. It is important to control for leaching of the ligand from the affinity gel. This can be done by incubating the gel as described, but in buffer alone, then spinning down the gel and adding the supernatant to a fresh aliquot of receptor solution, and assaying binding after an appropriate incubation period. It is advisable to perform this assay with two concentrations of radioligand, one below the K_d, and one at 10 times the K_d, and then to make a rough estimate of K_d and Bmax. Competition, due to the presence of leached ligand, will reduce K_d but not Bmax, while genuine uptake, measured at *step 4* will reduce Bmax but

not Kd. The measurement of ligand-binding curves is described in reference 1, with a summary in the appendix.

Ideally, the results should be 100% recoveries of prebound receptor and protein, and 0% recovery (i.e. 100% uptake) of the free receptor in the supernatant. Gels which show far from ideal behaviour in this respect can still be useful in purification. However, it is essential that there should be a clear difference between the uptake of free and preliganded receptors if true affinity chromatography is to be demonstrated.

We have used the following protocol for testing the elution of receptors from gels selected via *Protocol 2*.

Protocol 3. Eluting receptors from an affinity gel

1. Mix solubilized receptors (e.g. 0.3−1.0 ml) and gel (e.g. 0.1 ml) in a small plastic disposable column (0.5 × 5 cm, e.g. Kontes Disposaflex), and shake for 4 h at 4°C (or 60 min at 30°C). Alternatively, carry out this incubation in a microfuge tube.

2. Let the solution pass through the gel, and collect the eluate; or remove the gel by centrifugation, and remove the supernatant.

3. Rapidly wash the gel with 10 volumes of eluting buffer, without the specific eluting ligand.

4. Add the eluting solution, supplemented with the eluting ligand to the gel (1 ml per 0.1 ml gel). Shake vigorously for 4−15 h at 4°C or 10−30 min at room temperature. Let the solution pass through the gel, or remove the supernatant from the gel, and collect the eluate.

5. Repeat *step 4* with different solutions, e.g. containing different concentrations of NaCl, or different kinds and concentration of eluting ligands.

6. If the eluting solution contains ligands, apply 0.2 ml of eluate to a 2 ml column of Sephadex G50 F (0.7 × 5 cm), pre-equilibrated with detergent buffer, wash in with two successive 0.2 ml aliquots of detergent buffer, taking care not to disturb the gel bed, and then elute the receptor with 0.4 ml of detergent buffer.

7. Determine both the protein concentration and the ligand-binding activity of the eluted fraction. See comments in *Section 5.2* for determination of low concentrations of protein.

A ligand with an intermediate affinity for the receptor is ideal for elution. Ligands with too high an affinity are difficult to remove after elution. Ligands with low affinity have to be used at high concentrations, and may increase the ionic strength of the solution. In the case of muscarinic receptors, no binding activity could be detected after elution with *N*-methylscopolamine (NMS, K_d = 1.2 nM), presumably because the receptor−NMS complex could not be readily dissociated. In contrast, a very high concentration of carbamoylcholine (K_d = 180 μM) was needed for elution from some of the gels. This tended to cause elution of the bulk proteins as well as the receptor. However, 100 mM carbamoylcholine has been used successfully to elute mAChRs from ABT-Sepharose.

In our experience, it is well worth while to take trouble to optimize the uptake of receptor with respect to bulk contaminating proteins onto the affinity gel. It is very important to avoid adsorption of the bulk contaminants in the first place. Once adsorbed, it is difficult to wash them off without losing the receptor itself. It may be desirable to sacrifice a proportion of the uptake in the interests of minimizing the uptake of contaminating proteins. When eluting an affinity column, on the preparative scale, it is advisable:

(a) to perform the wash at low temperature (4°C) to slow down the kinetics of receptor dissociation;

(b) to follow the process of elution of bulk proteins, by connecting a UV monitor in series with the column. You can then continue to wash until there is no further release of protein, measured by absorbance at 280 nM. This often takes more than 10 column volumes of detergent buffer. Only at this point should the specific eluting ligand be added.

4.5 Affinity chromatography with polypeptide ligands

Affinity gels made with polypeptide ligands, such as hormones, venoms, and antibodies, appear to be different from those made with low molecular weight ligands in several respects. In particular, polypeptides have a number of functional groups, e.g. amino, carboxyl, or thiol groups. Some of these are necessary for binding to the receptor, but others are not. Thus, it is reasonable to expect that coupling of the polypeptide to the gel via these non-essential groups should not affect receptor binding, because polypeptides are macromolecules, and the functional groups are distributed over their surfaces.

In consequence, it is to be expected that affinity gels retaining a high affinity for their cognate receptors will be more easily prepared from polypeptide ligands than from low molecular weight non-peptide ligands. In reflection of this, antibodies are routinely purified by affinity chromatography on gels with immobilized peptide antigens as ligands, and the reverse is clearly feasible if specific antibodies are available. The problem is how to prepare specific anti-receptor antibodies before purification of the receptor. Another problem is that the affinity of receptors for gel-immobilized polypeptide ligands or antibodies may be so high that the receptors cannot be eluted biospecifically with the free ligands (see Chapter 9 for an instance of this problem).

4.5.1 Preparation of monoclonal antibodies

It is possible, in principle, to obtain cell clones secreting antibodies against receptors from hybridomas of myeloma cells with the spleen cells of mice into which crude receptor preparations have been injected as antigens. An introduction to these techniques, as applied to membrane proteins, is given in reference 62, Chapter 3. On the simplest assumption, that the fraction of hybridomas expressing anti-receptor antibodies is proportional to the molar abundance of the receptors in the crude preparation used as the antigen, it is expected that we will have to screen more than 10 000 cells to obtain a clone, if cell homogenates, or crude membrane preparations are used as a starting point. Somewhat in contrast to this, monoclonal antibodies recognizing the insulin R (89a) or the EGFR (89b) were obtained after the screening of only 1000 hybridomas, after the injection of intact cells bearing these receptors.

In the case of EGFRs, 10^7 cells containing $2-3 \times 10^{13}$ receptors ($6-9$ μg) were twice injected intraperitoneally, followed by an intravenous booster injection of 2×10^6 cells. This suggests that the above simplifying assumption is not necessarily valid, and that the antigenicity of these type 1 receptors on intact cells is greater than that of the average protein. It remains to be seen whether such strong antigenicity is restricted to receptors of this type, or whether group 2 and 3 receptors behave similarly.

Monoclonal antibodies against insulin Rs and IGF$_1$Rs have been obtained using partially purified receptors as the immunogens (16). In this particular case, the monoclonal antibodies were used to discriminate between these two receptor subtypes, which had previously been co-purified by affinity chromatography on an insulin column. In general, partially purified receptor preparations can be used as antigens for raising monoclonal anti-receptor antibodies, although, as far as the group 3 receptors are concerned, this strategy has not been outstandingly successful.

Several attempts have been reported at raising antibodies against ligands, particularly against irreversible ligands. Such antibodies can, in principle, be used for the purification of receptor−ligand complexes. In practice, however, this approach has not been successful, largely because the ligand, once immobilized within the ligand-binding pocket, is not accessible to the antibody. However, an interesting example of a small molecule being used as an antigen is provided by the case of antiphosphotyrosine antibodies, which have, for instance, been used in the purification of the PDGFR (19; the antiphosphotyrosine antibody technique is further discussed in reference 2, Chapter 8).

4.6 Elution of receptors from affinity gels made from immobilized polypeptide ligands

In most instances, specific ligands cannot be used to elute receptors from affinity gels made from immobilized monoclonal antibodies, or even immobilized polypeptide ligands. Instead, extremes of pH, or denaturing agents may need to be employed. For instance, IGF$_1$Rs were eluted from monoclonal antibody affinity columns using solutions of pH 11 or 2.2 (16), whereas the elution of insulin Rs from an insulin affinity column required the use of solutions containing 4.5 M urea in addition to insulin. Since there is no general solution to this problem, different approaches will be needed in different cases, as in the case of the insulin R. However, it should be noted that there is an instance of insulin Rs being eluted from a monoclonal antibody affinity column by means of insulin. In this case, there was a directly competitive interaction between the antibody and the polypeptide ligand.

5. Molecular characterization of receptors

5.1 The ligand-binding activity of purified receptors

The absolute protein concentrations present in solutions of purified receptors are usually very low. Because of this, and because of their hydrophobic natures, receptors tend to be lost by adsorption to the walls of glass tubes, particularly in the absence of detergents. For example, the binding assay for crude mAChRs solubilized with digitonin can be performed using small columns of Sephadex G50 pre-washed with a detergent-free buffer solution (e.g. 20 mM phosphate, pH 7). In contrast, the binding activity of the purified receptors is hardly measurable using this protocol. However, the binding activity can be assayed by replacing glass columns and tubes with items made of polypropylene, and by adding 0.04% digitonin to the pre-washing medium.

The binding activity of purified receptors in solution, or after reconstitution into phospholipid vesicles, can be assayed by the usual techniques employed for solubilized or membrane-bound receptors, namely membrane filtration or centrifugation assays of various kinds for reconstituted membrane-bound receptors, and gel-filtration, precipitation, or adsorption assays for soluble receptors (see Appendix). Both the theoretical and the practical aspects are discussed in detail in reference 1. Appropriate assay protocols for different receptor types are contained in the remaining chapters of this book.

Other parameters that can be measured include molecular size by SDS-PAGE (Chapters 2, 8, 10), isoelectric point by isoelectric focusing (see Chapter 8) and estimation of the molecular size, composition, and relative contributions of receptor and detergent to the receptor−detergent complex (see Chapter 2 for a detailed protocol).

5.2 Protein estimation and the polypeptide composition of purified preparations

The protein concentration of purified receptor preparations is usually too low to be estimated accurately by means of standard methods such as the Lowry method (see Chapter 3, Section 2) or the Bradford dye-binding method (dye-binding kits are supplied by Pierce Chemical Co.). However, there are more sensitive methods, including:

(a) Fluorescence assays based on the use of fluorescamine (Fluram, Sigma) or orthophthaldehyde. A suitable protocol is given in Chapter 2, Section 4.

(b) A dye-binding method base on the use of amido black is useful and sensitive (see Chapter 6, Section 6). An excellent and exhaustive discussion of colorimetric protein assay methods is given by Peterson (90), who also reviews interference with different assay methods, for instance by detergents.

(c) It is always desirable to have some insight into the polypeptide composition of the purified preparation. To discover this, concentrate the material using a Centricon 10, radioiodinate it, and analyse it by SDS-PAGE followed by autoradiography. Appropriate protocols are given in Chapters 2, 8, and 10. This method is very sensitive, if long-winded, and can be made semi-quantitative by densitometric scanning of the autoradiographs developed for different periods of time, and comparison with standards. Protein may be recovered by chloroform/methanol precipitation before SDS-PAGE. An alternative to radioiodination is silver staining (cf. Chapters 3, 6, 10).

(d) An alternative method is to inject the concentrated eluate ($100-200$ μl) onto a high-resolution gel-filtration column (e.g. a Superose 12 FPLC gel-filtration column) equilibrated in detergent buffer (e.g. 50 mM Tris-Cl, 0.1% SDS, pH 7.5, eluted at 0.3 ml/min). Measurement of the elution profile at 214 nm is primarily due to the absorbance of peptide bonds, and is very sensitive. This method separates proteins from interfering contaminants, such as detergents. A molecular weight-resolved profile is produced, from whose area the protein content of the original sample can be estimated reasonably accurately by comparison with injections of standard polypeptides (e.g. $0.3-10$ μg of BSA, ribonuclease, and ovalbumin). Microgram amounts of protein are readily quantitated in this way. Recoveries are typically greater than 60%, and the method is non-destructive in the sense that the sample can be recovered for further manipulations.

(e) It is worth emphasizing that problems of cross-contamination and adsorption, particularly to glass surfaces, can be acute when handling tiny amounts of protein. Handling should be performed, as far as possible in clean disposable plasticware (e.g. ELKAY tubes, microfuge tubes). Tubes can be siliconized by filling them with 5% dichlorodimethylsilane in chloroform for a few minutes, draining, rinsing several times with deionized water, and then heating them at 60°C for 10 min. High-quality deionized or double-distilled water must be used to make up reagents and buffers. If necessary, HPLC-grade water should be purchased. It is worthwhile to maintain a separate stock of clean, acid-washed glassware (50% nitric acid, 24 h, rinse with deionized water) for making up and storing buffers, and to make up buffers fresh. Items can be cleansed of surface contamination by filling them up with clean buffer or deionized water containing detergents as appropriate, sonicating them in a cleaning bath and rinsing. Disposable plastic gloves should be worn to avoid cross-contamination with skin keratins.

(f) The most sensitive method of protein determination is complete acid hydrolysis followed by amino-acid analysis. A protocol is given below:

Protocol 4. Acid hydrolysis

1. Take a chromic acid-cleaned Pyrex test tube, 12 mm × 100 mm.

2. Place the sample in the bottom of the tube. At least 20 pmoles of sample needs to be analysed. A buffer blank should be included in a separate tube.

3. With an oxygen-gas flame, heat the neck of the tube *c.* 1.5 cm from the top, and draw it down into a capillary through which a Pasteur pipette will pass.

4. Dilute Aristar grade concentrated HCl (BDH chemicals) 1 : 1 with HPLC-grade water to give a 6 N solution.

5. Bubble oxygen-free nitrogen through the HCl for about 30 min.

6. Add a crystal of phenol to the sample in the hydrolysis tube, to prevent destruction of tyrosine.

7. Add 1.0 ml of deoxygenated 6 N HCl to the sample.

8. Displace the air in the test tube with N_2. Seal off the constriction in the tube by melting it.

9. Heat the tube in an oven at 110°C for 18 h.

10. Cool and open the tube.

11. Remove all the HCl by evaporating in a vacuum desiccator over NaOH.

12. Take up the residue in solvent for amino-acid analysis.

5.3 Determination of amino-acid and oligonucleotide sequences of receptors

One obvious use for purified receptors is the determination of partial amino-acid sequences. The simplest procedure is to elute receptors from the bands obtained on SDS-PAGE (see Chapter 10, Section 6) and then determine the sequence of the amino terminal. A good example is that of the insulin R (12). In this case, 60−300 pmol of purified receptors was used and amino-terminal sequences of about 20 residues were determined. The amino-terminal sequences of the nAChR subunits, α-subunit of the glyR, and group 1 receptors have also been determined by this means. However, no free amino groups were detected when the $GABA_AR$ or any of the group 3 receptors were subjected to amino-terminal sequencing, suggesting that their amino terminals are blocked.

When free amino terminals are not available, the purified receptors must be subjected to partial hydrolysis. The specific cleavage of peptide bonds with endoproteases, such as V8 protease or trypsin (TPCK-treated chymotrypsin-free preparation), or chemical reagents, such as cyanogen bromide (CNBr), is commonly attempted. Protocols are given in Chapter 10. When purified receptors are denatured and then exhaustively digested with trypsin all of the susceptible bonds are expected to be cleaved at least in the hydrophilic sequences, which results in the formation of a large number of short peptides. Short hydrophobic peptides are best separated from each other by reverse-phase HPLC. Longer peptides are obtained by CNBr treatment because the methionine content of receptors is usually low. The resulting peptides can be separated by SDS-PAGE when they are long enough (e.g. more than 50 amino-acid residues) (25), or by reverse-phase HPLC. The peptides thus isolated (e.g. 100 – 500 pmol) are subjected to amino-acid sequence analysis with a gas phase or solid-phase sequencer. Further details are given in Chapter 10.

At this stage, the raising of antibodies against peptides can be attempted. Peptides conjugated to limpet haemocyanin, trypsin inhibitor, BSA, or other proteins have been used as antigens (cf. Chapter 8) and the resultant antibodies used to confirm that the peptide used as the antigen is a part of the receptors by immunoprecipitation or Western blots (Chapter 10). In order to determine the full sequence of the receptor gene, oligodeoxynucleotide probes synthesized on the basis of peptide sequences are used to screen an appropriate cDNA library. Either a mixture of relatively short oligonucleotides with all possible sequences or a relatively long oligonucleotide with a sequence based on codon usage frequency is used. For example, a mixture of 32 different oligonucleotides with a length of 14 bases was used for the isolation of cDNA clones of the α-subunit of the nAChR (23), but a single oligonucleotide with a length of 51 bases in the case of EGFR (17). Methods of screening and sequencing of cDNA are common to all proteins and detailed descriptions are given in *DNA cloning, a practical aproach* (61).

Genomic libraries as well as cDNA libraries have also been screened to obtain genes for receptors and to deduce their amino-acid sequences. This was particularly useful for βRs, α_2Rs, and mAChRs owing to the fact that the genes for these receptors were found not to contain introns. A similar approach should be useful for genes for other group 3 receptors lacking introns. Amongst the group 3 receptors examined so far, introns have been found in the rhodopsins, the $5HT_{IC}$, D_2 dopamine, and substance K receptors. The assumption that the deduced sequence is that of the purified receptor may be tested by examining whether the deduced sequence contains the sequences of other peptides which are determined but not used as probes, or if the relevant receptor reacts with antibodies raised against partial peptides in the deduced sequence. These tests provide indirect evidence that the deduced sequence is that of the purified receptor. However, direct evidence can be obtained by expressing the cDNA or genes and examining their function (see Chapter 11).

The amino-acid sequence deduced from the nucleotide sequence is compared to the known sequences of other receptors by means of computer programs. Amongst the useful structural information derived from the sequence is the hydropathy profile. The presence of hydrophobic domains with a length of approximately 20 amino-acid residues suggests that the regions are transmembrane segments. It is mostly by this criterion that group 1, 2, and 3 receptors have been assumed to have one, at least four, and seven transmembrane segments, respectively, although categorical direct evidence is still lacking. The interpretation of receptor sequences is discussed in Chapter 12.

5.4 Expression of cDNA

Genomic or cDNA clones ligated into expression vectors are used to transform suitable cell lines. Alternatively, mRNA transcribed from these clones is injected into recipient cells such as *Xenopus* oocytes. The expression of receptors in the recipient cells is confirmed by measuring the ligand-binding activity of the receptors, the reactivity of cell homogenates or membrane preparations with antibodies, or the responses induced by agonists in recipient cells. These include changes in membrane potentials or metabolites such as cAMP. The measurement of receptor—response coupling is discussed in *Receptor-effector coupling, a practical approach* (2).

There are several conditions that need to be satisfied before expression experiments can be performed:

(a) the recipient cells should express little if any endogenous receptor;

(b) a mixture of mRNAs encoding all subunits has to be injected if the receptor is composed of hetero-oligomers;

(c) the components (e.g. G-proteins), with which receptors interact, have to exist in the recipient cells.

The conditions are sometimes difficult to satisfy. As to condition (c), the activities of group 3 receptors expressed in a given cell line might be limited by the species of G-proteins present in the recipient cells, although the responses of mAChRs (34) and substance K receptors (42) have been well expressed in *Xenopus* oocytes. In initial studies on the glyR, condition (b) was not satisfied because only the cDNA corresponding to one of at least two subunits was available. In this case, the following strategy was used. The glycine-induced changes in Cl^- ion permeability were observed in *Xenopus* oocytes into which a mixture of mRNAs derived from the medulla had been injected, and this expression of glyR was shown to be attenuated by the addition of antisense mRNA corresponding to part of the deduced sequence for one subunit. As the activities of many receptors are known to be expressed in oocytes, into which a mixture of mRNA derived from brain has been injected, this strategy may be applicable to other receptors.

Recently the ion-channel activity of the $GABA_A$ receptor was shown to be expressed in *Xenopus* oocytes which received only α-subunit-encoding mRNA but not β-subunit-encoding mRNA. This result indicates that caution is necessary in the interpretation of expression experiments and also that the negative experiment using antisense mRNA may be useful. The expression experiments generally do not tell us about the direct function of receptors, because we do not know whether the responses are due exclusively to transformation by the genomic or cDNA clones or injected mRNA, or are also in part due to the collaboration of endogenous components in recipient cells. Group 2 receptors correspond to the former category and group 3 receptors to the latter, but this has to be confirmed by further lines of evidence.

5.5 Reconstitution of receptors and identification of their functions

The direct function of receptors is elucidated by reconstituting purified receptors in phospholipid vesicles with or without other components. The reconstitution of purified nAChRs into phospholipid vesicles and subsequent measurements of ion fluxes induced by acetylcholine have provided evidence that the hetero-oligomer, $\alpha_2\beta\gamma\delta$, functions as an ion channel as well as bearing the acetylcholine-binding sites, and that no other protein components are necessary for these functions (see reference 62, Chapter 5). Further characterization, for instance, determination of the structure−activity relationships, the functional role of each subunit, and a detailed description of ion-channel properties, has been carried out using oocytes into which the nAChR mRNA or site-directed mutants had been injected. Amongst the other group 2 receptors, the $GABA_AR$ and glyR, are known to function as Cl channels but they are not as well characterized as the nAChR.

A large number of results show that the functions of a number of group 3 receptors are mediated through the activation of G-proteins. Evidence that receptors directly interact with G-proteins has been obtained by reconstituting purified βARs or mAChRs with purified G-proteins into phospholipid vesicles and observing their interactions. Relevant examples are described in reference 2. Group 3 receptors may also have other functions besides their interaction with G-proteins, and the interactions between many of these receptors or receptor subtypes and their G proteins may be regulated by as yet unidentified third components. The arrestin−rhodopsin interaction is a case in point. In addition, the effector systems with which each individual G-protein interacts have not been fully identified or well characterized in each case.

Many group 1 receptors are known to have tyrosine kinase activity. However, their endogenous substrates are not known. In other cases, even the immediate mechanism of receptor−response coupling is unidentified. Generally speaking, identification of the *function* of receptors requires identification of the *components* with which the receptors interact, exceptions being those such as the group 2 receptors which incorporate the effector mechanism within the structure. Ideally the final identification of the component can be performed by reconstitution of purified receptors with purified effector components.

5.6 New strategies for the molecular characterization of receptors

Traditionally, the determination of receptor sequence has followed the isolation of the protein. In some cases, however, difficult and laborious procedures of solubilization and purification have been avoided, owing to recent advances in recombinant DNA technologies. The strategy is based on the isolation of cDNA or genomic clones for receptors without using purified receptors. The structure of receptors and the cellular responses to them can be examined by using these clones, and it should be possible to produce substantial amounts of receptors from these clones in the near future.

One way to dispense with the purification procedure is to utilize monoclonal antibodies should these be available. Clones for receptors can be screened with monoclonal antibodies against receptors from cDNA libraries constructed with expression vectors. This strategy has been used to screen a cDNA library for transducin, a G-protein in the retina, although peptide sequences were also used for confirmation in this case. It should be noted that this approach often produces false positives.

Different receptors in the same group have been shown to have common sequences, and therefore it is possible to isolate cDNA or genomic clones for another receptor in the same group by using oligonucleotides corresponding to the common sequences. Two different genomic clones were isolated by this strategy using partial sequences of a subtype of the mAChRs and were shown to be genes for previously unidentified subtypes of mAChRs (46). In another instance, a gene that is now known to encode a receptor, was isolated by homology screening with a probe based on the $\beta_2 AR$ even though the ligand for the putative receptor had not been identified at the time (91). This line of approach has been taken more frequently as the number of receptors with known sequences increases, and the sequences of dopamine D_2 and variants of β_1 and β_2 receptors have been determined in this way (48-50).

Very recently, the structures of the substance K and the $5HT_{1C}$ receptors have been deduced from cDNA clones that had been isolated by using functional assays without knowing or assuming the partial amino-acid sequences (42,43). The activities of the receptors were monitored by measuring changes in membrane potentials induced in *Xenopus* oocytes, into which mRNAs transcribed from successively fractionated cDNA clones had been injected. It is essential in this strategy that the cDNA clone contains the full length that is necessary for expression of the receptor activity, and therefore the length of the mRNA may be a limiting factor. In the case of the substance K receptor, the cDNA is composed of 2.5 kbp and the receptor of 384 amino-acid residues. A different strategy has also been used for isolation of cDNA clones for serotonin receptors, again based on electrophysiological assays of mRNA-injected oocytes. The single-stranded DNA derived from a cDNA library was used to deplete the mRNA encoding the receptor, and then the attenuation of the expression of receptor activity was monitored (92). This strategy (negative selection) appears to be less restricted by the length of mRNA than the other strategy (positive selection), although the latter is more straightforward and simpler than the former. Both of these strategies are expected to be used for characterization of receptors, especially those composed of a single homogeneous protein. The application to receptors composed of hetero-oligomers is also possible theoretically, although the strategy would become more complicated.

Acknowledgements

We would like to thank Dr T. Oda for this helpful comment on a part of this chapter (Section 4).

References

1. Hulme, E. C. (ed.) (1990). *Receptor—Ligand Interactions: A Practical Approach*, IRL Press, Oxford, in press.

T. Haga, K. Haga, and E. C. Hulme

2. Hulme, E. C. (ed.) (1990). *Receptor—Effector Coupling: A Practical Approach*, IRL Press, Oxford, in press.
3. Peralta, E. G., Winslow, J. W., Peterson, G. L., Smith, D. H., Ashkenazi, A., Ramachandran, J., Schimerlik, M. I., and Capon, D. J. (1987). *Science*, **236**, 600.
4. Greenfield, C., Patel, G., Clark, S., Jones, N., and Waterfield, M. D. (1988). *EMBO J.*, **7**, 139.
5. Turner, A. J. and Bachelard, H. S. (eds) (1987). *Neurochemistry, A Practical Approach*, IRL Press, Oxford.
6. Blank, U., Ra, C., Miller, L., White, K., Metzger, H., and Kinet, J.-P. (1989). *Nature*, **337**, 187.
7. Leonard, W. J., Depper, J. M., Crabtree, G. R., Rudikoff, S., Pumphrey, J., Robb, R. J., Kronke, M., Svetlik, P. B., Peffer, N. J., Walmann, T. A., and Greene, W. C. (1984). *Nature*, **311**, 626.
8. Morgan, D. O., Edman, J. C., Standring, D. N., Fried, V. A., Smith, M. C., Roth, R. A., and Rutter, W. J. (1987). *Nature*, **329**, 310.
9. Radeke, M. J., Misko, T. P., Hsu, C., Herzenberg, L. A., and Shooter, E. M. (1987). *Nature*, **325**, 593.
10. Yamamoto, T., Davis, C. G., Brown, M. S., Schneider, W. J., Casey, M. L., Goldstein, J. L., and Russell, D. W. *Cell*, **39**, 27.
11. Ebina, Y., Ellis, L., Jarnagin, K., Edery, M., Graf, L., Clauser, E., Ou, J.-H., Masiarz, F., Kan, Y. W., Goldfine, I. D., Roth, R. A., and Rutter, W. J. (1985). *Cell*, **40**, 747.
12. Ullrich, A., Bell, J. R., Chen, E. Y., Herrera, R., Petruzelli, L. M., Dull, T. J., Gray, A., Coussens, L., Liao, Y.-C., Tsubokawa, M., Mason, A., Seeburg, P. H., Grunfeld, C., Rosen, O. M., and Ramachandran, J. (1985). *Nature*, **313**, 756.
13. Roth, R. A. and Cassell, D. J. (1983). *Science*, **219**, 299.
14. Fujita-Yamaguchi, Y., Choi, Y., Sakamoto, Y., and Itakura, K. (1983). *J. Biol. Chem.*, **258**, 5045.
15. Ullrich, A., Gray, A., Tam, A. W., Yang-Feng, T., Tsubokawa, M., Collins, C., Henzel, W., LeBon, T., Kathuria, S., Chen, E., Jacobs, S., Francke, U., Ramachandran, J., and Fujita-Yamaguchi, Y. (1986). *EMBO J.*, **5**, 2053.
16. LeBon, T. R., Jacobs, S., Cuatrecasas, P., Kathuria, S., and Fujita-Yamaguchi, Y. (1986). *J. Biol. Chem.*, **261**, 7685.
17. Ullrich, A., Coussens, L., Hayflick, J. S., Dull, T. J., Gray, A., Tam, A. W., Lee, J., Yarden, Y., Liberman, T. A., Schlessinger, J., Downward, J., Mayes, E. L. V., Whittle, N., Waterfield, M. D., and Seeburg, J. (1984). *Nature*, **309**, 418.
18. Yarden, Y., Escobedo, J. A., Kuang, W. J., Yang-Feng, T. L., Daniel, T. O., Tremble, P. M., Chen, E. Y., Ando, M. E., Harkins, R. N., Francke, U., Fried, V. A., Ullrich, A., and Williams, L. T. (1986). *Nature*, **323**, 226.
19. Daniel, T., Tremble, P. M., Frackelton, A. R., and Williams, L. T. (1985). *Proc. Natl Acad. Sci. USA*, **82**, 2684.
20. Takayanagi, R., Inagami, T., Snajdar, R. M., Imada, T., Tamura, M., and Misono, K. S. (1987). *J. Biol. Chem.*, **262**, 12104.
21. Chinkers, M., Garbers, D. L., Chang, M.-S., Lowe, D. G., Chin, H., Goeddel, D. V., and Schultz, S. (1989). *Nature*, **338**, 78.
22. Carpenter, G. (1987). *Ann. Rev. Biochem.*, **56**, 881.
23. Noda, M., Takahashi, H., Tanabe, T., Toyosato, M., Furutani, Y., Hirose, T., Asai, M., Inayama, S., Miyata, T., and Numa, S. (1982). *Nature*, **299**, 793.

24. Wu, W. C.-S. and Raftery, M. A. (1981). *Biochemistry*, **20**, 694.
25. Schofield, P. R., Darlison, M. G., Fujita, N., Burt, D. R., Stephenson, F. A., Rodriguez, H., Rhee, L. M., Ramachandran, J., Reale, V., Glencorse, T. A., Seeburg, P. H., and Barnard, E. A. (1987). *Nature*, **328**, 221.
26. Siegel, E. and Barnard, E. A. (1984). *J. Biol. Chem.*, **259**, 7219.
27. Grenningloh, G., Rienitz, A., Schmitt, B., Methfessel, C., Zensen, M., Beyreuther, K., Gundelfinger, E. D., and Betz, H. (1987) *Nature*, **328**, 215.
28. Graham, D., Pfeiffer, F., Simler, R., and Betz, H. (1985). *Biochemistry*, **24**, 990.
29. Popot, J.-L. and Changeux, J.-P. (1984). *Physiol. Rev.*, **64**, 1162.
30. Ovchinnikov, Y. A. (1982). *FEBS Lett.*, **148**, 179.
31. Dixon, R. A. F., Kobilka, B. K., Strader, D. J., Benovic, J. L., Dohlman, H. J., Frielle, T., Bolanowski, M. A., Bennett, C. D., Rands, E., Diehl, R. E., Mumford, R. A., Slater, E. E., Siegel, I. S., Caron, M. G., Lefkowitz, R. J., and Strader, C. D. (1986). *Nature*, **321**, 75.
32. Yarden, Y., Rodriguez, H., Wong, S. K.-F., Brand, D. R., May, D. C., Burnier, J., Harkins, R. N., Chen, E. Y., Ramachandran, J., Ullrich, A., and Ross, E. M. (1986). *Proc. Natl Acad. Sci. USA*, **83**, 6795.
33. Benovic, J. L., Shorr, R. G. L., Caron, M. G., and Lefkowitz, R. J. (1984). *Biochemistry*, **23**, 4510.
34. Kubo, T., Fukuda, K., Mikami, A., Maeda, A., Takahashi, H., Mishina, M., Haga, T., Haga, K., Ichiyama, A., Kangawa, K., Kojima, M., Matsuo, H., Hirose, T., and Numa, S. (1986). *Nature*, **323**, 411.
35. Peralta, E. G., Winslow, J. W., Peterson, G. L., Smith, D. H., Ashkenazi, A., Ramachandran, J., Schimerlik, M. I., and Capon, D. J. (1987). *Science*, **236**, 600.
36. Peterson, G. L., Herron, G. S., Yamaki, M., Fullerton, D. S., and Schimerlik, M. I. (1984). *Proc. Natl Acad. Sci. USA*, **81**, 4993.
37. Haga, K. and Haga, T. (1985). *J. Biol. Chem.*, **260**, 7927.
38. Kobilka, B. K., Matsui, H., Kobilka, T. S., Yang-Feng, T. L., Francke, U., Caron, M. G., Lefkowitz, R. J., and Regan, J. W. (1987). *Science*, **238**, 650.
39. Regan, J. W., Nakata, H., DeMarinis, R. M., Caron, M. G., and Lefkowitz, R. J. (1986). *J. Biol. Chem.*, **261**, 3894.
40. Cotecchia, S., Schwinn, D. A., Randall, R. R., Lefkowitz, R. J., Caron, M. G., and Kobilka, B. K. (1989). *Proc. Natl Acad. Sci. USA*, **85**, 7159.
41. Lomasney, J. W., Leeb-Lundberg, L. M., Cotecchia, S., Regan, J. W., DeBernardis, J. F., Caron, M. G., and Lefkowitz, R. J. (1986). *J. Biol. Chem.*, **261**, 7710.
42. Masu, Y., Nakayama, K., Tamaki, H., Harada, Y., Kuno, M., and Nakanishi, S. (1987). *Nature*, **329**, 836.
43. Julius, D., Dermott, A. B., Axel, R., and Jessell, T. M. (1988). *Science*, **241**, 558.
44. Jackson, T. R., Blair, A. C., Marshall, J., Goedert, M., and Hanley, M. R. (1988). *Nature*, **335**, 437.
45. Wada, K., Ballivet, M., Boulter, J., Connolly, J., Wada, E., Deneris, E. S., Swanson, L. W., Heneman, S., and Patrick, J. (1988). *Science*, **240**, 330.
46. Bonner, T. I., Buckley, N. J., Young, A. C., and Brann, M. R. (1987). *Science*, **237**, 527.
47. Frielle, T. M., Collins, S., Daniel, K. W., Caron, M. G., Lefkowitz, R. J., and Kobilka, B. K. (1987). *Proc. Natl Acad. Sci. USA*, **84**, 7920.

48. Regan, J. W., Kobilka, T. S., Yang-Feng, T. L., Caron, M. G., Lefkowitz, R. J., and Kobilka, B. K. (1988). *Proc. Natl Acad. Sci. USA*, **85**, 6301.

49. Bunzow, J. R., Van Tol, H. H. M., Grandy, D. K., Albert, P., Salom, J., Christie, M., Machida, C. A., Neve, K. A., and Civelli, O. (1988). *Nature*, **336**, 783.

50. Fargin, A., Raymond, J. R., Lohse, M. J., Kobilka, B. K., Caron, M. G., and Lefkowitz, R. J. (1988). *Nature*, **335**, 358.

51. Dohlman, H. G., Caron, M. G., and Lefkowitz, R. J. (1987). *Biochemistry*, **26**, 2657.

52. Gilman, A. G. (1987). *Ann. Rev. Biochem.*, **56**, 615.

53. Berridge, M. J. and Irvine, R. E. (1984). *Nature*, **312**, 315.

54. Yatani, A., Mattera, R., Codina, J., Graf, R., Okabe, K., Padrell, E., Iyengar, R., Brown, A. M., and Birnbaumer, L. (1988). *Nature*, **336**, 680.

55. Yatani, A., Codina, J., Imoto, Y., Reeves, J. P., Birnbaumer, L., and Brown, A. M. (1987). *Science*, **238**, 1288.

56. Kim, Donghee, Lewis, D. L., Graziadei, L., Neer, E. J., Bar-Sagi, D., and Clapham, D. E. (1989). *Nature*, **337**, 557.

57. Matsuoka, I., Sakuma, H., Syuto, B., Morishi, K., Kubo, S., and Kurihara, K. (1989). *J. Biol. Chem.*, **264**, 706.

58. Helenius A. and Simons, K. (1975). *Biochim. Biophys. Acta.*, **415**, 29.

59. Hjelmeland, L. M. and Chrambach, A. (1984). In *Receptor Biochemistry and Methodology*, (ed. J. C. Venter and L. C. Harrison), Vol. 1, p. 35. Alan R. Liss, New York.

60. Lowe, C. R. (1979). *An Introduction to Affinity Chromatography*, North-Holland, Amsterdam.

61. Glover, D. M. (ed.) (1985). *DNA Cloning, A Practical Approach*, IRL Press, Oxford, Vols 1–3.

62. Findlay, J. B. C. and Evans, W. H. (eds) (1987). *Biological Membranes: A Practical Approach*, IRL Press, Oxford.

63. Dean, P. D. G., Johnson, W. S., and Middle, F. A. (eds) (1985). *Affinity Chromatography: A Practical Approach*, IRL Press, Oxford.

64. Peterson, G. L., Rosenbaum, L. C., Broderick, D. J., and Schimerlik, M. I. (1986). *Biochemistry*, **25**, 3189.

65. Ashani, Y. and Catravas, G. N. (1980). *Anal. Biochem.*, **109**, 55.

66. Rivnay, B. and Metzger, H. (1982). *J. Biol. Chem.*, **257**, 12800.

67. Peterson, G. L., Rosenbaum, L. C., and Schimerlik, M. I. (1988). *Biochem. J.*, **255**, 553.

68. Anholt, R., Lindstrom, J., and Montal, M. (1981). *J. Biol. Chem.*, **356**, 4377.

69. Criado, M., Eibl, H., and Barrantes, F. J. (1982). *Biochemistry*, **21**, 3622.

70. Janski, A. M. and Cornell, N. W. (1980). *Biochem. J.*, **186**, 423.

71. Repke, H. (1987). *Biochim. Biophys. Acta.*, **929**, 47.

72. Haga, T., Berstein, G., Nishiyama, T., Uchiyama, H., Ohara, K., and Haga, K. (1988). In *Neuroreceptors and Signal Tansduction*. (ed. S. Kito, T. Segawa, K. Kuriyama, M. Tohyama and, R. W. Olsen), p. 239. Plenum Publishing, New York.

73. Kito, Y., Naito, T., and Nashima, K. (1982). In *Methods in Enzymology*, Academic Press, New York, Vol. 81, p. 167.

74. Hekman, M., Feder, D., Keenan, A. K., Gal, A., Klein, H. W., Pfeuffer, T., Levitzki, A., and Helmreich, E. J. M. (1984). *EMBO J.*, **3**, 3339.

75. Laduron, P. and Ilien, B. (1982). *Biochem. Pharmacol.*, **31**, 2145.

76. Poyner, D. R., Birdsall, N. J. M., Curtis, C., Eveleigh, P., Hulme, E. C., Pedder, E. K. and Wheatley, M. (1989). *Mol. Pharmacol.*, **36**, 420.

77. Berrie, C. P., Birdsall, N. J. M., Haga, K., Haga, T., and Hulme, E. C. (1984). *Br. J. Pharmac.*, **82**, 839.

78. Harmon, J. T., Nielsen, T. B., and Kempner, E. S. (19). In *Methods in Enzymology*. (ed. C. H. Hirs and S. W. Timasheff), Vol. 117, p. 65. Academic Press, New York.

79. Florio, V. A. and Sternweis, . (1985). *J. Biol. Chem.*, **260**, 3477.

80. Umezawa, K. and Aoyagi, T. (1984). In *Receptor Biochemistry and Methodology*. (ed. J. C. Venter and L. C. Harrison), Vol. 2, p. 139. Alan R. Liss, New York.

81. Senogles, S. E., Amlaiky, N., Falardeau, P., and Caron, M. G. (1988). *J. Biol. Chem.*, **34**, 18966.

82. Cervantes-Olivier, P., Durieu-Trautman, O., Delavier-Klutchko, C., and Strosberg, A. D. (1985). *Biochemistry*, **24**, 3765.

83. Berrie, C. P., Birdsall, N. J. M., Dadi, H. K., Hulme, E. C., Morris, R. J., and Stockton, J. M. (1985). *Biochem. Soc. Trans.*, **13**, 1101.

84. Pfeiffer, F., Graham, D., and Betz, H. (1982). *J. Biol. Chem.*, **257**, 9389.

85. Sigel, E., Mamalaki, C., and Bernard, E. A. (1982). *FEBS Lett.*, **147**, 45.

86. Caron, M. G., Srinivasan, Y., Pitha, J., Kociolek, K., and Lefkowitz, R. J. (1979). *J. Biol. Chem.*, **254**, 2923.

87. Regan, J. W., Barden, N. M., Lefkowitz, R. J., Caron, M. G., DeMarinis, R. M., Krog, A. J., Holden, K. G., Mathews, W. D., and Hieble, J. P. (1982). *Proc. Natl Acad. Sci. USA*, **79**, 7223.

88. Haga, K. and Haga, T. (1983). *J. Biol. Chem.*, **258**, 13575.

89a. Roth, R. A., Cassell, D. J., Wong, K. Y., Maddux, B. A., and Goldfinr, I. D. (1982). *Proc. Natl Acad. Sci. USA*, **79**, 7312.

89b. Waterfield, M. D., Mayes, E. L. V., Stroobant, P., Bennet, P. L. P., Young, S., Goodfellow, P. N., Banting, G. S., and Ozanne, B. (1982). *J. Cellular Biochem.*, **20**, 149.

90. Peterson, G. L. (1983). In *Methods in Enzymology*, (ed. C. H. Hirs and S. N. Timasheff), Vol. 91, p. 95. Academic Press, New York.

91. Kobilka, B. K., Frielle, T., Collins, S., Yang-Feng, T., Kobilka, T. S., Francke, U., Lefkowitz, R. J., and Caron, M. G. (1987). *Nature*, **329**, 75.

92. Lubbert, H., Hoffman, B., Snutch, T. P., Van Dyke, T., Levine, A. J., Hartig, P. R., Lester, H. A., and Davidson, N. (1987). *Proc. Natl Acad. Sci. USA*, **84**, 4332.

93. Hjemland, L. M. (1986). In *Methods in Enzymology*, (ed. P. M. Conn), Vol. 124, p. 135. Academic Press, New York.

94. Davis, A. (1984). In *Receptor Biochemistry and Methodology*, (ed. J. C. Venter and L. C. Harrison), Vol. 3, p. 161. Alan R. Liss, New York.

95. Hjelmland, L. M. and Chrambach, A. (1984). In *Methods in Enzymology*, (ed. W. B. Jakoby), Vol. 104, p. 315. Academic Press, New York.

96. Hari, J., Pierce, S. B., Morgan, D. O., Sara, V., Smith, M. C., and Roth, R. A. (1987). *EMBO J.*, **6**, 3367.

97. Petruzelli, L., Herrera, R., and Rosen, O. M. (1984). *Proc. Natl Acad. Sci. USA*, **81**, 3327.

98. Shorr, R. G., Lyddiatt, A., Lo, M. M. S., Dolly, J. O., and Barnard, E. A. (1981). *Eur. J. Biochem.*, **116**, 143.

2

Solubilization, purification and molecular characterization of muscarinic acetylcholine receptors

T. HAGA, K. HAGA and E. C. HULME

1. Introduction

Here we present the procedures developed in our laboratory for the solubilization, purification, and characterization of mAChRs from porcine brain and heart. At least five different mAChR subtypes in mammalian tissues are known at present and their amino-acid sequences have been determined (1−3,26). However, they have not been fully characterized as regards their pharmacological properties and localization, and methods of sufficient specificity to permit the purification of a given subtype from the mixture often present in the tissue of interest have not been reported. The procedures described here can be applied to purify any of the mAChR subtypes.

2. Preparation of the affinity gel

mAChRs were purified initially from calf brain by affinity chromatography with dexetimide as the immobilized ligand (4), and then from porcine heart (5), porcine cerebrum (6), and rat forebrain (7) by affinity chromatography with ABT [3-(2′-aminobenzhydryloxy)tropane] (8) as the ligand. For unknown reasons, the ligand-binding activity was lost on purification by dexetimide affinity chromatography. However, mAChRs purified on the ABT-agarose gel exhibit ligand-binding activity. ABT-agarose is not available commercially, hence the preparation of ABT and ABT-agarose is described here (see *Figure 1*).

Protocol 1. Synthesis of 2-aminobenzhydrol (8)

1. Dissolve 2-aminobenzophenone[a] (30 g, 152 mmol) in methanol/benzene (300/100 ml) in a 500 ml round-bottom flask.

2. Add NaBH$_4$ (5.7 g, 150 mmol) little by little (e.g. in 0.5 g portions) to the 2-aminobenzophenone solution, which is chilled in an ice bath and stirred with a magnetic stirrer.

3. Stir the solution for 2 h at room temperature, and then check the reaction mixture by TLC (see Section 2.3). If the reactant is detected, add NaBH$_4$ (e.g. 0.3−1 g), stir (e.g. for 10 min) and check once more by TLC.
4. After confirming that the reaction is complete, evaporate the solvent *in vacuo*, add a slurry of ice and water (~500 ml) to the oily residue, and then stir the mixture with a glass rod.
5. Collect the resulting white solid (2-aminobenzhydrol) on a filter paper on a Büchner funnel, wash it with water (~1 l) and then dry it *in vacuo* over P$_2$O$_5$. The yield is approximately 80%.

a Obtainable from Aldrich.

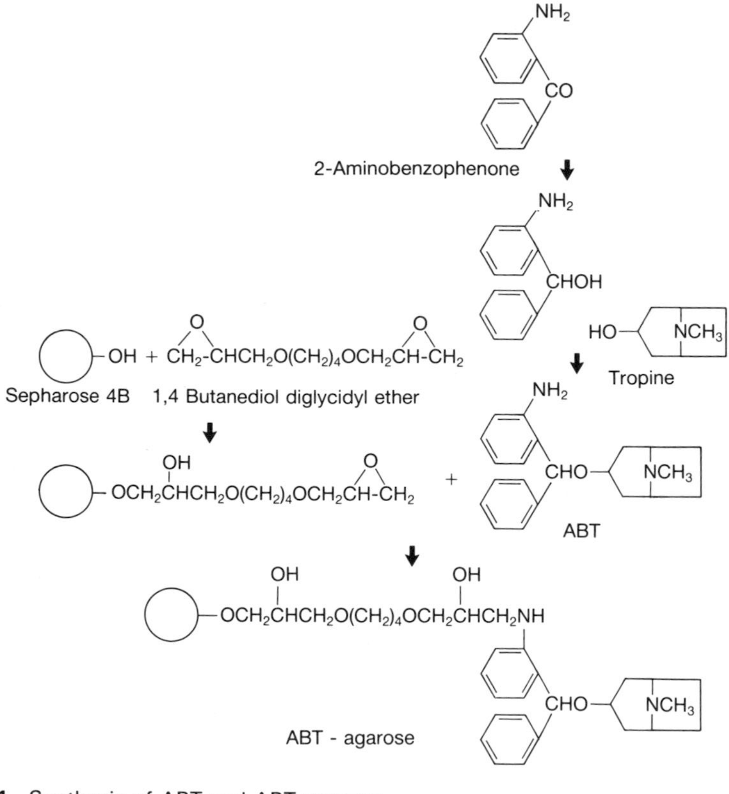

Figure 1. Synthesis of ABT and ABT-agarose.

Protocol 2. Synthesis of ABT (8)

1. Put concentrated sulphuric acid (46 ml) into a 300 ml two-necked round-bottom flask with a thermometer in an ice−salt bath.
2. Powder the tropine*a* (40 g, 280 mmol) and then add it to the chilled sulphuric

acid little by little, taking special care with the initial several grams, so that the temperature of the reaction mixture remains below 30°C. Stir the reaction mixture with a magnetic stirrer at first, and then with a thick glass rod by hand, because the reaction mixture becomes too viscous to be stirred with a magnetic stirrer as the amount of added tropine increases. After the addition of tropine is complete, stir the reaction mixture at room temperature and make sure that the reaction mixture is homogeneous.

3. Add well-dried 2-aminobenzhydrol (18.6 g, 94 mmol) to the reaction mixture, stir the mixture with a glass rod by hand so that it becomes homogeneous, and then incubate it in an oil bath at 70−80°C.

4. After 30 min, cool the reaction mixture and then pour it into a slurry of ice and water (∼1/l) in a 2 l beaker.

5. Add 5 M NaOH to the mixture, stirring with a magnetic stirrer, until the pH of the mixture becomes 7: measure the pH with a test paper.

6. Remove the resulting yellowish pellet by extraction with benzene. Repeat twice (∼500 ml each).

7. Add 5 M NaOH to the aqueous solution in an ice-cold bath, stirring with a magnetic stirrer, and confirm that an oily white precipitate is formed. Stop the addition of NaOH when no more precipitate is formed and the pH of the solution is in the region of 11.

8. Extract the oily precipitate with benzene twice (∼500 ml each) and wash the benzene extract with water (∼500 ml each) repeatedly, more than 10 times, until the pH of the water phase is neutral.

9. Add anhydrous sodium sulphate (∼200 g) to the benzene extract, filter the solution the next day, and wash the sodium sulphate with benzene twice (∼150 ml each). Evaporate the combined filtrates under reduced pressure. The resulting oil (ABT) weighs approximately 15 g and the yield is 50% of the theoretical value.

10. Dissolve the oil in a small volume of hot methanol, stand the solution at room temperature overnight, collect the resulting crystals on a filter paper, dry them under reduced pressure, and store them in a desiccator over silica gel.

a Obtainable from Aldrich.

2.1 Identification of ABT

The Rfs of 2-aminobenzophenone, 2-aminobenzhydrol, tropine, and ABT on TLC (Art 5554 (Merck), silica gel 60 F254; solvent, chloroform/methanol (9/1, v/v)) are 0.83, 0.54, 0.04, and 0.09, respectively. The spots corresponding to 2-aminobenzophenone, 2-aminobenzhydrol, and ABT can be detected on the basis of their UV absorption, and those of tropine and ABT by exposing the TLC plate to iodine vapour. Only the spot corresponding to ABT is detected by both methods. The ABT is dissolved in ethanol to a final concentration of 25 mM and then diluted 100-fold

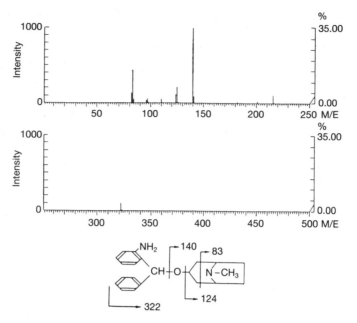

Figure 2. Mass spectrogram of ABT.

with water. The resulting 0.25 mM solution has an absorption of 0.5 at 285 nm. The results of mass spectroscopic analysis are shown in *Figure 2*. The IR and NMR data for ABT are given in reference 8.

Protocol 3. Preparation of epoxy-activated agarose (6)

1. Wash approximately 400 ml of packed Sepharose 4B on a sintered glass funnel with 4 l of distilled water and then with 400 ml of 0.3 M NaOH, and then suspend it in 700 ml of 0.3 M NaOH and pour the suspension into a 1 l round-bottom flask.

2. Add 1,4-butanediol diglycidyl ether[a] (100 g, 0.49 mol) to the gel suspension and then stir the contents of the flask at 20−25°C overnight, using a rotary evaporator. Do not evacuate, but tape the flask to the rotary evaporator to keep it in place.

3. Collect the gel on a sintered glass funnel and then wash it with 50% aqueous dioxane (3 l) to remove any excess 1,4-butanediol diglycidyl ether.

[a] Obtainable from Aldrich.

Protocol 4. Preparation of ABT-agarose (6)

1. Wash the epoxy-activated agarose gel (400 ml) with dioxane/0.2 M NaOH

(400/400 ml), suspend it in dioxane/0.2 M NaCl (300/400 ml), and then put the suspension into a 2 l flask.

2. Add ABT (10 g, 31 mmol) in 100 ml of dioxane to the gel and then stir the gel suspension for 24 h at $35-40°C$ using a rotary evaporator. Add 50 ml of 2 M $KHCO_3$ and 20 ml of 5 M KOH to the gel suspension and then incubate the suspension at $35-40°C$ for another 15 h.

3. Collect the reacted gel on a sintered glass funnel, and then wash it with 500 ml of 50% aqueous dioxane, and save the filtrate to recover unreacted ABT (more than 90% of the ABT can be extracted from the filtrate with benzene). Wash the gel with 3 l of aqueous 50% dioxane, and then suspend it in 400 ml of absolute ethanol and leave it for at least 30 min, to block unreacted epoxide groups.

4. Wash the gel successively with 3 l each of distilled water, 0.5 M NaCl/0.1 M potassium acetate buffer (pH 4.0), 0.5 M NaCl/0.1 M $KHCO_3$, and distilled water.

5. Determine the content of bound ABT by measuring the UV absorption spectrum of the gel suspension after 1:20 dilution of the gel with ethylene glycol. The absorption maximum of ABT is shifted from 285 to 293 nm by binding to the gel. The density of bound ABT should be approximately 4 μmol/ml of gel.

After the gel has been used for affinity chromatography, the final washing procedure (*step 4*) can be employed for regeneration of the gel, and the gel can be used repeatedly, although the recovery of mAChR decreases gradually.

3. The preparation and solubilization of membranes

The brain is one of the richest sources of mAChRs and is known to contain all five mAChR subtypes, particularly m1, m3, and m4. On the other hand, the heart is thought to mostly contain a single mAChR subtype (mAChR II or m2). The following are the protocols used for the preparation of membranes from porcine cerebra and atria, and for the solubilization of these membranes with a mixture of digitonin and sodium cholate.

3.1 Synaptic membranes from porcine cerebra

The following procedure is a modification of the method described by Jones and Matus (9), modified so as to allow the preparation of a large amount of synaptic membranes (6). Six hundred grams of porcine cerebra are used as the starting material and membrane preparations containing approximately 20 nmol of mAChR are obtained. The procedure takes a whole day and as described needs three centrifuges and three ultracentrifuges. However, it can be scaled down as appropriate.

Table 1. Stock solutions for porcine cerebral membrane preparation

(a) 2 M sucrose, 2 litres: add 600 ml of water to 1368 g of sucrose, warm and stir the suspension, and then adjust the final volume to 2 litres.

(b) 1.14 M sucrose, 300 ml: mix 2 M sucrose (171 ml) and water to give a final volume of 300 ml.

(c) Medium A (0.32 M sucrose/10 mM potassium phosphate buffer (pH 7.0) (KPB)/1 mM EDTA), 5 litres: mix 2 M sucrose (800 ml), 1 M KPB (50 ml), 0.4 M EDTA (pH 7) (12.5 ml) and water to give a final volume of 5 litres.

(d) Medium B (10 mM KPB/1 mM EDTA), 5 litres: the same as medium A but without the sucrose.

(e) 0.5 M phenylmethylsulphonyl fluoride (PMSF, Sigma), 20 ml: dissolve 1.75 g of PMSF in 20 ml of isopropanol in a hot water bath and store at 1 ml per tube at −20°C. Note: PMSF is toxic.

(f) 5 mg/ml pepstatin, 20 ml: dissolve 100 mg of pepstatin (Sigma) in 20 ml of dimethyl-sulphoxide in a hot water bath and store at 1 ml per tube at −20°C.

Protocol 5. mAChR-enriched membranes from procine brain

1. Prepare the solutions shown in *Table 1* the previous day and keep them in a cold room. Solutions (v) and (vi) should be kept in a freezer. Use water that has been passed through an ion-exchange column and then distilled in a glass vessel. All procedures should be carried out at 0−5°C unless otherwise stated.

2. Thaw the PMSF and pepstatin solutions in a hot water bath, and then add 1 ml each to medium A and medium B just before use.

3. Remove the cerebrum and caudate nucleus (600 g wet weight) from nine porcine brains: we usually use brains from a local abattoir within an hour after the death of the animals, put them in ice and start homogenization within another hour.

4. Add 2 l of medium A to the tissue and then homogenize the mixture in a Waring blender for 10 sec twice and then with a Physcotron (or Silverson or Polytron) for 1 min twice. Add 2.8 l of medium A to the suspension and then centrifuge the mixture at 8500 r.p.m. (11 000 *g*) for 90 min; we usually use three centrifuges for this.

5. Add 2 l of medium B to the pellet, homogenize the suspension with a Physcotron, and add 2.5 l of medium B to the suspension and then centrifuge the mixture at 11 000 *g* for 90 min.

6. Add a small amount of medium B (*c*. 100 ml) to the pellet and suspend it with a glass rod, then mix the suspension with 2 M sucrose (720 ml) and medium B to give a final volume of 1.8 l (sucrose concentration, 0.8 M), and then homogenize the mixture with a Physcotron.

7. Add 10 ml of 1.14 M sucrose to each of 18 Beckman type 35 tubes and 5 ml to each of 18 Beckman type SW 27 tubes. Place the membrane suspension (approximately 70 or 30 ml) on the 1.14 M sucrose layer with a bent-tipped Pasteur pipette until the tubes are full, and then centrifuge them at 34 000 r.p.m. or 25 000 r.p.m. for 2 h.

8. Discard the 0.8 M sucrose layer, collect and pool the interface between the 0.8 and 1.14 M sucrose layers and the 1.14 M sucrose layer, adjust the volume to 640 ml, and then homogenize the mixture with a Physcotron. Store the suspension in 40 ml tubes at −80°C.

3.2 Membranes from porcine atria

The following procedure was adapted from that of Peterson and Schimerlik (10).

Protocol 6. mAChR-enriched membranes from porcine atria

1. Cut 40 porcine atria (approximately 1 kg wet weight) into small pieces and then homogenize the pieces in 2 l of medium A (25 mM imidazole buffer (pH 7.4)/1 mM EDTA/0.02% sodium azide/0.1 mM PMSF/1 μg/ml pepstatin) at 4°C for 1 min with a Waring blender and then for 1 min with a Physcotron (or Polytron) homogenizer.

2. Centrifuge the homogenate at 2000 r.p.m. (600 *g*) for 15 min and then filter the supernatant through two layers of cheesecloth.

3. Centrifuge the filtrate at 30 000 r.p.m. (70 000 *g*) in Beckman type 35 rotors for 1 h and then suspend the pellet in buffer A (final volume, 160 ml).

4. Apply the suspension (16 ml per tube) to a sucrose gradient, consisting of 10 ml of 28% and 9 ml of 13% (w/v) sucrose in buffer A, and then centrifuge the gradient at 25 000 r.p.m. in a Beckman SW 27 rotor for 1 h.

5. Collect the 13% and 28% sucrose layers above the pellet, and store them at −80°C.

3.3 Yield of membrane mAChRs

The protein concentration and ligand {[[3H]quinuclidinyl benzilate, [3H]QNB (Amersham; NEN; see reference 31, Chapter 1)} binding activity of the cerebral membrane preparation are approximately 20 mg/ml and 30 pmol/ml, respectively (see Appendix for an outline of the assay protocols, and reference 31 for full details). The membranes should be diluted to give a mAChR concentration of ~1 pmol/ml and assayed with ~5 pmol/ml [3H]QNB in KPB buffer, 90 min, 30°C. The yield of [3H]QNB binding sites (approximately 20 nmol) corresponds to 50% of the sites initially present in the homogenate and the specific activity (1.5 pmol/mg protein) is increased twofold compared to that of the homogenate. A preparation with a higher specific activity (e.g. 4 pmol/mg of protein) can be obtained from rat forebrain on a smaller scale by means of the original method (9); for this purpose, it is necessary to apply smaller amounts of membrane to the sucrose gradient than those mentioned here.

The specific [^{3}H]QNB binding activity of the atrial membrane perparation is 1.5 − 1.9 pmol/mg of protein, and a membrane preparation containing approximately 1.2 nmol of [^{3}H]QNB binding sites is obtained from 1 kg of tissue. Both the yield and the specific activity are lower than those reported originally (10); a preparation with 350 pmol mAChR and a specific activity of 4.5 pmol/mg of protein has been reported to be obtained from 200 g of tissue.

3.4 Solubilization

Protocol 7. Solubilization of mAChRs

1. Add 10 g of digitonin to 500 ml of water in a 1 l beaker, stir and warm the suspension at approximately 90°C for 10 − 20 min, and then add 5 ml of 20% sodium cholate and cool the suspension at 4°C; it is not necessary to warm the suspension if water-soluble digitonin is used, but water-soluble digitonin is more expensive and is not necessarily required for this process.

2. Thaw the membrane suspension derived from 150 g of cerebra or 4 kg of atria (approximately 5 nmol [^{3}H]QNB binding sites and 3.5 g protein) and mix it with a solution containing 10 ml of 1 M KPB (pH 7.0; see *Table 1*), 10 ml of 5 M NaCl, and 2.5 ml of 0.4 M EDTA (total volume, 495 ml) at 4°C.

3. Add 0.2 ml each of 0.5 M PMSF and 5 mg/ml of pepstatin to the membrane suspension and then 505 ml of the digitonin−cholate mixture (*1*) with stirring. Stir the suspension for 1 h at 4°C and then centrifuge it at 34 000 r.p.m. (90 000 *g*) for 1 h, and collect the supernatant. The supernatant is immediately subjected to the affinity chromatography, as described in the next section.

3.5 Assay for ligand-binding activity of solubilized mAChRs

Protocol 8. Assay for ligand-binding activity of solubilized mAChRs

1. Prepare small columns of Sephadex G50 fine (2 ml, 0.75 × 4.6 cm). While Sephadex G25 fine or other kinds of permeation gel may be used, the use of Sephadex G50 medium or coarse is not recommended for assays using [^{3}H]-QNB because of the lower resolution and lower buffer retention of these gels. However, these problems are less acute if a hydrophilic ligand such as (−)[^{3}H]methylscopolamine (Amersham, NEN) is used to assay the receptors

(31). When 0.2 ml of sample is applied, macromolecules such as the receptor-[³H]QNB complexes are eluted betwen 0.8 and 1.2 ml, and free [³H]-QNB is eluted after 1.3 ml: the exact values may differ depending on the shape of the column and so should be determined beforehand. For example, apply a mixture of blue dextran and dinitrophenyl-alanine in 20 mM KPB (1 mg/ml and 0.1 mg/ml, respectively, 0.2 ml), collect the eluate (0.1 ml per tube, 35 tubes), dilute it with 1 ml of water and then measure the absorbance at 620 and 370 nm.

2. A [³H]QNB solution obtained from Amersham (1 mCi/ml of ethanol) is diluted with ethanol ten-fold and then stored at −80°C. The stock [³H]QNB solution (0.1 mCi/ml, approximately 2.5−3 μM) is diluted with 20 mM KPB or 20 mM KPB/0.1% digitonin just before use.

3. Mix the solubilized mAChR preparation, [³H]QNB solution (final concentration, 1−5 nM), buffer (20 mM KPB/0.1% digitonin), and atropine (final concentration, 0 or 1 μM) (total volume, 0.5 ml). Incubate the mixture for 60−90 min at 30°C. A portion of the incubation mixture (0.2 ml in duplicate) is used for assaying the bound [³H]QNB (see *step 4*). The difference in the bound [³H]QNB in the absence and presence of atropine is defined as the specific binding.

4. Wash the Sephadex columns with 20 mM KPB or 20 mM KPB/0.04% digitonin (5 ml). Add 0.2 ml of the incubation mixture to a column and then 0.1 ml followed by 1.0 ml of 20 mM KPB. Let each addition run in completely before the next addition. Collect the whole eluate (1.3 ml) in a scintillation vial. Add 10 ml of a Triton−toluene cocktail {40 g of 2,5-diphenyloxazole (PPO) and 1 g of dimethyl-1,4-bis[2-(5-phenyloxazoylyl)]benzene (POPOP) in 7 l of toluene and 3 l of Triton X-100} to the eluate from the Sephadex G50 column, mix and then measure ³H with a liquid scintillation counter. The use of commercial scintillators for aqueous solutions is more convenient (see, for example Beckman, Packard). Wash the Sephadex columns with 10 ml of 0.02% azide and store them in the same solution.

3.6 Yield of solubilized mAChRs

Approximately 50% of the [³H]QNB-binding activity in cerebral or atrial membrane preparations is recovered in the supernatant after extraction with a mixture of digitonin and sodium cholate.

4. Purification of mAChRs from solubilized extracts

4.1 Affinity chromatography

The affinity chromatographic procedure is schematically shown in *Figure 3*. All procedures are performed at 4°C, unless otherwise specified.

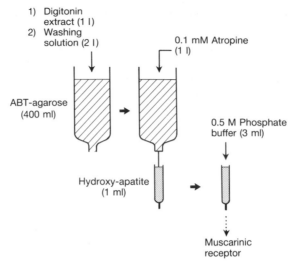

Figure 3. Flow diagram for the purification of mAChRs.

Protocol 9. Affinity chromatography of mAChRs

1. Wash the ABT-gel (400 ml), with 400 ml of 0.1% digitonin/20 mM KPB at a flow rate of 50 ml/h using a peristaltic pump. Apply the solubilized mAChR preparation (approximately 2.5 nmol of [³H]QNB-binding sites in 950 ml, see Section 3.4) to the column at the same flow rate. The solutions used for the ABT-gel or hydroxyapatite column should be made with water-soluble digitonin (Wako Chemicals or Sigma D1407) or a digitonin solution from which the precipitate has been removed (see Chapter 1, Section 2.4).

2. Wash the ABT-gel with 1.2 l of 0.1% digitonin/20 mM KPB/0.15 M NaCl at the same flow rate. It is advisable to monitor the A_{280} nm of the eluate.

3. Connect a small polypropylene column of hydroxyapatite (HA, e.g. Bio-Rad HTP) (1 ml, 0.75 × 2.3 cm) to one end of the ABT-gel column and then apply 1.2 l of 0.1 mM atropine/0.1% digitonin/20 mM KPB/0.15 M NaCl.

4. Disconnect the HA column from the ABT-gel column and then apply the following solutions successively to the HA column: (i) 20 ml of 0.1 M KPB/0.1% digitonin/10 mM carbachol; (ii) 0.5 ml of 0.5 M KPB/0.1% digitonin/10 mM carbachol; (iii) 3 ml of 0.5 M KPB/0.1 % digitonin/10 mM carbachol; and (iv) 3 ml of 1 M KPB/0.1 % digitonin/10 mM carbachol. Scarcely any mAChRs are eluted with the first two solutions. Most of the mAChRs are eluted with the third solution and some with the fourth solution. The latter two solutions are frozen in liquid N_2 and stored at −80°C.

4.2 Further purification

Further purification is carried out by gel permeation HPLC, or by rechromatography on ABT-Sepharose (6). Alternatively, purification procedures involving wheat-germ agglutinin or DEAE cellulose column chromatography may be performed prior to the affinity chromatography step (5; Chapter 1, Section 3.4). A preparation obtained by a single round of affinity chromatography is pure enough for most purposes because a single major band corresponding to pure mAChR is observed on sodium dodecyl sulphate polyacrylamide gel electrophoresis (SDS-PAGE).

4.3 Determination of protein concentration with fluorescamine

The protein concentrations of purified receptor preparations are usually too low to be measured with everyday reagents such as the Folin−Lowry reagent. As little as 0.1 μg of protein in 0.2 ml of solution can be detected by a method employing fluorescamine. The following procedure was adapted from that in reference 11. Digitonin at less than 1% does not interfere with the assay, but compounds with free amino groups cannot be included in the samples.

Protocol 10. Fluorescamine assay for protein

1. Put 0.2 ml of sample into a polypropylene tube and then add 0.2 ml of 0.2 M potassium borate buffer (pH 8.0). Dilute the sample with 0.1% digitonin/20 mM KPB when the protein concentration of the sample is expected to be higher than 50 μg/ml.

2. Add 0.2 ml of 0.03% fluorescamine (Fluram, Roche, also available from Sigma) in acetone to the protein solution, while agitating vigorously with a Vortex Mixer, and then add 2 ml of water.

3. Measure the fluorescence with excitation at 390 nm and emission at 475 nm.

4. Assay control samples containing bovine serum albumin (BSA) simultaneously. Dilute a stock aqueous BSA solution (1 mg/ml) ten- or 100-fold with 0.1% digitonin/20 mM KPB or other solutions containing detergents just before doing the assays. Dilute BSA solutions cannot be stored, even in polypropylene tubes because of adsorption. We use 0, 0.1, 0.25, 0.5, 1, and 2 μg of BSA as control samples; the fluorescence is linear with respect to the amount of BSA.

5. In typical experiments, the protein concentration of the eluate from the hydroxy-apatite column with 0.5 M KPB/0.1% digitonin is around 0.1 mg/ml, which can be estimated by using 5 or 10 μl of the solution.

4.4 Yield and specific binding activity of purified mAChR

The [^{3}H]QNB-binding activity of the purified mAChR preparation is assayed as described in Section 3.5, except that it is necessary to wash the Sephadex G50 columns with 0.04% digitonin/20 mM KPB, rather than with 20 mM KPB alone, and to use polypropylene tubes not glass ones. In typical experiments, the concentration of [³H]QNB-binding sites in the eluate with 0.5 M KPB/0.1% digitonin from the hydroxyapatite column is 200−500 pmol/ml and the total yield is 0.6−1.5 nmol (25−50% of the solubilized sites). The yield appears to depend more on the batch of ABT-gel than on the species of tissue, i.e. cerebra or atria.

The molecular weights of the protein component of the mAChRs are known to be 50 000−60 000 (1−3), and therefore the specific ligand-binding activity should be 17−20 nmol/mg of protein if the mAChRs are purified to homogeneity and every purified mAChR molecule binds a single ligand molecule. In practice the highest specific activity determined was 14 nmol/mg of protein for a preparation purified by a combination of affinity chromatography and other methods (5−7). The specific activity of a preparation purified by a single affinity chromatographic step usually ranges from 2 to 5 nmol/mg of protein, although the major protein band on SDS-PAGE corresponds to mAChRs. It is possible that a substantial portion of the purified mAChR is inactivated during purification and is not capable of binding the ligand.

5. Electrophoretic analysis

SDS-PAGE is used routinely to check the degree of purity of protein preparations and to estimate the apparent molecular sizes of the dissociated forms. mAChRs have also been examined by SDS-PAGE. A single major band is observed for mAChR preparations purified from porcine cerebra or atria, and the apparent molecular weight is around 70 000, that of atrial mAChR being a little higher than that of cerebral mAChR.

When the concentration of purified receptors is not high enough for detection on protein staining, purified samples may be [¹²⁵I]iodinated and then visualized by autoradiography. Furthermore, SDS-PAGE may be used for estimation of molecular size of mAChR in unpurified preparations; for this purpose, the mAChRs are labelled with an irreversible ligand, [³H]propylbenzilylcholine mustard ([³H]PrBCM; NEN), and the mobility of the [³H]PrBCM−mAChR complex on SDS-PAGE is estimated.

5.1 Iodination of mAChRs (6)

Protocol 11. Labelling of mAChRs with ¹²⁵I (6)

1. Take 0.1 mCi of ¹²⁵I (10 µl, 2 mCi/0.2 ml of 0.1 M NaOH; NEN, Amersham) in a 1 ml polypropylene tube, add 20 µl of 0.5 M KPB and then 100 µl of the mAChR sample in KPB/0.1% digitonin; the amount of mAChR may be as little as 1 pmol or less, but the procedure works best on amounts of ∼10 pmol.

2. Add 10 µl of freshly prepared 1 mg/ml chloramine-T (Pierce) and then vortex the solution.

3. After 1 min, add 10 μl of a freshly prepared solution of 3 mg/ml sodium metabisulphite and then vortex the mixture.

4. Apply the reaction mixture (150 μl) to a small column (e.g. Kontes Disposaflex) of Sephadex G50 fine (0.75 × 4.6 cm) equilibrated with 20 mM KPB/0.1% digitonin. Wash in with KPB/digitonin buffer (650 μl) and discard the eluate (800 μl).

5. Add 400 μl of KPB/digitonin buffer to the column and collect the eluate.

5.2 Affinity labelling of purified mAChRs with [³H]propyl-benzilylcholine mustard

Protocol 12. Affinity labelling of purified mAChRs with [³H]propyl-benzilylcholine mustard

1. Remove the protective ligands from an 0.2 ml aliquot of purified mAChR by gel-filtration on a small column (2 ml) of Sephadex G50 F contained in a plastic column (e.g. Kontes Disposaflex) equilibrated at 4°C in 20 mM Na-Hepes, 0.1% digitonin, pH 7.5 (HD buffer). Wash the aliquot of receptor into the gel with 2 × 0.2 ml of ice-cold HD, and elute the receptor with 0.5 ml of HD. Collect the eluate in a clean polypropylene container (e.g. a microfuge tube) cooled on ice. This whole procedure can be scaled up as desired.

2. Determine the concentration of binding sites in the gel-filtered eluate, if this is unknown. To do this, dilute the eluate *c.* tenfold, and assay it with [³H]-*N*-methylscopolamine (10^{-8} M) for 60 min at 15°C, terminating the binding reaction by gel-filtration on a 2 ml column of Sephadex G50 F (see Chapter 1, Section 2.5 and Appendix 1). Adjust the volume of the gel-filtered receptor to obtain a concentration of $10-20$ pmol/ml. Dilute the receptor with HD buffer if necessary. Note that the gel-filtration procedure dilutes the original receptor stock 2.5-fold.

3. Meanwhile cyclize an aliquot of [³H]PrBCM (NEN). Allow the stock to warm to room temperature before opening the bottle, to avoid introducing water by condensation. Dilute an aliquot of the ethanolic stock solution *c.* thirtyfold from a concentration of *c.* 10^{-5} M to *c.* 3×10^{-7} M in 10 mM sodium phosphate buffer, pH 7.5. Incubate the solution for 60 min at 25°C to allow cyclization to take place. The cyclized solution keeps for several hours at 0°C, but should be used as fresh as possible. NEN provide an information leaflet on the use of this reagent.

4. Add cyclized [³H]PrBCM to the gel-filtered receptor solution in HD buffer, aiming to achieve a [³H]PrBMC concentration approximately twice the concentration of receptor binding sites, i.e. $20-40$ pmol/ml. Incubate for 90 min at 15°C. The use of this temperature provides a useful compromise between receptor stability, which is reduced at higher temperatures, and the [³H]PrBCM alkylation rate constant, which is reduced at lower temperatures.

Table 2. Stock solutions for SDS-PAGE

(a) Lower gel buffer (L): 36.3 g Tris, 0.8 g SDS and water; adjust pH to 8.8 with 5 M HCl; final volume, 200 ml.

(b) Upper gel buffer (U): 6.05 g Tris, 0.4 g SDS with water; adjust pH to 6.8 with 5 M HCl; final volume, 100 ml.

(c) Acrylamide solution (A): 116.8 g acrylamide, 3.2 g methylene bis acrylamide and water; final volume, 400 ml.

(d) Sample diluting solution (S): 10 g glycerol, 5 ml of 2-mercaptoethanol, 2.5 g SDS, 6.25 ml of 1 M Tris, 0.5 ml of 1% bromophenol blue and water; adjust pH to 6.8 with 5 M HCl; final volume, 50 ml.

(e) Running buffer (10 times concentrated): 90.9 g Tris, 432 g glycine, 30 g SDS and water; final volume, 3 litres.

5. Terminate the incubation by gel-filtering the reaction mixture on a 5 ml column of Sephadex G50 F in ice-cold HD buffer, by scaling up the procedure described in *step 1*. Alternatively, concentrate the reaction mixture using a Centricon 10 centrifugal concentrator (Amicon) at $4°C$, either before or after gel-filtration. These procedures remove the free [^{3}H]PrBCM and its hydrolysis products, and permit further analysis of the labelled receptor. Sample the reaction mixture to assess the extent of labelling of the mAChR-binding site. Provided that the mAChRs are ligand-free before labelling, this procedure allows [^{3}H]PrBCM incorporation to a level of *c.* $75-100\%$ of the total binding site concentration measurable with a reversible ligand such as [^{3}H]-NMS.

5.3 SDS-PAGE by the Laemmli method (12)

Protocol 13. SDS-PAGE by the Laemmli method (12)

1. Make up stock solutions as shown in *Table 2*. Use the highest possible reagent quality, e.g. BDH electran grade.

2. Prepare the separating gel [10% acrylamide; 6.7 ml (A), 5 ml (L), 8.3 ml water, 50 μl freshly prepared solution of 100 mg/ml ammonium persulphate (APS), and 20 μl tetramethyl ethylene diamine (TEMED)] and then the stacking gel [4.5% acrylamide; 1.5 ml (A), 2.5 ml (U), 6 ml water, 30 μl APS, and 10 μl TEMED] between glass plates for slab electrophoresis ($14 \times 14 \times 0.1$ cm).

3. Mix a mAChR sample with an equal volume of the sample diluting solution (S) and then apply the mixture to the gel (40 μl per lane). Do not heat the sample because the band of mAChR undergoes aggregation when heated.

4. Perform electrophoresis for about 4 h at a constant current of 20 mA.

A fuller description of the practicalities of this method is given in Chapter 10, Section 3.

5.4 Detection of the mAChR band

5.4.1 Silver staining

The mAChR band can be detected by silver staining when 5−10 pmol of mAChR is applied. This amount corresponds to 250−500 ng of protein and is higher than the amount necessary for the detection of most standard proteins (50−100 ng), probably because the mAChR band is much broader than those of typical protein standards. The eluate from the hydroxyapatite column (200−500 pmol/ml) with a 0.5 M KPB/0.1% digitonin should be analysed by SDS-PAGE after the KPB has been removed by using dialysis or by means of a small Sephadex column.

We have used the Bio-Rad kit and followed the procedure recommended by the manufacturer. The mAChR bands are sometimes detected more clearly after restaining as follows (13). The gel is stained and then destained with a mixture of A:B:water (1:1:5, v/v) (A, 18.5 g NaCl, 18.5 g $CuSO_4$, 425 ml water and concentrated ammonium hydroxide, which is added until the first precipitate is completely dissolved; B, 218 g sodium thiosulphate and 500 ml water), and then stained once more from the beginning.

5.4.2 Autoradiography of ^{125}I-iodinated samples

Protocol 14. Autoradiography of ^{125}I-labelled mAChRs

1. Dry the gel immediately after electrophoresis or after silver staining.

2. Store it with an X-ray film and an intensifying screen (DuPont Lightening Plus) at −80°C for various periods; for example, 2−3 days for the ^{125}I-iodinated sample with 50 000 d.p.m. Pre-flashing the film speeds up this process, see Chapter 10, Section 3.

3. Develop the film.

Electrophoresis of molecular weight markers ([14C]methylated proteins; Amersham CFA, 626) in the lane next to a sample may be useful for identification of the band of mAChR and for estimation of its molecular size. Approximately 20 nCi of [14C]protein should be applied when developed after 2−3 days without an enhancing reagent. See Chapter 10, Section 3, for further practical details of these methods.

5.4.3 Autoradiography of [3H]PrBCM-labelled mAChRs

Protocol 15. Fluorography of 3H-PrBCM-labelled mAChRs

1. Dip the gel in an autoradiography-enhancing reagent according to the procedure recommended by the manufacturer (e.g. EN3HANCE, NEN).

2. Expose the treated gel to an X-ray film at $-80°C$ for various periods; for example, a week for [³H]PrBCM-labelled receptors containing 20 000 dpm. In this case, $1-2$ nCi of [¹⁴C]methylated marker is enough and appropriate. High sensitivity X-ray films, e.g. hyperfilm, are available from Amersham, which also supplies a very useful information booklet. See Chapter 10 for further details.

3. Develop the film.

Alternatively, a lower amount of [³H]PrBCM-labelled mAChR (e.g. 2000 dpm) may be detected by cutting the gel into slices and then counting as follows.

Protocol 16. Slicing and counting polyacrylamide gels

1. Cut the gel into 2.5 mm thick slices immediately after electrophoresis and place them in 20 ml (or 6 ml) glass scintillation vials.

2. Add 0.5 ml of 30% H_2O_2 to each slice, cap the tubes loosely and then incubate the slices at $60°C$ overnight. The slices should be dissolved. If not, persevere until they do dissolve.

3. Add 20 ml (or 6 ml) of a Triton−toluene cocktail and then count the radio-activity. It is advisable to store the vials in the dark for >2 h before counting, to allow chemiluminescence to decay.

---

6. Estimation of the molecular size of mAChRs

The molecular sizes of mAChRs have been estimated on the basis of different principles such as amino-acid sequences, SDS-PAGE analysis, and hydrodynamic measurements.

6.1 Estimation from amino-acid sequences

The amino-acid sequences of the mAChRs have been deduced from the nucleotide sequences of the corresponding cDNA clones as described in Section 7. The molecular weights of the protein components of the receptors are estimated to be ~52 kDa for three of the subtypes (m1, m2, m4) and 60 and 64 kDa for the m5 and m3 subtypes (1−3,26). The estimates are the sum of amino-acid residue weights from the assumed initiation codon to the assumed termination codon. The amino- and carboxy-terminal sequences have not been determined directly by protein chemical means for any of the subtypes and thus some amino-acid residues in the deduced amino-terminal sequences may be lacking in the expressed proteins, although this would be expected to exert little effect on the estimates of molecular size.

6.2 Estimation by SDS-PAGE

It is common to estimate the apparent molecular sizes of membrane proteins, as well as soluble proteins, by the comparison of their mobilities on SDS-PAGE with those of soluble protein standards. The reported sizes of mAChR range between 70 and

90 kDa, and are higher than those estimated from the amino-acid sequences. The difference has been suggested to be due to the contribution of sugars. Liang *et al.* (14) reported that the apparent molecular sizes of mAChR, which were estimated from the mobilities of [³H]PrBCM-labelled components solubilized from two different types of cultured cells, decreased from 92 to 77 kDa and from 67 to 45 kDa, respectively, on treatment with endoglycosidase F. A similar reduction is produced by digestion with *N*-glycanase (Chapter 10, Section 5.2). Peterson *et al.* (15) determined the amount of glucosamine in a mAChR preparation purified from porcine atria and estimated the proportion of carbohydrate to be 26.5% by weight on the assumption that the ratio of glucosamine to total carbohydrate is 40%, a ratio which was estimated from the data for three other membrane proteins. This proportion is consistent with molecular weights of 70 000 and 52 000 for the total receptor and the protein component of the m2 mAChR respectively.

Extensive studies on the behaviour of mAChR on SDS-PAGE have been reported both for purified (15) and for unpurified (16−18) preparations. These include estimation of molecular size by electrophoresis at different concentrations of acrylamide by the construction of Ferguson plots, effects of proteolysis, and reduction or oxidation of cysteine residues. The most detailed account of the methodology in this field is given by Peterson *et al.* (15).

6.3 Estimation from hydrodynamic properties

In the case of oligomeric proteins, the molecular size estimated from mobility on SDS-PAGE is that of the subunits, because oligomers are known to be dissociated into their subunits in the presence of SDS. On the other hand, oligomeric proteins generally remain associated in the presence of mild detergents such as Triton X-405, dodecyl maltoside, Lubrol PX, and sodium cholate. Thus the comparison of molecular weights of mAChRs in SDS and in mild detergents may give information on the quaternary structure of the receptors.

The behaviour of the mAChRs ([³H]QNB binding activity, [³H]QNB−mAChR complex, or [³H]PrBCM−mAChR complex) on sucrose density gradient centrifugation in H_2O and D_2O and on gel filtration chromatography is measured, and these data are combined to estimate the molecular size of mAChR. The estimated molecular weights of mAChRs in several mild detergents for both purified (15) and unpurified preparations (19,20) were found to be similar to or little higher than those estimated from receptor mobility on SDS-PAGE, indicating that mAChRs exist largely in a monomeric state. The principle of this kind of estimate is reviewed briefly in the following (see reference 21). Experimental details will be found in references 15 and 19−22, and a detailed protocol is given below (see Section 6.3.1).

6.3.1 Sucrose gradient centrifugation of detergent-solubilized receptors

---

Protocol 17. Sucrose gradient centrifugation of detergent-solubilized receptors

1. Sucrose gradient centrifugation is carried out in a gradient of 5−20% buffered sucrose containing the appropriate concentration of detergent in either H_2O

or D_2O. In order to permit separate evaluation of the relative contributions of detergent and protein to the molecular weight of the receptor−detergent complex, the partial specific volume of the detergent should be significantly different from the average value for proteins, which is $0.71−0.76$ cm^3/g. Suitable detergents include Triton X-405, dodecyl maltoside, octyl glucoside, and Lubrol PX. Cholate, deoxycholate, CHAPS, and CHAPSO are of marginal utility in this regard because of the similarities of their PSVs to that of protein (see Chapter 1, Section 2). However, the molecular weight of the receptor−detergent complex can still be calculated. The partial specific volume of digitonin is only very marginally different from that of proteins, and no separate estimate of the protein and detergent contributions is possible.

To run gradients, we routinely use a $4.5−4.8$ ml gradient volume in a Beckman SW55 Ti rotor, and a $0.2−0.5$ ml sample volume. These volumes can be scaled up as necessary for larger volume gradients. To pour low-volume gradients accurately, it is necessary to have a gradient maker with tapered mixing chambers (see *Figure 4*). A suitable device is available from ISCO.

2. Connect a peristaltic pump to the gradient mixer. Deliver the gradient via a thin stainless-steel needle into gradient tubes held in an upright position.

3. Make up 5% and 20% sucrose solutions in buffer plus detergent plus additives as appropriate. For a 4.8 ml gradient in Beckman SW55 Ti tubes, place 2.4 ml of the sucrose solutions in the mixer chambers, with central tap A turned off, and the outlet tap B closed. Place the denser sucrose solution in the chamber closest to the pump. Stir this chamber with an overhead stirrer, e.g. a plastic paddle, or microspatula. Make sure that there are no bubbles in the tap connecting the two chambers. Open the taps A and B and pump the gradient at *c.* 1 ml/min, keeping the outlet tube just above the meniscus, and moving it up by hand. When the gradient is complete, transfer it to a refrigerator, and age it for 3 h at 4°C. Wash out the gradient maker with 5% sucrose and pump it dry before starting to pour the next gradient. Note that gradients in D_2O can freeze if they are kept too long on ice.

    A better alternative, if available, is to use stepper-motor activated syringe pumps such as the Pharmacia P 500 pumps, with a gradient controller. Beckman market a purpose-built syringe-pump gradient former, which is capable of pouring several gradients at once.

4. Make up the molecular weight standards in buffer, e.g. *E. coli* β-galactosidase (Sigma grade IX, 1 mg/ml), bovine catalase (1 mg/ml) and pig heart lactate dehydrogenase (1 mg/ml) and cytochrome C mixed together (see *Table 3*). Dialyse the standards v. several changes of buffer at 0°C, to remove residual ammonium sulphate. A more extensive list is provided by Peterson *et al.* (15). The assay of these standards is described in *Table 3*.

5. Mix 0.15 ml of solubilized receptor preparation with 0.05 ml of the standard mixture and layer it onto the gradient with an Eppendorf tip, or adjust the volumes as desired to give the appropriate final gradient volume, following

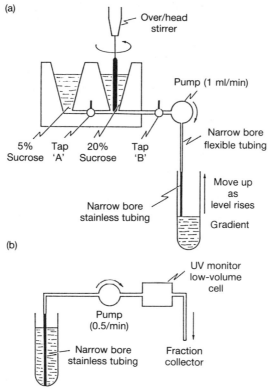

Figure 4. (a) Pouring a sucrose gradient. It is also possible to do this by flotation, delivering the light sucrose first to the bottom of the tube, and displacing it upwards with dense sucrose. In this variant, the delivery tube remains at the bottom of the gradient tube, and the light sucrose solution is in the chamber of the gradient mixer next to the pump. (b) Fractionating a sucrose gradient from the bottom.

the manufacturer's instructions for the individual rotor. Place the tubes in the precooled rotor (cold-room temperature).

6. Spin the gradients at 100 000 *g* for *c*. 18 h at 5°C in a swing-out rotor. Leave the brake off, so that the run coasts gently to a halt. Using tweezers, carefully remove the gradients from the rotor, and clamp them in an upright position in a cold-room.

7. Fractionate the gradients by pumping them out from the bottom via the piece of stainless steel tubing, positioned just above the bottom of the tube (*Figure 4b*). Prime the pump with 20% buffered sucrose containing detergent before commencing. Collect 20 × 0.25 ml fractions in microcentrifuge tubes (vary as appropriate). Keep tubing lengths and diameters as small as possible, to minimize dead space. Measure the dead volume, and subtract it from the total volume collected, when analysing the data. If using a UV monitor, use a low-

Table 3. Calibration enzymes and assays for hydrodynamic techniques

Enzyme	Partial specific volume, $\bar{V}$ (ml/g)	Sedimentation coefficient, $S_{20,w}$ $(S)^a$	Stokes radius (nm)
β-Galactosidase	0.76	15.93	6.84
Catalase	0.73	11.3	5.21
Fumarase	0.74	9.1	5.27
Lactate dehydrogenase	0.74	7.3	4.75
Malate dehydrogenase	0.74	4.32	3.69
Cytochrome C	0.73	1.71	1.87

[a] Values taken from reference 22. Note that $1S = 10^{-13}$ sec.

Enzymes may be purchased from Sigma: β-Galactosidase from *E.coli*, grade IX; bovine catalase; pig heart LDH; porcine heart MDH. Standards should be prepared by mixing them together at 1 mg/ml, and then dialysing for several hours against 20 mM Tris, pH 8.0, to remove ammonium sulphate.

Assays

β-Galactosidase: Add 20 μl of sample to 100 mM phosphate buffer pH 7.4 (0.115 g Na_2HPO_4 + 0.0297 g NaH_2PO_4/100 ml), 10 mM *ortho*-nitrophenyl-β-galactopyranoside (Sigma, 4 mg/ml). A yellow colour will develop within two minutes. Stop the reaction by adding 0.5 M Na_2CO_3, and measure OD at 410 nm. Samples may be stored in the dark on ice for an hour.

Catalase: Dilute H_2O_2 (200 volumes, BDH) 1 in 1000 in 10 mM phosphate buffer pH 7.4. Add 20 μl of sample to 2 ml of this buffer, and record the rate of change of OD at 230 nm.

Fumarase: Add 50 μl of samples to 2 ml of 10 mM phosphate pH 7.4, 50 mM L-malate and measure rate of change of OD at 240 nm.

Lactate dehydrogenase: Add 50 μl of sample to 2 ml of 0.1 M Tris (0.121 g/100 ml), 0.1 M lactic acid (0.9 g/100 ml), 1 mM NAD (60 mg/100 ml) pH 9.0 (adjusted with NaOH). Record rate of change of OD at 340 nm.

Malate dehydrogenase: Add 25 μl of sample to 2 ml of 10 mM phosphate pH 7.4, 2.5 mM *cis*-oxaloacetate (33 mg/100 ml), 0.17 mM NADH (12 mg/100 ml), and record rate of change of OD at 340 nm.

Cytochrome C: Measure absorbance of sample at 410 nm.

volume flow cell. Alternatively, and better, use a commercial gradient fractionator (e.g. ISCO, Buchler Densi-flow) which displaces the gradient upwards through a UV monitor by means of a dense layer of 50% sucrose pumped in from the bottom. Between runs wash out the pump and delivery system with 20% buffered sucrose containing detergent.

8. Usually the receptor will be prelabelled with a radiolabelled ligand before centrifugation. It is not usually necessary to separate bound from free ligand before centrifugation, because the free ligand will remain at the top of the gradient. Indeed, it is often advisable to include a concentration of radiolabelled ligand in the gradient which is sufficient to maintain a reasonable degree of receptor occupancy, and thus protect the binding sites against denaturation during the run.

9. Assay each fraction for receptor binding using the Sephadex G50 F assay to separate bound from free ligand. Assay the individual fractions for the activity of molecular weight standards, as shown in *Table 3*.

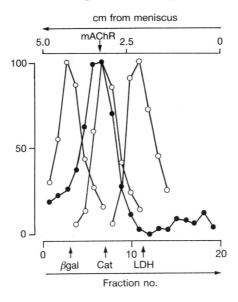

Figure 5. Fractionation of forebrain mAChRs on a 5 – 20% sucrose gradient in Hepes/digitonin buffer, pH 7.5.

10. Plot the distribution of receptor-binding activity, and of enzymatic activity or A_{280} nm of molecular weight standards in the gradient, as shown in *Figure 5*. From the positions of maximal activity, calculate the distances migrated by receptor and standards. Also, note the peak widths at half-height. Receptor peaks are often wider than those of standards because of heterogeneous glycosylation, or the formation of specific or non-specific complexes of the receptor with other components. Plot the distance migrated against the $s_{20,w}$ for the standards. The relationship should be linear, and this can be tested by linear regression analysis [see Petersen *et al.* (15)]. However, because of the lack of equality between the partial specific volumes of the receptor – detergent complex, and the molecular weight standards, the $s_{20,w}$ of the receptor – detergent complex cannot be simply read off from the standard curve. Instead, correction factors must be applied. The calculation is very clearly explained by Peterson *et al.* (15), and this reference should be consulted for the details. An alternative approach is given by Poyner in *Receptor – effector coupling, a practical approach*, Chapter 1 (27). Reference 33 should be read carefully by anyone wishing to understand the finer points and pitfalls. Firstly, the partial specific volume of the receptor – detergent complex is calculated as follows,

$$\bar{v}_R = \frac{(1 - \bar{v}_s \varrho_D{}^S) - x(1 - \bar{v}_s \varrho_H{}^S)}{\varrho_H{}^R(1 - \bar{v}_s \varrho_D{}^S) - x \varrho_D{}^R (1 - \bar{v}_s \varrho_H{}^s)} \tag{1}$$

71

where $\bar{v}_R$ and $\bar{v}_S$ are the partial specific volumes of the receptor and a closely matched standard, respectively, and

$$x = \frac{y_D{}^S \, y_H{}^R}{y_H{}^S \, y_D{}^R} \tag{2}$$

where y is the distance migrated by the receptor (R) or standard (S) in H_2O (H) and D_2O (D). ϱ_H and ϱ_D represent the average densities taken at $0.5y$ in the gradient for H_2O and D_2O. Tables of densities of sucrose solutions are given in reference 28.

The $s_{20,w}$ value of the receptor—detergent complex is then calculated as follows:

$$s_{20,w}^R = \frac{y^R \, (1 - \bar{v}_R \varrho_{20,w}) \, (1 - \bar{v}_S \varrho^S{}_{T,m})}{y_S \, (1 - \bar{v}_S \varrho_{20,w}) \, (1 - \bar{v}_R \varrho^R{}_{T,m})} \times s_{20,w}^S \tag{3}$$

where $\varrho_{20,w}$ is the density of H_2O at $20°C$, $\varrho_{T,m}$ is the average density at $0.5y$ at $5°C$ in the H_2O gradient, and $\bar{v}_R$ is as calculated above.

6.3.2 Gel-filtration of detergent-solubilized receptors

To estimate a value for the molecular weight of the receptor—detergent complex, it is necessary to determine a value for its Stokes' radius. This is done by gel-filtration on a suitable column, for instance agarose gels (such as Sepharose, Sepharose-Cl, Superose), agarose—polyacrylamide composites (such as Sephacryl, Ultrogel), or the TSK silica-based gel-fitlration media (see Chapter 10). The Pharmacia/LKB booklet, *Gel filtration, theory and practice*, can usefully be consulted on the properties of different gel-filtration media. Sephacryl S300 is a popular choice. A Superose 6 column connected to an FPLC pump is also likely to be very useful for the gel-filtration of receptor—detergent complexes, and offers the possibility of relatively short run times.

Protocol 18. Gel-filtration of detergent-solubilized receptors

1. Use a column with an adjustable top support. Pour a column of Sephacryl S300 of approximately 0.5×50 cm. Make sure that the packing is uniform. To achieve this, use a column extension tube (see the Pharmacia booklet), or the set-up shown in *Figure 6*. Before starting, ensure that the gel suspension is free of fines, by suspending it 1:1 in water, and allowing it to settle until a well-defined bed is formed. Carefully decant off any supernatant, and resuspend the gel once more to 1:1 in water.

2. Put a layer of water on the bottom support of the column, and allow some of it to run through, to sweep out any bubbles that may be trapped in the bed support.

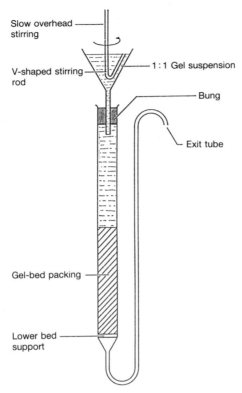

Figure 6. Pouring a gel-filtration column.

3. Introduce the gel suspension until the column is full, and the requisite amount of gel is in the funnel. Stir slowly, and allow the column to run until packing is complete, and the desired bed height has been achieved.

4. Remove the funnel and bung. Put a layer of water over the top of the gel bed, and carefully insert the endpiece. Ensure that all bubbles are expelled. To avoid the introduction of bubbles, use degassed buffers. Lower the bed support onto the top of the gel bed so that there is as little dead space as possible, and tighten it into place.

5. To discover a reasonable flow rate for the column, initially run it under gravity, with the exit tube at its final level. Measure the flow rate. Connect up a peristaltic pump to the column inlet, and pump the column at a proportion of the gravity flow rate (e.g. 50%). With a column of the above dimensions, a flow rate of $30-100$ μl/min is reasonable, and gives run times of $2-5$ h. Slower flow rates will give better resolution (within limits).

6. Working at 4°C, equilibrate the column with detergent buffer. Pass through three column volumes to achieve equilibration. Re-adjust the top bed support

73

as appropriate, if shrinkage occurs. Note that some buffers will have a higher viscosity, and so require a lower flow rate.

7. Calibrate the column using molecular weight markers. Pharmacia provide both low (13.7−67 kDa) and high (158−669 kDa) molecular weight marker kits for this purpose. Standards can be detected by A_{280}, A_{214}, or by enzymic activity measurements, as shown in *Table 3*. The excluded (void) volume v_0 is measured with blue dextran 2000 (supplied with the molecular weight marker kit), which can be detected by measuring A_{650} nm. The dissolution of blue dextran is hastened by warming to *c.* 50°C for a few minutes. It is advisable to centrifuge samples for several minutes in a microfuge at *c.* 12 000 *g* to remove particles, before applying them to the column. The volume applied should be 100−200 *μ*l, or 2−5% of the bed volume. The sample should be carefully pumped onto the bed at the standard flow rate. This is best done with the aid of an injection valve, e.g. the Pharmacia valve V7. The included volume, v_t should be measured with a low molecular weight radioligand, or a coloured compound such as DNP-lysine, or chromate ions. The aim to collect *c.* 50 timed fractions per run, preferably in microfuge tubes. Fractions can be weighed to increase accuracy. Determine the partition coefficient for each standard from the following equation:

$$KD = (v_e - v_0)/(v_t - v_0)$$

where v_e is the volume to the peak of the elution profile of the standard and v_t and v_0 are the volumes to the peaks of the elution profiles of the included and excluded markers.

8. Prelabel the receptor with the radioligand before gel-filtration, as for centrifugation. Unless the ligand dissociation rate constant is extremely slow, it is advisable to include radioligand in the column equilibration solution, to protect the receptor−ligand complex during chromatography. The concentration of the receptor−ligand complex in the fractions is measured by means of the Sephadex G50 F gel-filtration assay. Ligand dissociation during the run will lead to peak trailing. There is an obvious trade-off between this and the run time. This is not a problem if the receptor is labelled with an irreversible ligand.

9. The Stokes' radius (a^R) of the receptor−ligand−detergent complex is calculated using the relation:

$$a^R = A + B \cdot KD^R$$

where A and B are constants for the particular gel used and are determined from the calibration runs. KD^R is the partition coefficient of the receptor.

The molecular weight of the receptor−ligand−detergent complex is calculated from the equation:

$$M_R = \frac{6\pi\eta_{20,w}\ Na^R\ s^R_{20,w}}{1 - \bar{v}_R\ \varrho_{20,w}} \tag{4}$$

where $\eta_{20,w}$ is the viscosity of water at 20°C (1.00×10^{-3} kg/m/s), $s^R_{20,w}$ is calculated from density gradient centrifugation, a^R is measured by gel-filtration, $\varrho_{20,w}$ is the density of water at 20°C (0.998 g/ml) and N is Avogadro's number (6.022×10^{23}).

The number of grams of detergent bound per gram of receptor is calculated from the relationship:

$$\delta_D = \frac{\bar{v}_R - \bar{v}_{Protein}}{\bar{v}_{Detergent} - \bar{v}_R} \tag{5}$$

The molecular weight of the uncomplexed receptor is then given by

$$M = \frac{M_R}{1 + \delta_D} \tag{6}$$

and the frictional coefficient of the receptor-detergent complex is given by

$$f/f_0 = \frac{a^R}{(3\bar{v}_R M_R/4\pi N)^{1/3}} \tag{7}$$

The reader is referred to references 15 and 22 for further insight into these matters, for references to the primary literature, and for more tables of the physical properties of calibrating proteins.

A note on units: with $s_{20,w}$ expressed in seconds ($1S = 10^{-13}$ sec) and a in metres, equation (4) gives molecular weight in kg/mol. $\bar{v}$ is measured in ml/g and ϱ in g/ml. To use equation (7), convert a^R to units of cm and M_R to the usual units of g/mol.

In the case of mAChRs in Lubrol PX, the molecular weight (i.e. that of the mAChR−detergent complex) and $\bar{v}$ were estimated to be 133 000 and 0.814 ml/g, respectively (19). The value of δ_D was estimated to be 0.35 by using values of 0.735 and 0.935 ml/g for $\bar{v}$(protein) and $\bar{v}$(Lubrol PX), respectively, and therefore the molecular weight of mAChR to be 86 000.

7. Cloning, sequencing, and expression of cDNAs for mAChRs

7.1 mAChR I and mAChR II subtypes

The amino-acid sequences of the mAChRs have been determined by the following procedure (1,2).

(a) Purification of mAChR (1.1 nmol) from 1.4 kg of porcine cerebra by affinity chromatography (ABT-agarose gel) followed by gel-permeation HPLC.

(b) Determination of the amino-acid sequences of five peptides obtained on partial hydrolysis of purified mAChR with TPCK-trypsin followed by separation by reverse-phase HPLC (Chemcosorb 3 μ C_{18}–H column cf. Chapter 10). The determined sequences of the five peptides comprised 13, 13, 6, 16, and 15 residues, respectively.

(c) Chemical synthesis of oligonucleotides corresponding to the sequences of portions of the above peptides.

(d) Screening to detect cDNAs corresponding to mAChRs from a library of cDNA clones derived from mRNA from porcine cerebra (or porcine heart) by the use of the ^{32}P-labelled synthetic oligonucleotides (see reference 29).

(e) Determination of the nucleotide sequence of the cDNA clones and deduction of the amino-acid sequence from the nucleotide sequence.

The amino-acid sequence deduced from the nucleotide sequence of a cDNA cloned from a cerebral library contained the sequences of only three of the five peptides which had been determined by protein sequencing. Another cDNA was cloned from a library derived from porcine cardiac mRNA, and the amino-acid sequence deduced from the cDNA sequence was found to contain the sequences of the other two peptides. These results indicate that there are at least two different mAChRs, which were designated as mAChR I and II (23). Northern blot analysis indicates that mAChR I is more abundant in the cerebrum than in the heart, the reverse being the case for mAChR II.

7.2 Other subtypes

cDNAs or genomic DNAs for three additional subtypes of mAChRs have been cloned by screening with probes based on the nucleotide sequences of the above two cDNAs. They were termed m3, m4, and m5 (3,26), or M3 and M4 (24), whereas mAChR I and mAChR II has been designated m1 and m2 or M1 and M2, by different groups. The generally accepted nomenclature is now that of Bonner *et al.* (3,26). The genes for the mAChR subtypes were found to be free from introns (3,24), as in the case of the genes for β-adrenergic receptors.

7.3 Expression of cDNA clones for mAChRs

Muscarinic responses and muscarinic ligand-binding activity were expressed in *Xenopus* oocytes into which mRNA transcribed from cDNA clones had been injected (1,23,30), or in cultured (e.g. CHO) cells into which cDNA clones had been transfected (3,24). These results constitute evidence that the amino-acid sequences deduced from the nucleotide sequences of these cDNA clones represent those of mAChR and that there are at least five different mAChR subtypes.

The affinities of different mAChR clones for the selective antagonist, pirenzepine, have been found to be 25- to 50-fold higher for m1 and m4 than for m2; the apparent dissociation constants were 10−20 nM for the former and 500 nM for the latter (1,3,23). Thus, m1 and m4 belong to the original M1 subclass, and m2 to the original

M2 subclass, these initially having been determined pharmacologically by their relative affinities for selective ligands such as pirenzepine and its congeners (32). The reader is referred to the original literature for further discussion of this rapidly developing and still incompletely resolved area (see references 23−26).

8. Ligand-binding activity of purified mAChRs

The ligand-binding properties of solubilized and purified mAChRs are essentially the same as those of membrane-bound mAChRs as far as specificity for muscarinic ligands is concerned; for example, the same stereoselective binding and the same orders of potency for various ligands are observed for both purified and membrane-bound receptors. There are, however, several differences between purified and membrane receptors. First, the affinities of mAChRs for any given ligand are generally decreased by solubilization and purification, the decrease being about the same in magnitude for most ligands. Secondly, the agonist binding of membrane receptors is affected by guanine nucleotides but that of purified receptors is not. This reflects the interaction of membrane receptors with GTP-binding proteins (G-proteins) and is reproduced by reconstituting purified receptors with G-proteins in phospholipids, as described in *Receptor−effector coupling, a practical approach*, Chapter 5 (27). The third difference concerns the subtype-specific ligand, pirenzepine. The affinity for pirenzepine is dependent on source for membrane-bound mAChRs but not for purified mAChRs; mAChRs of cerebral membranes exhibit 50 times as high an affinity for pirenzepine as mAChRs of cardiac membranes, but the difference in affinity is only twofold for mAChRs solubilized and purified from the two tissues (25). On the other hand, the m1 and m2 subtypes expressed in *Xenopus* oocytes show 50-fold differences in their affinities for pirenzepine. The subtype-specific conformation of mAChR which is measured by the affinity of pirenzepine may not be determined only by the mAChR sequence but may also be affected by the interaction of the mAChRs with lipids such as cholesterol, or other membrane components (34).

References

1. Kubo, T., Fukuda, K., Mikami, A., Maeda, A., Takahashi, H., Mishina, M., Haga, T., Haga, K., Ichiyama, A., Kangawa, K., Kojima, M., Matsuo, H., and Numa, S. (1986). *Nature,* **323**, 411.
2. Kubo, T., Maeda, A., Sugimoto, K., Akiba, I., Mikami, A., Takahashi, H., Haga, T., Haga, K., Ichiyama, A., Kangawa, K., Matsuo, H., Hirose, T., and Numa, S. (1986). *FEBS Lett.,* **209**, 367.
3. Bonner, T. I., Buckley, N. J., Young, A. C., and Brann, M. R. (1987). *Science,* **237**, 527.
4. André, C., Backer, J.-P. De, Guillet, J. G., Vanderleyden, P., Vanquelin, G., and Strosberg, A. D. (1984). *EMBO J.,* **2**, 495.
5. Peterson, G. L., Herron, G. S., Yamaki, M., Fullerton, D. S., and Schimerlik, M. I. (1984). *Proc. Natl. Acad. Sci. (USA),* **81**, 4993.
6. Haga, K. and Haga, T. (1985). *J. Biol. Chem.,* **260**, 7927.

7. Wheatley, M., Birdsall, N. J. M., Curtis, C., Eveleigh, P., Pedder, E. K., Poyner, D., Stockton, J. M., and Hulme, E. C. (1987). *Trans. Biochem. Soc.,* **15**, 113.
8. Haga, K. and Haga, T. (1983). *J. Biol. Chem.,* **258**, 13575.
9. Jones, D. H. and Matus, A. L. (1974). *Biochim. Biophys Acta,* **356**, 276.
10. Peterson, G. L. and Schimerlik, M. I. (1984). *Prep. Biochem.,* **14**, 33.
11. Castell, J. V., Cervera, M., and Macro, R. (1979). *Anal. Biochem.,* **99**, 379.
12. Laemmli, U. K. (1970). *Nature,* **227**, 680.
13. Switzer, R. C., Merrill, C. R., and Shifrin, S. (1979). *Anal. Biochem.,* **98**, 231.
14. Liang, M., Martin, M. W., and Harden, T. K. (1987). *Mol. Pharmacol.,* **32**, 443.
15. Peterson, G. L., Rosenbaum, L. C., Broderick, D. J., and Schimerlik, M. I. (1986). *Biochemistry,* **25**, 3189.
16. Venter, J. C. (1983). *J. Biol. Chem.,* **258**, 4842.
17. Dadi, H. K. and Morris, R. J. (1984). *Eur. J. Biochem.,* **144**, 617.
18. Repke, H. and Schmitt, M. (1987). *Biochim. Biophys. Acta,* **929**, 62.
19. Haga, T. (1980). *FEBS Lett,* **113**, 68.
20. Berrie, C. P., Birdsall, N. J. M., Haga, K., Haga, K., Haga, T., and Hulme, E. C. (1984). *Br. J. Pharmacol.,* **82**, 839.
21. Clark, S. (1975). *J. Biol. Chem.,* **250**, 5459.
22. Haga, T., Haga, K., and Gilman, A. G. (1977). *J. Biol. Chem.,* **252**, 5776.
23. Fukuda, K., Kubo, T., Akiba, I., Maeda, A., Mishina, M., and Numa, S. (1987). *Nature,* **327**, 623.
24. Peralta, E. G., Ashkenazi, A., Winslow, J. W., Smith, D. H., Ramachandran, J., and Capon, D. J. (1987). *EMBO J.,* **6**, 3923.
25. Haga, T., Haga, K., Berstein, G., Nishiyama, T., Uchiyama, H., and Ichiyama, H. (1988). *Trends Pharmac. Sci.,* (Suppl. III), 12.
26. Bonner, T. I., Young, A. C., Brann, M. R., and Buckley, N. J. (1988). *Neuron,* **1**, 403.
27. Hulme, E. C. (ed.) *Receptor—effector coupling, a practical approach,* IRL Press, Oxford, in press.
28. Findlay, J. B. C. and Evans, W. H. (eds) (1987). *Biological membranes, a practial approach,* IRL Press, Oxford.
29. Glover, D. M. (ed.) (1987). *DNA cloning, a practical approach,* IRL Press, Oxford, Vols. 1−3.
30. Turner, A. J. and Bachelard, H. S. (eds) (1987). *Neurochemistry, a practical approach,* IRL Press, Oxford.
31. Hulme, E. C. (ed.) *Receptor—ligand interactions, a practical approach,* IRL Press, Oxford, in press.
32. Birdsall, N. J. M., Curtis, C. A. M., Eveleight, P., Hulme, E. C., Pedder, E. K., Poyner, D., and Wheatley, M. (1988). *Pharmacology,* (Suppl. 1), 22.
33. O'Brien, R. D., Timpore, C. A., and Gibson, R. E. (1978). *Anal. Biochem.,* **86**, 602.
34. Berstein, G., Haga, T., and Ichiyama,A. (1989). *Mol. Pharmacol.,* **36**, 601.

3

Dopamine receptors: isolation and molecular characterization

PHILIP G. STRANGE and RICHARD A. WILLIAMSON

1. Introduction

The actions of the catecholamine neurotransmitter dopamine can be accounted for by the binding of dopamine to D_1 and D_2 dopamine receptors. The properties of the two receptor classes have been discussed in detail elsewhere (1,2) and they appear to be distinct proteins with different pharmacological properties linked to different biochemical and physiological events. The isolation and purification of the receptor proteins is essential in understanding how the two receptor systems function. These methods will further enable the composition of the two proteins to be determined, will enable studies on receptor reconstitution to be performed, and will assist in the cloning of the genes for the receptors. Even when the genes have been cloned and the primary structure of the receptors has been obtained, studies on the isolated proteins will be required in order to locate key regions of the receptor, e.g. the ligand-binding site. This underlines the need for methods of purification.

As the dopamine receptors are integral membrane proteins the first stage in their isolation will be solubilization from the membrane by detergent. Once a suitable solubilized preparation has been obtained the receptor may be purified using protein purification procedures. This chapter discusses these methods in relation to dopamine receptors.

2. Solubilization of dopamine receptors

Solubilization of a receptor by detergent consists of releasing the protein from the membrane into a monodisperse form. It is most important that a true solubilization be achieved rather than the breaking of the membrane into small fragments, and a set of criteria may be laid down to establish true solubilization. These have been discussed in detail elsewhere (1) and soluble preparations of D_1 and D_2 receptors have been obtained that satisfy these criteria. The methods used will be discussed below.

2.1 D_1 dopamine receptor

Work on the isolation of this receptor has been much slower than for the D_2 dopamine receptor, mainly owing to the lack of a good radioligand. Radiolabelled benzazepines, e.g. [³H]SCH 23390 (Amersham or NEN Dupont) became available recently and this has allowed successful solubilization of this receptor. D_1 receptors (labelled by [^{125}I]SCH 23982, related to SCH 23390) have been solubilized using cholate (3) but the preparation requires reconstitution into vesicles before it can be assayed. Solubilization of D_1 receptors (labelled by [³H]SCH 23390) has been reported (4,5,6) using digitonin, the preparations have been characterized and a report has appeared of the partial purification of this receptor (see below).

2.2 D_2 dopamine receptor

Initial reports on solubilization of this receptor from brain using digitonin appeared in 1978 (7) and 1979 (8) and since then many different detergents have been used successfully to provide well-characterized soluble preparations of D_2 dopamine receptors (reviewed in reference 1). Currently, the highest yield of solubilization of D_2 receptors from brain is achieved with cholate (35−53% yield depending on species). We have found that a suitable preparation of bovine brain for solubilization is a mixed mitochondrial/microsomal membrane preparation. This contains a high proportion of the D_2 dopamine receptors present in the starting brain region but large membrane fragments have been removed. The method for preparing membranes is given in *Protocols 1* and *2*. D_2 dopamine receptors may then be solubilized from this membrane preparation using 0.3% cholate in high yield, and a method for this is given in *Protocols 3* and *4*.

Protocol 1. Preparation of a mixed mitochondrial − microsomal membrane preparation from bovine caudate nuclei

1. Dissect out caudate nuclei from bovine brains, this should be done within 1 h of death and caudate nuclei should be stored at −80°C until required (we have stored caudate nuclei for up to 3 months without deterioration in [³H]spiperone binding). The tissue should be stored in a closed container to avoid freeze drying.

2. Prepare the following solutions:
   (a) *Homogenization buffer*, Hepes (20 mm), EDTA (10 mm), EGTA (1 mM), sucrose (300 mM), adjusted to pH 7.4.
   (b) *Resuspension buffer*, Hepes (20 mM), EDTA (10 mM), EGTA (1 mM) adjusted to pH 7.4.
   (c) *Phenylmethylsulphonyl fluoride (PMSF) solution.* Dissolve 34.8 mg PMSF in 1 ml ethanol. This should be prepared fresh on the day of use.

3. All solutions and equipment should be cooled to 4°C before use, all manipulations should be carried out at 4°C. Excess amounts of buffers (a and b) can be stored at 4°C for future use.

4. Thaw caudate nuclei at room temperature and weigh the tissue; we generally use about 25 g (from about six brains) in a typical membrane preparation. The caudate nuclei should be used as soon as they are thawed and should not be refrozen.

5. Homogenize the caudate nuclei in 3 volumes buffer (a) supplemented with 1.5 μl PMSF solution/ml. PMSF is subject to hydrolysis and should be added to the buffer immediately before homogenization. 10 strokes (up and down = 1 stroke) of a Teflon/steel or Teflon/glass Potter−Elvejhm homogenizer of 0.18 mm radial clearance at 850 r.p.m. provides sufficient homogenization.

6. Add a further 6 volumes of buffer (a) to the homogenate (giving 9 volumes total).

7. Centrifuge the homogenate at 1000 g for 10 min.

8. Carefully decant and save the supernatant. Combine the pellets and resuspend them in half the volume of buffer (a) used in *steps 5* and *6* (equivalent to 4.5 volumes) using 5 strokes of the homogenizer.

9. Centrifuge the resuspended pellets at 1000 g for 10 min.

10. Decant the supernatant and combine it with the supernatant obtained in *Step 8*.

11. Centrifuge the combined supernatants (*8* and *10*) at 200 000 g for 30 min.

12. Discard the supernatants and resuspend the pellets in buffer (b) (2 ml buffer per gram of original wet tissue) using 5 strokes of the homogenizer.

13. Assay a small portion (100 μl) of the membrane preparation for protein as described in Protocol 2. Store the remainder of the membranes at −80°C until required. The membranes must not be exposed to repeated freeze/thaw cycles so for convenience they should be stored in small aliquots (e.g. 3 ml) in tubes or in vials, which can be frozen by immersion in a dry ice/isopropanol freezing mixture.

Protocol 2. Protein assay for a mixed mitochondrial − microsomal membrane preparation

The technique used is a slightly modified version of that in (30).

1. Prepare the following solutions:

   (a) *Diluting buffer* (same as the resuspension buffer in Protocol 1), Hepes (20 mM), EDTA (10 mM), EGTA (1 mM), adjusted to pH 7.4.

   (b) *Membrane preparation*. Add 100 μl of the mixed mitochondrial−microsomal membrane preparation (produced as in Protocol 1) to 500 μl buffer (a).

   (c) *Trichloroacetic acid (TCA) solution*, TCA (10% w/v) in water.

   (d) *Lowry solution A*, anhydrous sodium carbonate (2% w/v), sodium hydroxide (0.1 M).

   (e) *Lowry solution B*, copper sulphate pentahydrate (0.5% w/v), sodium potassium tartrate (1% w/v), adjusted to pH 7.4.

   (f) *Lowry solution C*, Folin−Ciocalteau's phenol reagent (Fisons), diluted 1:1 with distilled water on the day of use.

(g) *Bovine serum albumin (BSA) solution*, BSA (4 mg/ml) in water. Aliquots of this solution can be stored at −20°C for future use.

2. Prepare quadruplicate tubes containing 0, 10, 20, 30, and 40 μl of BSA solution (g) and, in separate tubes, 30 and 40 μl of diluted membrane preparation (b).

3. Add 1 ml 10% TCA solution (c) to each tube, vortex and centrifuge tubes at 3600 r.p.m. (2200 g) (MSE, Centaur 2) for 10 min.

4. Decant supernatant and leave tubes inverted to drain for 10 min.

5. Add 1 ml Lowry solution B (e) to 50 ml Lowry solution A (d), mix and then add 1 ml of the A/B solution to each tube. Vortex all tubes and leave for 10 min.

6. Add 100 μl of solution Lowry C (f) to each tube. Incubate tubes for 30 min.

7. Add 2 ml distilled water to each tube and measure the absorbance at 750 nm.

8. Construct a standard curve relating absorbance to protein concentration and read off the protein level in the unknown samples.

Protocol 3. Solubilization of D_2 dopamine receptors

1. Prepare a mixed mitochondrial−microsomal membrane preparation from bovine caudate nucleus as shown in *Protocol 1*.

2. Determine the protein concentration of the membrane preparations as described in *Protocol 2*.

3. Prepare the following solutions:

   (a) *Membrane diluting buffer* (same as the resuspension buffer in *Protocol 1*), Hepes (20 mM), EDTA (10 mM), EGTA (1 mM), adjusted to pH 7.4.

   (b) *Detergent buffer*, Hepes (20 mM), EDTA (10 mM), EGTA (1 mM), NaCl (2 M), sodium cholate (0.6% w/v), pH 7.4. The pH of the buffer should be adjusted to above 6.5 before adding the sodium cholate which is insoluble below pH 6.5. When all the components have dissolved adjust pH to 7.4.

   (c) *Proteinase inhibitor solution*. Add 5 mg quantities of leupeptin, chymostatin aprotinin, pepstatin A, and antipain to 10 ml distilled water. The chymostatin and pepstatin A are insoluble and are used in suspension. The protease inhibitors can be obtained from Sigma.

   (d) *Phenylmethylsulphonyl fluoride (PMSF) solution*. Dissolve 34.8 mg PMSF in 1 ml ethanol. This should be prepared fresh on the day of solubilization.

4. Cool all solutions and equipment to 4°C before use and carry out all manipulations at 4°C. Excess quantities of buffers (a) and (b) and proteinase inhibitor solution (c) should be stored at 4°C for future use.

5. Dilute the membrane preparation (*1*) with diluting buffer (a) to give a final protein concentration of 8mg/ml.

6. Add proteinase inhibitor solution (c) to the diluted membranes (20 μl inhibitor solution/ml diluted membranes).

7. Add PMSF solution (d) to diluted membranes at a concentration of 1 μl/ml.

8. Add an equal volume of buffer (b) to the diluted membranes giving a final protein concentration of 4 mg/ml and final cholate and NaCl concentrations of 0.3% (w/v) and 1 M.

9. Stir the suspension gently at 4°C for 1 h. Stirring should be done with a magnetic stir bar and should maintain the membranes in suspension without causing any frothing.

10. Centrifuge the suspension at 200 000 g for 1.5 h.

11. Decant the upper part of the supernatant from the pellet with great care. The last few ml of supernatant should be discarded, preventing any contamination of the solubilized receptor with unsolubilized lipid and small membrane fragments which can be seen as a whispy trace coming from the pellet.

12. Measure [³H]spiperone binding as described in *Protocol 5* and protein concentrations in *Protocol 4*. The solubilized receptor preparation should be stored at 4°C until required. [³H]Spiperone-binding activity will decrease with time and so the preparation should be used within 24 h and preferably on the day of production.

Protocol 4. Protein assay for cholate/sodium chloride solubilized receptor preparations.

This technique is a slightly modified version of that used to determine protein in the mixed mitochondrial–microsomal membrane preparation (*Protocol 2*).

1. Prepare the following:

   (a) *Soluble receptor preparation*. Prepared as described in *Protocol 3*.

   (b) *Trichloroacetic acid (TCA) solution*, TCA (25% w/v) in water.

   (c) *Sodium dodecyl sulphate (SDS) solution*, SDS (10% w/v) in water.

   (d) *Lowry solutions A, B, and C*, see *Protocol 2, step 1, d, e and f*.

   (e) *BSA solution*, see *Protocol 2, step 1g*.

2. Prepare quadruplicate tubes containing 0, 10, 20, 30, and 40 μl of BSA solution (e) and, in separate tubes, 30 and 40 μl of soluble receptor preparation (a).

3. Add 1 ml 25% TCA solution (b) to each tube, vortex and centrifuge tubes at 3600 r.p.m. (2200 g) (MSE, Centaur 2) for 10 min.

4. Decant supernatant and leave tubes inverted to drain for 10 min.

5. Add 1 ml Lowry A/B solution (see *Protocol 2, step 5*) to each tube and vortex.

6. Add 1 ml 10% SDS solution (c) to each tube. Vortex and leave tubes overnight.

7. Add 100 μl Lowry solution C (d) to each tube. Vortex and then incubate all tubes for 30 min.

8. Add 1 ml distilled water to each tube and measure the absorbance at 750 nm.

9. Construct a standard curve from the values for BSA and read off the protein concentration in the solubilized preparation.

D$_2$ dopamine receptors in brain membranes may be assayed using the radioligand [^{3}H]spiperone (9) and this radioligand is very useful for assaying receptors after solubilization. A method for the ligand-binding assay of cholate-solubilized D$_2$ dopamine receptors with [^{3}H]spiperone is given in *Protocol 5*.

Protocol 5. Ligand-binding assay for solubilised D$_2$ dopamine receptor using [^{3}H]spiperone

1. Prepare the following:

   (a) *Soluble D$_2$ dopamine receptor preparation.* Solubilize D$_2$ dopamine receptors from bovine caudate nucleus as in *Protocol 3*.

   (b) *Assay buffer*, Hepes (20 mM), EDTA (1 mM), EGTA (1 mM), adjusted to pH 7.4 (cool to 4°C).

   (c) [^{3}H]Spiperone working solution. This contains [^{3}H]spiperone (Amersham or NEN Dupont) at a concentration of about 10 nM in assay buffer (b). The exact concentration is determined on the day of the assay by scintillation counting of aliquots. For convenience the [^{3}H]spiperone obtained from the supplier is diluted into ethanol to a concentration of 5 μM, which is divided into aliquots of 0.5 ml and held at −20°C. The 5 μM solution can then be further diluted on the day of the assay.

   (d) *Butaclamol solution.* (+)- or (−)-butaclamol (Research Biochemicals Inc. or SEMAT) (10 μM) in assay buffer (b). For convenience these compounds are held as a frozen (−20°C) stock at 100 μM in buffer (b) containing 0.25% acetic acid (used to dissolve the drug powder) and diluted on the day of the assay.

   (e) *Charcoal suspension.* Mix together 1 part (by weight) of charcoal (Norit GSX, from BDH), 1 part (by volume) of albumin solution (see below) and 9 parts (by volume) of assay buffer (b) and stir for 30 min at 4°C. Stirring should be with a small magnetic stir bar and be vigorous enough to keep the charcoal in suspension but not so vigorous as to cause frothing. The albumin solution is kept as a stock at 4°C for convenience and contains bovine serum albumin (RIA grade from Sigma) (22%), NaCl (0.85%), NaN$_3$ (0.1%) (all w/v).

2. All manipulations are performed at 4°C.

3. Set up triplicate tubes (LP3, Hughes and Hughes) containing [^{3}H]spiperone

working solution (10 nM) (50 μl) and 50 μl of (+)- or (−)-butaclamol (10 μM). Start assay by adding to each tube 400 μl of soluble D_2 dopamine receptor preparation, vortex mix, and leave at 4°C for 4−16 h.

4. Take tubes in groups of maximum size of 12 and add to each tube 100 μl of charcoal suspension. This can be done conveniently with a Flow Laboratories Syringe Phaser equipped with a 2 ml plastic disposable syringe and a wide-gauge needle. It is particularly important to add the same amount of charcoal to each tube and this can prove a problem as the charcoal tends to fall out of suspension. Therefore the charcoal working suspension should be stirred constantly, rapidly drawn up into the syringe and rapidly dispensed. With batches of 12 tubes this can be achieved successfully. Following addition of charcoal vortex-mix tubes and immediately centrifuge (12 000 g, 2.5 min, Damon IEC microcentrifuge).

5. Pipette 300 μl of supernatant from each tube directly into a scintillation counting vial (5 ml volume, Hughes and Hughes) and determine radioactivity after mixing with 3 ml of Optiphase (Fisons).

6. The difference in radioactivity between tubes containing (+)- and (−)-butaclamol represents specific [³H]spiperone binding.

Although [³H]spiperone is an excellent radioligand for the D_2 dopamine receptor it must be remembered that it can also label $5HT_2$ serotonin receptors, α_1-adrenergic receptors, and an acceptor site termed the spirodecanone site (2). [³H]Spiperone binding to spirodecanone sites does not show stereoselectivity so that definition of specific [³H]spiperone binding as the difference in [³H]spiperone binding in the presence of (+)- and (−)-butaclamol will eliminate that component. Spirodecanone binding of [³H]spiperone has been a problem, however, in increasing the apparent non-specific binding of [³H]spiperone in some solubilized preparations (10) but this is not the case for cholate-solubilized preparations (11).

[³H]Spiperone binding to $5HT_2$ serotonin and α_1-adrenergic receptors may be checked by competition with selective antagonists, e.g. mianserin ($5HT_2$) and prazosin (α_1) and where present these components may be eliminated using a suitable selective blocking concentration (9). In soluble preparations of bovine caudate nucleus obtained using 0.3% cholate, specific [³H]spiperone binding is solely to D_2 dopamine receptors (12) which simplifies analysis considerably. It is of interest that in a previous study using 0.2% cholate for solubilization, [³H]spiperone binding was to D_2 dopamine and $5HT_2$ serotonin receptors (11) so that the $5HT_2$ serotonin receptor is selectively unstable to raising the cholate concentration.

The yield of a solubilization procedure is important, particularly if the preparation is to be used for purification of receptor, and the method of *Protocols 3* and *4* gives high-yield solubilization. The stability of the preparation is also important in order to preserve active receptor during purification. In early studies on preparations of soluble D_2 receptors obtained in 0.2% cholate it was noted (11) that during fractionation on sucrose density gradients or gel-filtration columns it was necessary to include phospholipid in buffers to preserve receptor activity, and this is now done in all fractionation procedures (see below). Also although solubilization with 0.3% cholate gives a slightly higher yield of solubilization than 0.2% cholate, the former preparation is much less stable ($t_{1/2}$ at 4°C: 0.3% cholate 20 h, 0.2% cholate 62 h, V.A. Sessions and P.G. Strange, unpublished results). Dilution of the 0.3% cholate/1 M sodium chloride preparation to 0.225% cholate/0.75 M sodium chloride gives a much stabler preparation, with full binding of the radioligand [^{3}H]spiperone used for assays. Therefore solubilization with 0.3% cholate/1 M sodium chloride followed by dilution to 0.225% cholate/0.75 M sodium chloride, maximizes extraction and stability.

These comments serve to underline the problems we have encountered when working with soluble D_2 dopamine receptors and show that choice of solubilization conditions is not necessarily straightforward.

3. Fractionation of soluble dopamine receptors

Purification of these receptors requires fractionation of protein mixtures containing the receptors and methods for this will be considered in this section. Protocols for fractionation of soluble D_1 receptors on lectin affinity columns and HPLC size exclusion columns have been described (4). One report of the fractionation of digitonin-solubilized D_1 dopamine receptors by affinity chromatography on a benzazepine (SCH 39111, related to SCH 23390) affinity column has appeared (6). A 200−250-fold purification of the receptor was achieved and this column should provide the basis of a full purification protocol.

Solubilized D_2 dopamine receptors have been fractionated by gel filtration (11), isoelectric focusing (13), sucrose density gradient centrifugation (11), and HPLC (14). These methods yield only moderate degrees of enrichment (<tenfold) and more powerful techniques will be required for complete purification which requires an enrichment of about 50 000-fold (starting from the detergent-solubilized preparation). To achieve such a high degree of purification an affinity method based on a specific property of the receptors must be used. Lectin affinity chromatography and ligand affinity chromatography have been used in this context.

3.1 Lectin affinity chromatography

The D_2 dopamine receptor is a glycoprotein and will adsorb to and can be eluted from wheat-germ agglutinin-Sepharose (1,15). Elution is biospecific, is best achieved with N-acetyl-glucosamine and gives about tenfold purification of receptors. Although the degree of purification achieved is not great, providing phospholipid is included in the eluting buffers, we have found that this technique provides a high overall recovery of soluble receptor and so may be a valuable adjunct to a purification procedure.

We have found there to be some variability in the properties of commercially available wheat-germ agglutinin-Sepharose. Whereas in our early study (15) high-efficiency adsorption and elution (N-acetyl-glucosamine, 100 mM) of D_2 receptors solubilized in either 0.2% or 0.3% cholate/1 M sodium chloride could be achieved, more recently this has not been found to be the case. In more recent studies we have found that uptake of soluble D_2 receptors by wheat-germ agglutinin-Sepharose is efficient only if 0.3% cholate and 1 M sodium chloride are present (16). Also whereas previously purification was routinely in excess of tenfold, in more recent studies eluting with 100 mM sugar this has declined to about fourfold. However, by using a much lower eluting sugar concentration (10 mM) high recovery of receptor and a better degree of purification are achieved. We have no explanation for the variability but it emphasizes the need for rigorous definition of conditions for adsorption and elution in such experiments. The current protocol is now described.

Protocol 6. Wheat-germ agglutinin-Sepharose affinity chromatography of the D_2 dopamine receptor

1. Prepare the following:

    (a) *Solubilized receptor preparation.* Prepared on the day of use as described in Protocol 3 and kept at 4°C.

    (b) *Sodium chloride solution*, sodium chloride (2 M) in water.

    (c) *Washing buffer*, Hepes (20 mM), EDTA (10 mM), EGTA (1 mM), sodium cholate (0.3% w/v), sodium chloride (1 M), soya bean phosphatidylcholine (0.06% w/v, Sigma), pH 7.4. The pH of the buffer should be adjusted to >6.5 before adding the sodium cholate which is insoluble below this pH. The soya bean phosphatidylcholine should be added last and the buffer stirred with a magnetic stir bar for several hours at room temperature until the lipid has dissolved, the buffer should then be adjusted to pH 7.4.

    (d) *Eluting buffer.* As buffer (c) but containing 10 mM N-acetyl-D-glucosamine.

    (e) *Eluting buffer.* As buffer (c) but containing 100 mM N-acetyl-D-glucosamine.

    (f) *Storage buffer*, Hepes (20 mM), EDTA (10 mM), EGTA (1 mM), sodium azide (0.02% w/v) adjusted to pH 7.4.

2. Immobilized wheat-germ agglutinin (WGA) can be obtained commercially from a variety of sources. We have used WGA-Sepharose (Sigma) and WGA-agarose (Vector Laboratories) and found them to be similar in performance. The purification procedure is carried out batchwise and this can be done conveniently using the WGA-matrix in a column made from a 20 ml syringe. The column is made by cutting a large hole in the syringe bung on the end of the plunger. The bung is then wrapped in 25 μm nylon bolting cloth which can be secured with a short length of thin (~ 2 mm) visking tubing. The bung is then inserted into the syringe with the tied end at the bottom. The syringe should be washed with 100 ml ethanol, 100 ml sodium chloride solution (*step 1b*) and 100 ml distilled water. The matrix (1 ml) can then be added to the syringe and a second bung used to close the top of the column.

3. All solutions and equipment should be cooled to 4°C before use and all manipulations should be performed at 4°C. Excess quantities of washing buffer (c) can be stored at 4°C until required.

4. Wash the WGA-matrix with 5 × 10 ml aliquots of washing buffer (c) over 10 min, the flow can be regulated using a Luerlock syringe tap (American Pharmaseal Company).

5. Let the matrix run dry and carefully expel any further buffer from the syringe by gently pushing in the bung used to close the column.

6. Close the Luerlock tap and add 5 ml solubilized receptor preparation (*Step 1a*) to the column (larger volumes can be added but the amount of matrix should be increased accordingly). Close the column with the second bung; to allow air to escape a hypodermic needle should be held on the inside edge of the syringe when inserting the bung, the needle can then be removed sealing the column.

7. Rotate the column on its long axis (i.e. not end-over-end) for 1 h so that the liquid and matrix mix thoroughly but gently.

8. Remove the bung and collect the run-through fraction from the column. Gently expel any further liquid from the matrix.

9. Wash the column with 3 × 10 ml aliquots of washing buffer (c) over about 5 min, gently expel each aliquot from the gel between additions. Save the combined wash fractions.

10. To elute the column add 5 ml eluting buffer (d) to the matrix, close the column and rotate it for 45 min.

11. Collect the eluate from the column and repeat *step 10* using eluting buffer (e).

12. Wash the column with 5 × 10 ml washing buffer (c) and 2 × 10 ml storage buffer (f). Discard the washes and store the column at 4°C with 10 ml storage buffer (f).

13. Measure the [³H]spiperone binding (*Protocol 5*) and protein (*Protocol 4*) levels in all fractions. A measure of purification can then be obtained by dividing the specific activity of receptor (fmoles receptor/mg protein) in the eluate by the specific activity in the starting material. Purification figures together with values of percentage uptake of receptor by the column, percentage yield of applied receptor in the eluate, and overall recovery of receptor can be used to monitor column performance.

3.2 Ligand affinity chromatography

This technique should in principle be the most powerful one for purification of a receptor. Attachment of a ligand selective for the receptor to a matrix such as Sepharose should provide a highly selective means of purifying a receptor from a complex mixture of proteins such as is found in detergent-solubilized preparations of brain. We have developed such affinity matrices based on the butyrophenones haloperidol and spiperone, which have a high affinity for the D_2 dopamine receptor. Either compound can be coupled to sepharose via the carbonyl group, whereas haloperidol can also be coupled via its hydroxyl group (17). For routine work we use an affinity matrix with haloperidol coupled via its hydroxyl group with a succinic acid linker to amino hexyl Sepharose and the method for synthesis is given in *Protocol 7*. Although other modes of coupling and different length spacer arms also give workable affinity matrices we have found this one to perform best.

Protocol 7. Synthesis of the haloperidol-Sepharose affinity matrix

The technique is a modified version of that of reference 31.

1. Prepare the following:
   (a) *Sodium phosphate buffer*, sodium dihydrogen orthophosphate (100 mM), adjusted to pH 6.8.
   (b) *Sodium borate buffer*, sodium metaborate (100 mM), adjusted to pH 9.0.
   (c) *Sodium hydroxide solution*, sodium hydroxide (1 M) in water.
   (d) *Sodium chloride solution*, sodium chloride (0.5 M) in water.

2. Dissolve 113 mg haloperidol (Janssen Pharmaceutica) in 4 ml chloroform. Add 7 µl succinyldichloride with vigorous mixing and incubate for 30 min at room temperature.

3. Add 2 ml distilled water to stop the reaction. Add 3 ml sodium phosphate buffer (a) and adjust the pH to 6.8 with sodium hydroxide solution (c).

4. Add 15 ml chloroform to the mixture and stir very vigorously with a magnetic stir bar for about 20 seconds.

5. Remove the organic layer (lower layer) and repeat the chloroform extraction step (4) a further 4 times.

6. Combine the chloroform extracts and gradually add anhydrous sodium sulphate with swirling. The first few grams of the sodium sulphate will clump together as it absorbs the water in the extract; however, subsequent crystals will remain free and flow smoothly in the extract. At this point the extract should be swirled gently for about 5 min without any further addition of anhydrous sodium sulphate.

7. Filter the extract to remove the sodium sulphate and wash the crystals with a further 15 ml of chloroform. Combine the extract and wash fractions and evaporate off the chloroform in a rotary evaporator at about 30°C.

8. It is essential to remove any traces of water from the product (haloperidol hemisuccinate) before the next stage of the reaction. The residue is thus dissolved in about 20 ml ethanol and the evaporation procedure (*step 7*) repeated. The ethanol evaporation step should be repeated until the product becomes very viscous (this usually requires 2 or 3 ethanol evaporations).

9. During the extraction and drying procedure 10 ml (~2.5 g) AH Sepharose 4B (Pharmacia) should be incubated in sodium chloride solution (d). When fully swollen (~3 h) the matrix should be placed in a 50 ml syringe (adapted as described in *Protocol 6, step 2*) and washed with 200 ml sodium chloride solution (d) and 200 ml distilled water.

10. Gently expel any remaining liquid from matrix and close the column with a Luerlock tap. Add 3 ml dioxane and 5 ml sodium borate buffer (b) to the column and adjust the pH to 9.0.

11. Dissolve the residue from *step 8* in 4 ml dry dioxane (Aldrich, Gold label grade) and keep the solution at 8°C (the dioxane will freeze if kept at 4°C). Add 41 μl triethylamine and 39 μl isobutyl chloroformate and incubate with occasional gentle mixing at 8°C for 20 min.

12. Add the reaction mixture dropwise to the matrix with occasional gentle swirling. The pH of the reaction mixture should be maintained at 9.0 throughout the period of addition using sodium hydroxide solution (c). Incubate the mixture for 1 h at room temperature (maintaining the pH at 9.0 throughout) and then incubate overnight at 4°C.

13. The coupled gel, which is usually brown in colour, should be thoroughly washed sequentially with 200 ml 50% (v/v) dioxane solution, 2 litres 50% (v/v) ethanol solution, 3 litres sodium chloride solution (2 M) and 500 ml distilled water. The washing procedure removes much of the brown coloration from the gel, though it remains very light brown in colour.

14. A measure of the concentration of haloperidol immobilized on the gel can be gained by scanning its absorbance over a range from 450 to 200 nm. A small portion of the matrix (100 μl) is suspended in 0.9 ml ethanediol and the sample is scanned in a scanning spectrophotometer (Unican SP8000) against an ethanediol reference. The procedure is repeated with uncoupled AH Sepharose. The difference in absorbance values at 250 nm (absorbance peak for ligand) and 350 nm (baseline value) is then determined for each sample and the value for the uncoupled gel is subtracted from the value for the coupled gel (allows for absorbance at 250 nm due to AH Sepharose). The concentration of immobilized haloperidol is then determined using its extinction coefficient and the absorbance value according to the Beer Lambert equation (absorbance = extinction coefficient × concentration × path length; ϵ = 18 000 at 250 nm).

If a soluble preparation of D_2 receptors is applied to the haloperidol affinity matrix substantial (50−70%) amounts of receptor are taken up, although uptake at 4°C is slow, not reaching equilibrium until at least 20 h. Adsorption of bulk protein is very low, provided that the concentration of sodium chloride is kept above 500 mM. Receptor uptake is biospecific and can be prevented by drugs that bind to the D_2 receptor, whereas drugs selective for other receptors do not prevent uptake. Thus, it would seem that the affinity matrix is behaving as a true affinity support.

Elution of receptor attached to an affinity column can in principle be achieved with a high concentration of a competing ligand. We chose to use the compound metoclopramide for this purpose as it has selectivity for the D_2 receptor but only a moderate affinity. Thus it can be used at a high concentration to compete with the immobilized haloperidol but should dissociate quickly from the eluted receptor. Metoclopramide (1 mM) does indeed remove D_2 receptor from the affinity column, and the metoclopramide may be removed from the eluate either by gel-filtration, dialysis, or by lectin affinity chromatography. The material obtained after dialysis or gel-filtration shows an enrichment of D_2 receptors of about 400-fold whereas the combined ligand/lectin affinity chromatography gives about 20 000-fold purification based on specific activity data for [^{3}H]spiperone binding. The method for the ligand affinity chromatography is given in Protocols 8, 9 and 10. [^{3}H]Spiperone binding to the purified preparations shows a pharmacological profile (based on inhibition studies with receptor-selective drugs) consistent with the presence of D_2 dopamine receptors.

Protocol 8. Affinity chromatography of the D_2 dopamine receptor

1. Prepare the following:
   (a) *Diluting buffer*, Hepes (20 mM), EDTA (10 mM), EGTA (1 mM) sodium acetate (8 mM), adjusted to pH 7.4.
   (b) *Washing buffer*, Hepes (20 mM), EDTA (10 mM), EGTA (1 mM), sodium cholate (0.225% w/v), sodium acetate (2 mM), sodium chloride (0.75 M), soya bean phosphatidylcholine (0.045% w/v), pH 7.4
      The pH of the buffer should be adjusted to >6.5 before adding the sodium cholate which is insoluble below pH 6.5, the soya bean phosphatidylcholine should be added last and the buffer stirred with a magnetic stir bar for several hours at room temperature until the lipid has dissolved. The pH should then be adjusted to 7.4 with NaOH solution.
   (c) *Eluting buffer*, as buffer (b) but containing metoclopramide (1 mM) (from Beecham Pharmaceuticals)
   (d) *Solubilized receptor preparation.* Prepared on the day of use as described in *Protocol 3*, diluted (3:1 v/v) with diluting buffer (a) (giving sodium cholate and sodium chloride concentrations of 0.225% and 0.75 M), and stored at 4°C until required.

2. The haloperidol-C_6-Sepharose affinity matrix is prepared as described in *Protocol 7*. The affinity chromatography procedure is performed batchwise using 10 ml matrix in a 50 ml syringe (see *Protocol 6, step 2* for alterations to syringe).

3. All solutions and equipment should be cooled to 4°C before use and all manipulations should be performed at 4°C. Excess quantities of the diluting and washing buffers (a and b) can be stored at 4°C for future use.

4. Wash the affinity matrix with 5 × 40 ml aliquots of washing buffer (b) over 30 min, the flow can be regulated using a Luerlock syringe tap.

5. Let the matrix run dry and gently expel any further buffer from the syringe by gently pushing in the bung used to close the column.

6. Close the Luerlock tap and add 45 ml diluted solubilized receptor preparation (d) to the column (larger volumes can be added but the amount of matrix should be increased accordingly). Close the column with the second bung. A hypodermic needle should be held on the inside edge of the syringe to allow air to escape when inserting the bung, the needle is then removed, sealing the column.

7. Rotate the column on its long axis (i.e. not end-over-end) for 20 h so that the matrix and liquid mix thoroughly but gently.

8. Remove the bung and collect the run through fraction from the column. Gently expel any further liquid from the matrix.

9. Wash the column with 100 ml washing buffer (b) and then mix 50 ml washing buffer (b) and matrix together for 30 min as in *step 7* above. Wash the column with a further 400 ml buffer (b) over 2 h. Determine the level of [^{3}H]spiperone binding in the starting material, run through, and first two wash fractions (Protocol 5).

10. To elute the column add 10 ml eluting buffer (c) to the matrix, close the column and rotate for 24 h.

11. Collect the eluate and repeat *step 10* with a further 10 ml buffer (c). This removes additional receptor from the affinity matrix, although the yield in this second elution is not as great as in *step 10*. Eluates from *steps 10* and *11* may be combined or processed separately.

12. Dialyse the eluate against 100 volumes of washing buffer (b): five changes of buffer over 72 h.

13. Determine the level of [^{3}H]spiperone binding in the dialysed eluate (*Protocol 5*: the [^{3}H]spiperone concentration in the assay should be increased to 3 nM to overcome any remaining metoclopramide in the eluates). Measure protein levels in the starting material, run through, and washes as described in *Protocol 4*, and in the eluate as described in reference 32 and *Protocol 10*.

14. The level of purification and column performance can be monitored as described in *Protocol 6, step 13*.

15. The affinity column should be washed as described in *Protocol 9* to remove any non-specifically bound proteins.

16. Further fractionation of the receptor can be obtained by subjecting the metoclopramide eluate to wheat-germ agglutinin-Sepharose affinity chromatography as described in *Protocol 6*. The sodium cholate and sodium chloride concentrations should be increased to 0.3% (w/v) and 1 M before adding to the lectin column.

17. Analysis of the protein compositions of eluates is performed following SDS-PAGE as in reference 33 and *Protocol 10*.

Protocol 9. Affinity column washing procedure

1. Prepare the following:
   (a) *Washing solution*, Hepes (20 mM), sodium cholate (1%), sodium chloride (3 M), adjusted to pH 7.0.
   (b) *Storage buffer* (same as the storage buffer in *Protocol 6*), Hepes (20 mM, EDTA (10 mM), EGTA (1 mM), sodium azide (0.02% w/v), adjusted to pH 7.4.
2. Drain column and wash with 100 ml washing solution (a) over 5 min.
3. Drain column and mix with 50 ml washing solution (a) over 25 min as in *Protocol 8, step 7*.
4. Drain column and wash with 200 ml washing buffer (a), 500 ml distilled water, and 100 ml storage buffer (b). Store in buffer (b) at 4°C.

Protocol 10. Protein assays for eluates from affinity columns

For samples eluted from the haloperidol-Sepharose affinity column (*Protocol 8, step 13*) the protein concentration is determined as in reference 32, using [³H]-dinitrofluorobenzene (Dupont) labelling. For samples obtained after the combined haloperidol-Sepharose and wheat-germ agglutinin-Sepharose chromatography steps the protein is determined using quantitative silver staining. Protein is extracted from the eluate using $CHCl_3/MeOH$ as in reference 34, separated by discontinuous polyacrylamide gel electrophoresis, and silver stained as in reference 35. The techniques for extraction and staining are outlined below, whereas that for electrophoresis is given in reference 33.

1. Prepare the following:
   (a) *Ligand/lectin affinity chromatography eluate,* prepared as in *Protocol 8, step 16*.
   (b) *Hepes buffer*, Hepes (20 mM), EDTA (10 mM), EGTA (1 mM), sodium cholate (0.225% w/v), sodium acetate (2 mM), sodium chloride (0.75 M), soya bean phosphatidyl choline (0.045% w/v), adjusted to pH 7.4. Prepared as described in Protocol 8, *step 1b*.
   (c) *Bovine serum albumin (BSA) solutions*, 0.1, 0.25, 0.5, 0.75, 1, and 1.5 µg BSA in separate 10 ml aliquots of Hepes buffer (b).
   (d) *Fixing solution A*, methanol (50% v/v), acetic acid (10% v/v) in water.
   (e) *Fixing solution B*, methanol (10% v/v), acetic acid (7% v/v in water).
   (f) *Dithiothreitol (DTT) stock solution*, DTT (5 mg/ml) in water, stored at −20°C in 1 ml aliquots until required.
   (g) *Silver nitrate solution*, silver nitrate (0.1% w/v) in water, prepared immediately before use.

(h) *Developing solution*, sodium carbonate (3% w/v), formaldehyde (0.05% v/v) in water, prepared immediately before use.

Extraction procedure

2. Add 5 ml of each BSA standard (c) and eluate (a) to separate screw-capped polypropylene tubes (50 ml, Falcon).

3. Add 20 ml methanol to each tube.

4. Vortex mix each tube and centrifuge at 3200 r.p.m. ($\sim$2000 g) (MSE, Centaur 2) for 10 s, the tubes must be tightly capped at this stage.

5. Add 5 ml chloroform to each tube and repeat *step 4*.

6. Add 15 ml distilled water to each tube and repeat *step 4*.

7. Aspirate off most of the upper phase, taking care not to disturb protein which is present as a precipitate at the interface.

8. Add 5 ml methanol and vortex gently. Centrifuge at 3600 r.p.m. (2200 g) for 4 min.

9. Decant off the supernatant.

10. Resuspend the precipitated protein in a further 5 ml of standard or eluate sample.

11. Repeat *steps 3 − 10* until all the standard samples and eluate samples have been processed.

12. Dry the tubes under nitrogen and dissolve the precipitated protein in SDS-PAGE sample buffer and run the protein on a 5%/8% discontinuous gel as described in reference 33.

Silver staining procedure

The technique is a modified version of that in reference 35. All incubations should be carried out at room temperature with continuous gentle agitation (e.g. on an orbital shaker). The gel should be handled as little as possible and never without clean gloves.

13. Incubate the gel sequentially in fixing solution A (d) and B (e) for 30 min each.

14. Rinse with distilled water and incubate for 30 min with a 1:1000 dilution (in water) of DTT stock solution (f).

15. Decant off the DTT solution and incubate the gel with silver nitrate solution (g) for 30 min.

16. Decant off the silver nitrate solution and rinse the gel rapidly (5 − 10 seconds) once with distilled water and twice with developing solution (h). Care must be taken to remove the developing solution before it discolours the gel.

17. Add developing solution ($\sim$500 ml) and shake gently until the desired level of staining is achieved. Stop the reaction by neutralization of the solution with acetic acid.

18. After 5 − 10 min, decant off the solution and rinse and store the gel in distilled water.

19. Measure the strength of the stained BSA standard and eluate bands (D_2 dopamine receptor) using a scanning densitometer.

20. Plot band density against BSA concentration and calculate the concentration of protein in the eluate using this standard curve. The values calculated using this technique are consistent with those from quantitative Coomassie blue staining of eluate and BSA standards run on a mini-gel system.

Examination of the material purified by ligand/lectin affinity chromatography by SDS−PAGE shows a major band at M_r 95 000 which frequently appears as a doublet (12). Photoaffinity labelling of D_2 receptors in membrane preparations using [^{3}H]azidomethyl spiperone (Dupont) (18) followed by SDS-PAGE indicates a species of M_r 95 000, suggesting that the M_r 95 000 species from the affinity chromatography is the D_2 dopamine receptor. In early work where solubilization of receptors was carried out in the presence of the proteinase inhibitors PMSF, EDTA, and EGTA a major band at M_r 80 000 was seen, whereas the current procedure includes the additional proteinase inhibitors aprotinin, antipain, chymostatin, leupeptin, and pepstatin A. This may imply that proteolysis of the M_r 95 000 species was occurring to the M_r 80 000 species, and it emphasizes the need for maximal suppression of proteolytic activity in such studies.

Whereas the gel-electrophoresis patterns indicate that a fairly high degree of purification has been achieved after ligand/lectin affinity chromatography, the data from specific activity indicate that the preparation is about 25% pure. This may be a reflection of instability of receptors during purification and, indeed, we have found values for receptor ligand binding in purified preparations to be variable. In other studies (19) reconstitution has been required in order to achieve reproducible levels of receptor binding, but we have not found this to be of any help.

The method described here and in reference 12, using ligand/lectin affinity chromatography, provides a method for the purification of D_2 dopamine receptors that should be of great value in understanding structure−function relationships for this receptor. Other methods for the use of affinity chromatography have been reported (19−22), most of which are quite similar to that reported here, and two further reports of the purification of the receptor have appeared (23,24).

4. Molecular properties of D_1 and D_2 dopamine receptors

Despite intensive work on the topic, there is little firm information on the molecular nature of these two receptors. Both are glycoproteins that adsorb to wheat-germ agglutinin (1,4,15) and so have glycoprotein side chains with available sialic acid or N-acetyl-glucosamine residues and evidence has been presented for heterogeneous glycosylation of the D_2 receptor (16). Both are coupled to guanine nucleotide regulatory proteins (G-proteins). The D_1 receptor is presumably coupled to G_s

whereas the D_2 receptor is coupled to G_i/G_o or a related species (25). The D_1 receptor is a species of M_r 72 000 from photoaffinity labelling (26) whereas the D_2 receptor is generally found to be a species of M_r 85 000−95 000 from photoaffinity labelling and purification studies (1). There are reports of an M_r 120 000 species from photoaffinity labelling (27) and purification studies (24). Whether this represents a precursor to the M_r 85 000−95 000 species which would then be derived proteolytically from the larger species, or whether it represents a distinct receptor subtype remains to be established.

The amino-acid sequence of the rat D_2 dopamine receptor has recently become available from gene cloning work (28). The cloned sequence codes for a 415 amino-acid protein (M_r 47 064) which corresponds well with the M_r of the fully deglycosylated receptor (M_r 53 000, R.A. Williamson and P.G. Strange, unpublished results). The amino-acid sequence when analysed for hydrophathicity predicts a structure with seven transmembrane α-helices. Thus the D_2 receptor is a member of the superfamily of G-protein-linked receptors (29) and shows considerable sequence homology to several members of this family. The availability of the amino-acid sequence, together with the method for purification of the receptor outlined in this chapter, will enable careful structural studies to be performed in order to verify the predicted structure.

Acknowledgements

We thank Sue Davies for typing the manuscript and the MRC, SERC, and Wellcome Trust for financial support.

References

1. Strange, P. G. (1987). In *Dopamine Receptors*. (ed. I. Creese and C.M. Fraser), p. 29. A.R. Liss, New York.
2. Strange, P. G. (1987). *Neurochem. Intl.*, **10**, 27.
3. Sidhu, A. and Fishman, P. H. (1986). *Biochem. Biophys. Res. Commun.*, **137**, 943.
4. Niznik, H. B., Otsuka, N. Y., Dumbrille Ross A., Grigoriadis, D., Tirpak, A., and Seeman, P. (1986). *J. Biol. Chem.*, **261**, 8397.
5. Hollis, C. M. and Strange, P. G. (1989). *Biochem. Soc. Trans.*, **17**, 774−775.
6. Gingrich, J. A., Amlaiky, N., Senogles, S. E., Chang, W. K., McQuade, R. D., Berger, J. G., and Caron, M. G. (1988). *Biochemistry*, **27**, 3907.
7. Tam, S. and Seeman, P. (1978). *Eur. J. Pharmacol.*, **52**, 151.
8. Gorissen, H., Aerts, G., and Laduron, P. (1979). *FEBS Lett.*, **100**, 1526.
9. Withy, R. M., Mayer, R. J., and Strange, P. G. (1981). *J. Neurochem.*, **37**, 1144.
10. Wheatley, M., Hall, J. M., Frankham, P. A., and Strange, P. G. (1984). *J. Neurochem.*, **43**, 926.
11. Hall, J. M., Frankham, P. A., and Strange, P. G. (1983). *J. Neurochem.*, **41**, 1526.
12. Williamson, R. A., Worrall, S., Chazot, P. L., and Strange, P. G. (1988). *EMBO J.*, **13**, 4129.
13. Madras, B. and Seeman, P. (1985). *J. Neurochem.*, **44**, 856.

14. Lew, J. Y. and Goldstein, M. (1984). *J. Neurochem.*, **42**, 1298.
15. Abbott, W. M. and Strange, P. G. (1985). *Biosci. Rep.*, **5**, 303.
16. Leonard, M. N., Williamson, R. A., and Strange, P. G. (1988). *Biochem. J.*, **255**, 877.
17. Juszczak, R. and Strange, P. G. (1987). *Neurochem. Intl.*, **11**, 389.
18. Niznik, H. B., Grigoriadis, D. E., and Seeman, P. (1986). *FEBS Lett.*, **209**, 71.
19. Senogles, S. E., Amlaiky, N., Johnson, A. L., and Caron, M. G. (1986). *Biochemistry*, **25**, 749.
20. Ramwami, J. and Mishra, R. K. (1986). *J. Biol. Chem.*, **261**, 8894.
21. Antonian, L., Antonian, E., Murphy, R. B. and Schuster, D. I. (1986). *Life Sci.*, **38**, 1847.
22. Soskic, V. and Petrovic, J. (1986). *Jugoslav. Physiol. Pharmacol. Acta*, **22**, 329.
23. Elazar, Z., Kanety, H., David, C., and Fuchs, S. (1988). *Biochem. Biophys. Res. Comm.*, **156**, 602.
24. Senogles, S. E., Amlaiky, N., Falardeau, P., and Caron, M. G. (1988). *J. Biol. Chem.*, **263**, 18996.
25. Senogles, S., Benovic, J. L., Amlaiky, N., Unson, C., Milligan, G., Vinitsky, R., Spiegel, A. M., and Caron, M. G. (1987). *J. Biol. Chem.*, **262**, 4860.
26. Amlaiky, N., Berger, J. G., Chang, W., McQuade, J., and Caron, M. G. (1987). *Mol. Pharmacol.*, **31**, 129.
27. Amlaiky, N. and Caron, M. G. (1986). *J. Neurochem.*, **47**, 196.
28. Bunzow, J. R., van Tol, H. H. M., Grandy, D. K., Albert, P., Salon, J., Christie, M., Machida, C. A., Neve, K. A., and Civelli, O. (1988). *Nature*, **336**, 783.
29. Strange, P. G. (1988). *Biochem. J.*, **249**, 309.
30. Lowry, O. H., Rosebrough, N. J., Farr, A. L., and Randall, R. J. (1951). *J. Biol. Chem.*, **193**, 263.
31. Wurzburger, R. J., Miller, R. L., Marcum, E. A., Colburn, W. A., and Spector, S. A. (1981). *J. Pharmacol. Exp. Ther.*, **217**, 757.
32. Schultz, R. M., Bleil, J. D., and Wassarman, P. M. (1978). *Anal. Biochem.*, **91**, 354.
33. Laemmli, U. (1970). *Nature*, **227**, 680.
34. Wessel, D. and Flugge, U. I. (1984). *Anal. Biochem.*, **138**, 141.
35. Morrissey, J. H. (1981). *Anal. Biochem.*, **117**, 307.

4

Opioid receptors

CATHERINE D. DEMOLIOU-MASON and ERIC A. BARNARD

1. Introduction

Evidence for the heterogeneity of opioid receptor sites in the CNS has been provided by a variety of studies. At least three classes of binding sites have been identified and designated as μ (morphine), δ (enkephalin) and $\varkappa$ (dynorphin). The assignment of these sites as distinct receptor subtypes has been supported by a variety of studies (reviewed in reference 1): specific ligand competition, selective protection from irreversible inactivation, differential anatomical distributions, the differing pharmacological profiles of various opiate and opioid peptide ligands, and differential affinity cross-linking studies (2).

Ligand-binding studies using mammalian brain membrane preparations, or neuroblastoma $\times$ glioma (NG108-15) cells or homogenates thereof, have located the opioid receptor binding activity on the plasma membranes. Subcellular fractionation studies have shown this activity to be concentrated in both the P_2 and P_3 membrane fractions (3). This ligand binding has been shown to be stereoselective and sensitive to ionic strength, pH, and temperature (4). A phospholipid requirement for the stabilization of receptor activity has been suggested from studies with phospholipases and various lipids (5).

Cations (Na^+, Mg^{2+}, Mn^{2+}) and guanine nucleotides are known to act as regulators in the opioid-directed inhibition of basal and stimulated adenylate cyclase (6) via coupling to the pertussis toxin substrates, the G_i/G_o proteins (7,8). The effects exerted by these agents have been related to receptor conformational changes which are manifested as changes in the receptor affinity (9,10). It has been shown, for example, that Na^+ ions and guanine nucleotides decrease the receptor affinity for pure agonists and increase it for antagonists, whereas the divalent cations Mg^{2+} and Mn^{2+} appear to act in the opposite direction; alkaloids with mixed agonist–antagonist activity are affected to a lesser extent (1). Studies with reducing and alkylating reagents have indicated that the sites for the regulation of ligand binding by cations and guanine nucleotides and the conformational changes may involve receptor $-SH$ groups (11,12).

2. Solubilization

The sensitivity of these receptors to cations, guanine nucleotides, and thiol reagents, as well as the lipid requirements of the receptor, are properties which have to be

Table 1. Buffers used for brain membrane preparation and/or solubilization

	Reference
(a) 50 mM Tris-HCl (pH 7.4)	31
(b) 50 mM Tris-HCl (pH 7.4)/0.3 M sucrose	19
(c) 50 mM Tris-HCl (pH 7.4)/1 mM EDTA	33
(d) 50 mM Tris-HCl (pH 7.4)/200 μl/ml Trasylol/50 μg/ml bacitracin/10 mM EDTA	14
(e) 50 mM Tris-HCl (pH 7.4)/1 mM EDTA/20 μg/ml bacitracin/40 KIU/ml Trasylol/1 mM PMSF	24
(f) 50 mM K$^+$phosphate (pH 7.5)/1 mM PMSF	28
(g) 50 mM imidazole (pH 7.5)	28
(h) 10 mM TES-KOH (pH 7.5)/0.3 M sucrose/ 1 mM EDTA-Na$_2$ 1 mM DTT/1 mM benzamidine – HCl/0.01% bacitracin/0.002% soya bean trypsin inhibitor/1 mM PMSF	13

Abbreviations: TES, *N*-tris (hydroxymethyl)methyl-2-aminoethanesulphonic acid; Tris, Tris(hydroxymethyl)aminoethane; EDTA, ethylene diamine tetra-acetic acid; PMSF, phenylmethylsulphonyl fluoride; DTT, dithiothreitol.

taken into consideration during membrane preparation and solubilization. Soluble opioid activity has been obtained from brain tissue (rat, bovine, chicken, frog), the NG108-15 cells and the human placenta, in several detergents, discussed below.

2.1 Buffer systems for membrane preparation and solubilization

The choice of the buffer system to be used for membrane preparation and detergent solubilization should be dictated by the biochemical properties of the subtype to be studied, and in the case of affinity chromatography purification by the nature of the immobilized ligand (i.e. agonist v. antagonist; opioid peptide v. opiate alkaloid). The most common buffer system used to prepare membranes and soluble extracts is 50 mM Tris−HCl (pH 7.4−7.5) containing 0.32 M sucrose. Buffer systems containing protease inhibitors and other reagents are shown in *Table 1*. Solubilized receptor activity is usually assayed at 0.1−0.05% (w/v) final detergent concentration by the polyethylene glycol (PEG) precipitation method (13,76; see Section 2.6).

In the course of testing the various buffer systems in our laboratory, we have observed that although 50 mM Tris−HCl (pH 7.5) can substitute partially for a divalent cation such as Mg^{2+} which is required for high-affinity opioid-agonist binding, other buffer systems, and in particular those containing chelating reagents and no divalent cations, do not. Soluble extracts of membranes depleted in divalent cations in any buffer bind opioid ligand with the following order of decreasing affinity: etorphine > opiate antagonist > mixed opiate agonist−antagonists > pure opiate agonists > opioid peptides.

The omission of protease inhibitors and reducing agents during membrane preparation and solubilization can also result in losses of receptor activity. Such losses are greater when (as is common) membranes are pre-incubated at 30−37°C

(15 – 30 min) in order to remove endogenous ligands prior to solubilization. A combination of protease inhibitors and dithiothreitol (DTT, $0.1 – 1.0$ mM) in the Tes-KOH buffer system (*Table 1*) reduces the losses of opioid receptor activity. In this medium, membrane-bound or solubilized receptor activity is stabilized even at ambient temperature (up to 4 h) and during storage at $-70°C$ for several months. Maneckjee *et al.* (14) have reported that the addition of Trasylol (20 KIU/ml), Leupeptin (5 mg/ml), and EDTA-K$_2$ (1 mM) in diluted extracts made in CHAPs (1 mM) can likewise prevent the loss of activity with time (up to 7 days) during storage at $-20°C$. Our experience has shown that, irrespective of the protease inhibitors used, membranes and soluble extracts retain their activity better when stored at high protein concentrations ($5 – 10$ mg/ml) after freezing rapidly in liquid N$_2$. Rapid freezing as well as mixing during thawing prevent the partitioning of the various components in the medium.

Phospholipids have also been added to detergent extracts in order to stabilize the receptor conformation in an active state. Phosphatidylserine (0.5 M) was found to partially stabilize rat brain extracts in CHAPS upon storage at $-20°C$ and to enhance the activity of Triton X-100 extracts. Phosphatidylethanolamine or lecithin had either no effect or inhibited opiate binding activity (14,15). Recent studies with partially purified μ-opioid receptors from Triton X-100 bovine brain-extracts have shown that polyunsaturated fatty acids or acidic phospholipids with polyunsaturated fatty acids can restore lost receptor binding in a concentration-dependent manner (16).

2.2 Solubilization with ionic detergents

Puget *et al.* (17) originally reported that a complex containing bound [^{3}H]etorphine (Amersham) could be recovered in Na$^+$-cholate extracts (1% w/v, 15 min, 4°C) of rat brain membranes prelabelled with 2 nM [^{3}H]etorphine in 50 mM Tris – HCl (pH 7.4). This preparation was used to study the hydrodynamic properties of the prelabelled receptor species but was not tested for [^{3}H]opioid-ligand binding activity after dissociation.

Gioannini *et al.* (18) used glycodeoxycholate (glyDOC) to solubilize mammalian, chicken, and frog brain membranes.

Protocol 1. Brain membrane solubilization with glyDOC

1. Prepare crude membranes in 0.32 M sucrose/50 mM Tris-HCl (pH 7.4).
2. Centrifuge the membranes at 35 000 g (15 min, 4°C).
3. Resuspend the pellet in 1.5 volumes of 0.25% glyDOC in 50 mM Tris-HCl (pH 7.4) containing 1 M NaCl.
4. Centrifuge at 100 000 g (35 min, 4°C).
5. Dilute the glyDOC/NaCl extracts with ice-cold 50 mM Tris-HCl (pH 7.4) containing 1 mM EDTA-K$_2$ and 0.1% digitonin for assaying binding activity.

In our laboratory we have tested the ability of K^+-cholate and Na^+-deoxycholate (1%, w/v) to solubilize opioid receptors from rat brain membranes pretreated with Mg^{2+} (10 mM) (see below). Although both detergents solubilized membrane protein effectively, the preparations displayed only [^{3}H]etorphine and antagonist-binding activity (13).

2.3 Solubilization with the zwitterionic detergent, CHAPS

Solubilization of NG108-15 cell membranes, as well as rat and bovine brain membranes, with CHAPS (10 mM) was first described by Simmonds *et al.* (19) and subsequently modified by others (20,21).

Protocol 2. Membrane solubilization with CHAPS

1. Resuspend P_2-pelleted membranes in 2 volumes of 10 mM Tris-HCl (pH 7.5) (12−15 mg protein/ml).

2. Add CHAPS to a final concentration of 10 mM.

3. Homogenize the suspension in a ground-glass tissue homogenizer, and incubate for 15 min on ice.

4. Centrifuge the suspension at 105 000 g (60 min, 4°C). Remove the resulting clear supernatant from the solid pellet and the cloudy suspension floating above it. The cloudy material can be removed more efficiently by gel-filtration on Sepharose 4B or 6B in 10 mM Tris-HCl (pH 7.5)/1 mM CHAPS (22).

Opioid receptors in the clear supernatant have been partially purified by the PEG-precipitation method (20,21,23,24). Acidify the supernatant (5 ml) with 0.25 ml of 1 M K-acetate buffer (pH 5.6). Precipitate the membrane proteins with 2.65 ml of 50% PEG (M_r 8,000) in 0.1 M potassium-acetate (pH 5.6) (15 min, 4°C). Wash once with one volume of 10 mM Tris-HCl/1 mM CHAPS/1 mM DTT and resuspend it in 10 mM Tris-HCl (pH 7.4) (0.3 × original volume) by sonication for 20 sec.

In a variation, pre-incubation of P_2 membranes (30 min, 37°C) in 50 mM Tris-HCl (pH 7.4)/1 mM EDTA/20 μg/ml bacitracin/40 KIU/ml Trasylol/1 mM PMSF is followed by solubilization in the same buffer containing 10 mM CHAPS (1 h, 4°C) (24).

2.3.1 Crude synaptosome (P2) fraction from brain

Protocol 3. Preparation of crude synaptosome (P2) fraction

1. Potter-Elvejem homogenize brain (0.2 mm clearance, 500 c.p.m., 10 complete strokes) in 9 volumes (w/v) of ice cold 0.32 M (iso-osmolar) sucrose containing 20 mM buffer adjusted to pH 7.4 (e.g. sodium phosphate, or HEPES). Measure the volume (*V*).

2. Centrifuge the homogenate in *c.* 50 ml aliquots at 1000 g for 5 min at 4°C.

Carefully pour off the supernatant, taking care not to disturb the pellet (designated P_1) which is enriched in unbroken cells and brain microvessels.

3. Centrifuge the supernatant (S_1) in *c*. 50 ml aliquots at 10 000 *g* for 20 min at 4°C to pellet the crude synaptosome fraction, P_2. Pour off the supernatant (S_2 fraction).

4. The P_2 fraction can be resuspended for binding studies by gentle vortexing mixing in *c*. 20 ml of sucrose or buffer per pellet. The P_2 pellet consists of two layers; a brownish bottom layer enriched in heavier mitochondria, and a light-coloured upper layer enriched in synaptosomes (sometimes designated the buffy coat). With care, resuspension of the upper layer alone can be achieved. The upper layer contains *c*. 50% of the protein content of the original homogenate. Resuspension to half the original homogenate volume thus yields a preparation equivalent to 0.1 g wet weight tissue/ml, corresponding to a protein concentration of *c*. 10 mg/ml. Thus, pour off the resuspended P_2 fraction into a measuring cylinder, and resuspend to $V/2$ ml. Ensure even resuspension with a few strokes of the Potter−Elvejem homogenizer.

5. If resuspension in buffered iso-osmolar sucrose is used, the fraction can be used as a starting point for synaptosome preparations using sucrose, ficoll, or percoll gradients.

6. Use the P_2 fraction immediately in binding studies, or recover by recentrifugation (100 000 *g*, 30 min) for freezing and storage. Freeze by immersing the tube in a dry-ice/isopropanol slurry, and store at −70°C.

The P_2 fraction is typically *c*. 1.5-fold enriched in synaptic receptors compared to the original homogenate. The total preparation time is *c*. 1 h. Resuspension and recentrifugation of the S_1 fraction, which slightly enhances the yield, has been omitted in favour of speed.

A further fraction, the crude microsomal fraction, P_3, can be obtained by centrifuging the S_2 fraction for 60 min at 100 000 *g*. This produces a further small pellet, which is somewhat more enriched in receptors (e.g. two-fold in mAChRs on a protein basis).

2.4 Solubilization with non-ionic detergents

Simon *et al.* (25) originally reported that the non-ionic detergent Brij 36T [poly(oxyethylene)10-lauryl ether] could solubilize a complex containing bound [^{3}H]etorphine from rat brain membranes. Similar results were subsequently obtained in other laboratories (26−28) as well as using Triton X-100 (29,30). However, neither these complexes nor their corresponding unliganded soluble extracts displayed any opiate-binding activity after solubilization.

2.4.1 Triton X-100

Extraction of rat brain membranes with Triton X-100 was reported to yield active soluble receptor after removal of the excess detergent (31).

Protocol 4. Brain membrane solubilization with Triton X-100

1. Make P_2 pelleted membranes from rat brain (minus cerebellum) and resuspend them in 50 mM Tris-HCl (pH 7.5) at 10 mg protein/ml.
2. Solubilize with 1% Triton stir gently for 15 min, 4°C and then centrifuge at 100 000 g (30 min, 4°C).
3. Remove the detergent by stirring the supernatant gently with 0.4 g/ml Bio-beads SM-2 (Bio-Rad) (2 h, 4°C) which are prewashed with 50 mM Tris-HCl (pH 7.5).
4. Separate the supernatant from the Bio-beads and concentrate it on an Amicon PM10 membrane to 15−20 mg protein/ml.

2.4.2 Digitonin

Solubilization of active opiate receptors in this detergent from frog, chicken, and mammalian brain has been described by several groups (13,32−34). For some of these preparations the inclusion of a high salt concentration (0.5−1.0 M NaCl) in the detergent medium was found necessary for the extraction (32,33).

According to the method of Demoliou-Mason and Barnard for mammalian brain (13), the membranes are prepared from brain (minus cerebellum) in TES-KOH buffered medium (see *Table 1*).

Protocol 5. Brain membrane solubilization with digitonin/Mg^{2+}

1. Resuspend P_2 membranes to the original volume (1:20, wet weight per volume) in the same buffer (minus sucrose), stir on ice (30 min), centrifuge (40 000 g, 30 min, 4°C) and resuspend to a concentration of 10−12 mg protein/ml, for storage at −70°C.
2. Thaw the frozen membranes and resuspend them to 1.0−0.5 mg protein/ml in ice-cold 10 mM TES-KOH (pH 7.5)/1 mM EGTA-K^+/10 mM $MgSO_4$/0.1 mM DTT/1 mM benzamidine−HCl/0.01% bacitracin/0.002% soya bean trypsin inhibitor, referred to as 'Mg-buffer'.
3. Pellet the membranes at 40 000 g (30 min, 4°C), and resuspend them in Mg-buffer to 2−3 mg protein/ml.
4. Incubate at 30°C (60 min), dilute the membranes 1:4 with Mg-buffer and pellet them as above.
5. Solubilize the membranes in Mg-buffer with digitonin (2% w/v; 2:1, detergent to protein) by gentle agitation for 60 min at room temperature.
6. Centrifuge the suspension at 120 000 g (60 min, 4°C), collect the supernatant, freeze it in liquid N_2, and store it at −70°C.

2.4.3 Solubilization by sonication

This method is that of Cho *et al.* (35). Briefly, rat brain membranes (P_2) are resuspended in 0.32 M sucrose (1 : 8, wet weight per vol). This suspension (15 ml) is sonicated in an ice bath for 9 min using a Branson ultrasonifier (model W 14D). The sonicated suspension is first treated with 0.5% Triton X-100 (30 min, 4°C) and then centrifuged at 100 000 g (60 min, 4°C).

2.5 Evaluation of the detergents used

Compared to other neurotransmitter and hormone receptor systems, the progress in solubilizing and purifying the opioid receptors in an active form has been slow, as a result of the high sensitivity of this receptor to detergents. Even with detergents which can solubilize the opioid receptor in an active state, there are limitations with regard to the recovery of receptor-binding activity and the efficiency in solubilizing it from mammalian brain. The yields of soluble receptor activity obtained with the various detergents are shown in *Table 2*. Comparisons can not be made directly between these, since different buffer systems and radiolabelled ligands for assaying activity have been used. Furthermore, the yields are obtained relative to the activity present in membranes and this may differ depending on the method and conditions used during prepration. Several conclusions can be drawn, however: ionic detergents, for example, which can solubilize a large proportion of the membrane protein, give soluble extracts which are either inactive or bind only opiate antagonists, with the exception of [^{3}H]etorphine, a potent alkaloid agonist which binds with high affinity to all receptor subtypes in membranes and displays the characteristics of an antagonist in ligand-binding studies in the presence of Na$^+$. Although the addition of Na$^+$ may increase the efficiency of the detergent, the stabilization of the receptor subtypes in the antagonist state with low agonist affinity reduces the usefulness of the preparation, especially in affinity purification studies, since there are no subtype-specific antagonists available at present. However, some of the ionic detergents as well as the non-ionic ones can be used effectively in solubilizing covalently cross-linked receptor–radiolabelled ligand complexes for partial purification by conventional chromatographic methods.

Solubilization of opioid-binding sites by sonication and Triton X-100 treatment of rat brain membranes also yields solubilized extracts which bind antagonists preferentially; binding of μ and $\varkappa$ opiate agonists at affinities considerably lower than those seen in the membranes, is detected only after removal of the detergent and the addition of lipids (35) or exchange of the detergent (36). Removal of Triton X-100 by Bio-beads, for example, appears to increase the receptor affinity for μ-opiate agonists. Assaying receptor activity in Triton extracts is difficult, since removal of the free ligand by equilibrium dialysis (5 h, 37°C) (26) or by charcoal (0.7%) adsorption in the presence of excess bovine serum albumin (0.5%) is required (35). Recently, Ueda *et al.* (37) have reported the use of nitrocellulose membranes (BA85, Schleicher and Schuell) in assaying the activity of purified opioid receptors from Triton extracts, by filtration.

CHAPS-solubilized extracts of mammalian brain after partial purification by PEG precipitation, show (at 1 mM CHAPS) higher affinities (in the nM range) for agonists in addition to binding etorphine and antagonists (21). Opiate agonist (but not

Table 2. Percentage yield of solubilized opioid-receptor activity from brain tissue.

Detergent (%,w/v)	Bovine	Rat	Human	Guinea-pig	Chicken	Toad	References
K$^+$-cholate (1%)		6					13
Na$^+$-deoxycholate (1%)		52					13
gly-Doc (0.25%)/NaCl (1 M)	43	15			33		18
Triton X-100 (1%)		70	1.3				15
CHAPS (10 mM)		17–34				40–70	19–24
digitonin (1%)							32,34
digitonin (2%)/MgSO$_4$ (10 mM)	41	45					13
digitonin (0.5%)/NaCl (0.5 M)		29			56	53	18
digitonin (1%)/NaCl (0.5 M)	38	73					24
digitonin (0.5%)/NaCl (0.75 M)							57
digitonin (0.5%)/NaCl (1 M)				25–30			39

Abbreviations: Gly-Doc, glycodeoxycholate; CHAPS, 3-[(3-cholamidopropyl)dimethylammonio]-1-propanesulphonate.

antagonist) binding in CHAPS extracts of NG108-15 cell membranes is inhibited by guanine nucleotides and by sodium, suggesting that the receptor in CHAPS retains some of its physiological properties (22). Low opioid-receptor binding activity and yields, however, are also the limiting factors in CHAPS-solubilization of mammalian brain tissue. The addition of glycerol (50% v/v, final concentration) to the membranes, prior to solubilization with 10 mM CHAPS, has been suggested (23) to improve the recovery of the activity (>75%), as well as the use of purified synaptosomal membranes; this results in greater recoveries of receptor activity (>90%) and a greater efficiency of solubilization (>90%).

Good yields (30−40%) of soluble [³H]antagonist-binding activity from mammalian brain have been achieved by the addition of NaCl during solubilization with digitonin, with detergent dilution prior to assaying the receptor activity (38). However, these preparations contain only very low levels of opiate-agonist-binding activity. Higher activities have been obtained after sucrose gradient fractionation of the digitonin/NaCl extracts in the absence of Na^+ and lower concentrations of digitonin (38,39).

Pre-incubation of rat brain membranes with 10 mM Mg^{2+} (13) prior to solubilization with 2% digitonin (also in the presence of Mg^{2+}) appears to be the most effective procedure used so far in stabilizing the opioid receptor in the agonist state. This procedure gives higher yields of solubilized opioid peptide and opiate-agonist-binding activity with affinities close to those observed in the membrane. In addition, agonist binding (assayed at 0.1% digitonin) retains the same sensitivity to cations (Mg^{2+}, Na^+), guanine nucleotides, and pertussis toxin as observed for the membrane-bound receptor; this indicates that the opioid receptor is thus solubilized in a state functionally coupled to G-protein(s) (8,13). The recommended procedure is detailed in Section 2.4.2 above.

2.6 [³H]ligand-binding assay for solubilized opioid receptors

Radiolabelled-ligand binding is assayed in the presence of 10 mM Mg^{2+}. If the effects of other cations or reagents are to be studied they can be included in the assay medium (solution 5) at two times the required concentration. All chemical reagents used are available from Sigma.

Protocol 6. [³H]ligand-binding assay for digitonin/Mg^{2+} solubilized receptors

1. Prepare the following solutions:
   (a) Make up 1.0 l of 10 mM TES-KOH, pH 7.5; store at 4°C.
   (b) Using solution (a) make up 50 ml of 0.6% (w/v) γ-globulin.
   (c) Using solution (a) make up 50 ml of 36% (w/v) polyethylene glycol (mol. wt, 8000).
   (d) Using 250 ml of solution (a) and double-distilled H_2O make up 500 ml of 10% (w/v) polyethylene glycol and store at 4°C.

Solutions (b) and (c) can be aliquoted and frozen at $-20°C$. Solutions (b$-$d) should be chilled on ice before use.

(e) Prepare fresh 100 ml of assay medium containing 10 mM TES-KOH, pH 7.5; 2 mM EGTA-K$^+$; 20 mM MgSO$_4$; 2 mM benzamidine-HCl; 0.02% (w/v) bacitracin; and 0.002% (w/v) soya bean trypsin inhibitor. MgSO$_4$ should be added after adjusting to pH 7.5 with 2 N KOH and the pH should be checked and readjusted if required.

(f) Prepare radiolabelled and unlabelled ligand in 10 mM TES-KOH, pH 7.5 at 10 times the required concentration.

2. Pipette in Eppendorf tubes (1.5 ml):
   250 μl assay medium (e)
    50 μl 10 $\times$ concentrated radiolabelled ligand
    50 μl 10 mM TES-KOH, pH 7.5; or 10$\times$ concentrated unlabelled ligand
   125 μl 10 mM TES-KOH, pH 7.5
    25 μl solubilized receptor in 2% digitonin

3. Incubate at 30°C for 1 h or at room temperature for 2 h, then transfer the samples on ice to cool for 5$-$10 min.

4. Add to each sample 50 μl solution (b) followed by 200 μl solution (c). Vortex the samples for $\sim$30 s but avoid the formation of bubbles. Incubate on ice for 30 min.

5. Briefly vortex each sample, pipette out 600 μl aliquots and filter on Whatman GF/C filters which have been presoaked in double-distilled H$_2$O, using a filtration apparatus connected to a vacuum pump. Wash each filter rapidly with 2 $\times$ 3 ml solution (d), dry the filter under a lamp and transfer to scintillation vials for counting.

6. Count directly 25 or 50 μl of the remaining sample to calculate the radiolabelled-ligand concentration.

---

3. Purification studies

3.1 Molecular exclusion chromatography and sucrose gradient centrifugation

Sepharose CL-4B or 6B (Pharmacia), Ultrogel AcA22 (M_r range 100 000$-$1 200 000 Da) and Ultrogel AcA34 (LKB) (M_r range 20 000$-$350 000 Da) have been used for molecular fractionation studies of the opioid receptor species in detergent extracts of brain membranes. Briefly, the column (1$-$2 cm $\times$ 60$-$100 cm) is equilibrated and run at 4°C, in the same buffer medium used for solubilization but containing lower detergent concentrations (0.1$-$0.05% digitonin or 1 mM CHAPS). Crude solubilized extract (2$-$3 ml) is applied and fractions (3$-$4 ml) are collected (10$-$15 ml/h) and assayed for [^{3}H]ligand-binding activity, or aliquots are counted directly for radiolabelled ligand$-$receptor complexes.

Table 3. Solubilized macromolecules of brain opioid receptors.

Source	Detergent	M_r (kDa)	size in Å	$S_{20,w}$	References
rat*	1% Brij 36T	370	59		28
rat*	1% Brij 36T	380	48		52
rat*	1% Brij 36T				27
mouse*	1% Brij 36T	<20; 100 – 500	43; 55; 63		26
rat*	1% Triton X-100	120; 240; 440	$\leq$20; 50; 63; 70		30
rat*	1% Na$^+$-cholate	500			17
rat*	10 mM CHAPS	690	73	15.6	21
rat	10 mM CHAPS	110; 560	31; 76	8.0; 15.6	53
rat	10 mM CHAPS		70		24
rat, bovine	10 mM CHAPS	160; 390	45; 65	9.2; 13.2	19
rat	1% digitonin/0.5 M NaCl		23; 36; 42; 51; 64; 78		24
rat	2% digitonin/10 mM MgSO$_4$	400; 750 – 875		19; 34 – 49	54
guinea-pig	0.5% digitonin/1 M NaCl	350 – 400			39
frog	1% digitonin			10; 12	32
frog*	1% digitonin		42; 71	10.8; 15.7	55
frog	1% digitonin/0.5 M NaCl	180; 470			56

* Values were obtained with prebound [^{3}H]ligand – receptor complexes.

Linear sucrose gradients, 5−25% w/v, are prepared in solubilization buffer medium (10−12 ml) containing 0.1−0.05% detergent or 0.1 mM CHAPS. The crude extract (0.2−0.3 ml) is layered on top of the gradient and centrifuged at 100 000 g for 2.5−3 h or 16 h at 4°C. Fractions, 0.35−0.5 ml, are collected. Prolonged centrifugation has been reported to result in a 40−60% decrease of the activity (39). Details of the protocols that are followed are given in Chapter 2, Section 6.

The apparent sizes of opioid receptor species obtained in various detergents by gel permeation chromatography and sucrose density gradient centrifugation, are shown in *Table 3*.

3.2 Lectin-affinity chromatography

Lectin-affinity chromatography has been used for the partial purification of opioid receptors from digitonin/NaCl extracts (33). Opioid receptors bind reversibly to immobilized wheat-germ agglutinin. Since the proportion of receptor activity retained on the lectin is unaffected whether free or [^{3}H]ligand−prebound-receptor is applied, this method can be used for the removal of ligand after affinity purification.

Partially purified opioid-agonist-binding activity which is sensitive to guanine nucleotides has been obtained in our laboratory (40) using this approach.

Protocol 7. WGA-purification of digitonin/Mg^{2+} solubilized receptors

1. Pack a 2 ml column of WGA-Sepharose 6MB (Sigma) and equilibrate it with 25 volumes of 10 mM TES-KOH (pH 7.5) containing 1 mM benzamidine-HCl, 1 mM EGTA-K$^+$, 5 mM MgSO$_4$, 0.001% soya bean trypsin inhibitor, 0.5 mM DTT, and 0.1% digitonin.

2. Dilute Mg^{2+}-digitonin extracts (3 ml) (see Section 2.4.2) four-fold with the above buffer and apply twice to the column (10 ml/h, 25°C).

3. After washing with 50 ml of the equilibration buffer, elute the bound receptor with 0.2 M *N*-acetyl-glucosamine in the same buffer.

Under these conditions 40% of [^{3}H]D-Ala^2-D-Leu^5-enkephalin ([^{3}H]DADLE) binding activity is recovered, giving a 60-fold purification. A greater recovery can be achieved when *N*-acetylchitotriose (Sigma) (10 mM) is used in place of *N*-acetyl-glucosamine, but the cost is much greater.

3.3 Hydrophobic chromatography

Phenyl-Sepharose CL-4B (LKB) has been used to partially purify opioid receptors from rat brain (40) and to purify *x*-receptors from frog brain (41).

Protocol 8. Phenyl-Sepharose-purification of digitonin/Mg^{2+} solubilized receptors

1. Dilute Mg^{2+}-digitonin extracts of rat brain membranes (20 ml, 55 mg protein) 40-fold to 0.05% (w/v) concentration of digitonin with 10 mM K$^+$-phosphate buffer (pH 6.8)/1 mM DTT/1 mM benzamidine-HCl/1 M ammonium sulphate.

2. Apply the diluted extract at a rate of 50 ml/h to a 5 ml column of phenyl-Sepharose equilibrated with 10 volumes of the above buffer.

3. Elute loosely-bound proteins with a stepwise or linear gradient of decreasing concentrations of ammonium sulphate (1-0.2 M; 150 ml total volume) in K^+-phosphate buffer (pH 7.6)/0.05% digitonin.

4. Elute tightly bound hydrophobic proteins in the absence of salt with steps of increasing digitonin concentrations (40 ml of 0.05%, 80 ml of 0.25%, 40 ml of 0.5%, 20 ml of 1.0%, and 10 ml of 2.0% w/v). Collect fractions (4 ml). All steps are carried out at 25°C.

In total, 70% of the receptor activity applied is recovered in the absence of salt and is distributed in several peaks within the digitonin eluates, giving a purification in the range of 40−100-fold.

For the purification of $\varkappa$-receptors, frog brain membranes solubilized with 1% digitonin (41) are extracted with Bio-beads to remove excess detergent, diluted with ammonium sulphate (0.5 M final concentration) and applied to a 16 ml phenyl-Sepharose column. Proteins are eluted with a step gradient of decreasing salt concentrations (100 ml each of 0.5, 0.25, 0.125, and 0.06 M); receptor activity is eluted with 100 ml of a decreasing salt gradient (0.06 M to H_2O).

3.4 Affinity cross-linking studies

Covalently labelled opioid−receptor complexes from brain or NG108-15 cell membranes have been identified using (a) irreversible ligands, (b) photochemical analogues of opiates or opioid peptides, or (c) bifunctional cross-linking reagents.

3.4.1 Irreversible opiate ligands

Two requirements have to be satisfied for the selective covalent labelling of receptor sites, these are:

(a) The ligand should possess a high affinity for the receptor.

(b) The ligand should possess an electrophilic centre with intrinsic reactivity which can align correctly with a proximal nucleophilic site on the receptor.

The various alkylating ligands which have been synthesized and characterized are reviewed in reference 42. Their use is limited, however, since the majority of these ligands are not commercially available in a radiolabelled form [see *Receptor−ligand interactions, a practical approach* (76), Chapter 1 for a listing of available radioligands].

[^{3}H]chlornaltrexamine ([^{3}H]CNA), an irreversible narcotic antagonist, has been applied (43) at 1 nM concentration to label opioid receptors in mouse brain membranes.

Protocol 9. Covalent labelling of receptors with [^{3}H]CNA

1. Suspend the membranes (10 mg protein/ml) in 50 mM Tris-HCl (pH 7.7)/100 mM NaCl, divide into two equal parts and to each add a volume equal

to 10% of the total volume Tris/NaCl buffer containing 1×10^{-4} M of either dextrorphan or naltrexone. Incubate these mixtures for 10 min at 25°C.

2. Add [^{3}H]CNA to give a final concentration of 1×10^{-9} M, incubate the mixture for 1 min and stop the reaction by adding an equal volume of 0.1 M sodium thiosulphate.

3. Centrifuge the membranes at 49 000 g for 15 min, resuspend each pellet in 14 ml Tris/NaCl buffer, centrifuge again and resuspend in 8 ml 50 mM Tris-HCl (pH 7.7) containing 1% Triton X-100 (protein: Triton X-100 = 1.0). Solubilize for 45 min at 25°C and centrifuge at 100 000 g for 15 min. Remove the supernatant with a Pasteur pipette and centrifuge once more at 100 000 g for 15 min.

4. Dialyse the supernatants against 50 volumes of 50 mM Tris-HCl buffer containing 1% Brij 36T, 0.04% sodium azide for 5 h with buffer changes every hour.

5. Transfer the supernatants to centrifuge tubes and precipitate the proteins by adding 5 g of ammonium sulphate (92% solution). Centrifuge as before, discard the supernatant and resuspend each pellet in 3 ml Tris-Brij 36T buffer and centrifuge again. Use the supernatant for further characterization.

Bidlack *et al.* (44) used [^{125}I]bromoacetamidomorphine (a μ-agonist) and [^{125}I]chloroacetylnaltrexone (an antagonist) at 20 nM concentration to label specifically rat neuronal membranes in 50 mM Tris-HCl, pH 7.5 (2 h, 37°C). Unreacted ligand was removed by four washes of the membranes by resuspension and centrifugation. [^{125}I]bromoacetamidomorphine labelled three major proteins with M_rs of 43 000, 35 000, and 23 000. [^{125}I]chloroacetylnaltrexone labelled mainly the 23 000 Da protein.

Klee *et al.* (45) used [^{3}H]fentanyl isothiocyanate ([^{3}H]FIT) to label the opioid receptor from NG108-15 cell membranes.

Protocol 10. Covalent labelling of receptors with [^{3}H]FIT

1. Suspend plasma membranes in 10 mM K^+-phosphate (pH 8.0) at $0.5-1$ mg protein/ml and incubate with 10 nM [^{3}H]FIT ($30-120$ min, 37°C) with or without competing opioid ligands.

2. Chill the suspension on ice and wash and harvest the membranes by centrifugation.

3. Solubilize the [^{3}H]FIT-labelled membranes with 50 mM CHAPS in phosphate-buffered saline.

4. Partially purify the solubilized extracts first on a WGA-Sepharose column (2 ml) in 10 mM CHAPS/0.15 M NaCl/10 mM Tris-HCl (pH 7.5), and then on a 0.5×0.76 cm hydroxyapatite column with elution by 0.25 M K^+-phosphate.

[^{3}H]FIT labelled specifically a 58 000 Da subunit. Proteins with M_rs of 35 000, 25 000, 17 000, and 12 000 were found to be labelled non-specifically.

The alkylating enkephalin analogue [^{3}H]Tyr-D-Ala-Gly-Phe-LeuCH$_2$Cl ([^{3}H]DALECK; 47) has been used to label covalently a μ-specific subunit of M_r 58 000 in rat brain membranes (46,47). The method used then has now been modified in our laboratory to increase the labelling efficiency and reduce the non-specific alkylation of membrane proteins.

Protocol 11. Covalent labelling of receptors with [^{3}H]DALECK

1. Prepare rat brain membranes as described in Section 2.4.2.

2. Incubate the membranes at $1-2$ mg protein/ml with 20 nM DADLE (1 h, 30°C) in 10 mM TES-KOH (pH 7.5) buffer containing in addition to protease inhibitors 6 mM MgSO$_4$ and 1 mM DTT.

3. Alkylate the membranes with 0.1 M iodoacetamide (10 min, 25°C) and wash 4 times by centrifugation and resuspension in 40 volumes of 10 mM TES-KOH each time.

4. Incubate the membranes at $0.8-1.0$ mg protein/ml in the Mg-TES buffer (without DTT) (1 h, 25°C) containing 5 nM [^{3}H]DALECK with or without unlabelled competing ligands (10^{-6} M). At the end of the incubation readjust the pH to 8.0 with KOH.

5. After 10 min, inactivate unreacted [^{3}H]DALECK by the addition of 1 mM DTT and two-fold dilution with 20 mM Na$^+$-acetate (pH 4.5) buffer and harvest the membranes by centrifugation at 40 000 g.

6. Wash the membranes in 40 volume of sodium acetate buffer and then solubilize them with 1% Zwittergent B12 (Calbiochem) or Na$^+$-cholate (Serva) in Mg-buffer (1 h, 4°C).

7. Partially purify the 100 000 g soluble fraction on a WGA-column (Section 3.2).

A specifically labelled protein of M_r 58 000$-$60 000 is identified. Non-specific labelling represents $<5\%$ of total (*Figure 1*).

3.4.2 Photoreactive ligands

Carbene- and nitrene-forming photoaffinity analogues of opiates and enkephalin peptide analogues have been used to covalently label the opioid receptors. This method has both advantages and disadvantages. On the one hand, the photoreactive ligands have a much shorter reactive lifetime than that of the electrophilic alkylating analogues and can be activated within their targets upon irradiation, thus in theory minimizing non-specific labelling. On the other hand, opioid receptors have been shown to be inactivated by short wavelength (254 nm) ultraviolet light ($t_{1/2}$ <7 min), and opiates also are labile upon irradiation (48). Other limitations to the use of photoreactive opiate and opioid peptide analogues is the present unavailability of any of them

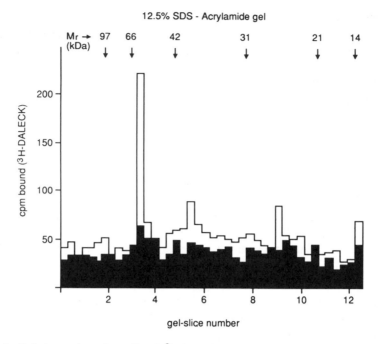

Figure 1. Gel-electrophoresis profile of [³H]DALECK cross-linked rat brain membranes. Brain membranes were pre-incubated with 20 nM DADLE (1h, 30°C) in the presence of 6 mM MgSO₄ and 1 mM DTT, alkylated with 0.1 M iodoacetamide (10 min, 24°C), washed and incubated with 5 mM [³H]DALECK (1 h at 25°C). [³H]DALECK-bound opioid receptors were cross-linked at pH 8.0 (10 min). Unreacted [³H]DALECK was inactivated with 1 mM DTT as described in Protocol 12, *step 5*. The membranes were subsequently solubilized with 1% Zwittergent B12; radiolabel-bound solubilized proteins were purified on WGA and elecrophoresed on an SDS 12.5% acrylamide gel. 3 mm slices were cut, dissolved in 200 µl soluene overnight and then counted in toluene-based scintillant. Non-specific labelling was measured in the presence of 10⁻⁶ M unlabelled DALECK. The profiles represent total and non-specific cpm recovered/slice. Arrows indicate the elution profile of molecular weight standards (kDa).

commercially in a radiolabelled form and the highly hydrophobic nature of the nitrophenylazido group usually added to the structure, which can result in high non-specific labelling.

Specific labelling of the δ-receptor subunit by this technique has been reported by Zajac *et al.* (49), who used the radiolabelled enkephalin analogue [³H]Tyr-ᴅ-Thr-Gly-Phe(pN₃)Leu-Thr ([³H]Aᵤ-DLET).

Protocol 12. Photoaffinity labelling of receptors

1. Incubate rat brain or NG108-15 cell membranes at 0.4 −0.8 mg protein/ml in 50 mM Tris-HCl (pH 7.4) with 10 nM [³H]Aᵤ-DLET (60 min, 35°C) with or without competing opioid ligands.

114

2. Centrifuge at 90 000 g (30 min, 4°C).

3. Resuspend 0.5 ml of the membranes to the original concentration with Tris and irradiate (5 min, 4°C) in a cylindrical polystyrene tube (1.1 cm diameter) at a distance of 40 cm with a 150 W lamp (Prolabo, France) (550 μm/cm^2 intensity at 254 nm).

4. Chill the irradiated membranes on ice, wash and harvest by centrifugation, resuspend in Tris buffer and solubilize either with 1% Na$^+$-cholate or 2% (w/v) SDS.

Fractionation of the Na$^+$-cholate extracts identified high molecular weight species ($\sim$200 000 Da), whereas gel-electrophoresis of the SDS extract showed [^{3}H]label incorporation into proteins of M_r 45 000$-$50 000 and 90 000. Irradiation of Na$^+$-cholate extracts of membranes pretreated with [^{3}H]A$_2$-DLET showed that only the protein species of M_r 45 000$-$50 000 were labelled specifically. The risk exists that these were derived from the longer polypeptide identified by other covalent labelling methods.

3.4.3 Bifunctional cross-linking reagents

This method has employed the cross-linking of a radiolabelled opioid peptide to receptors in membranes or soluble extracts by means of a bifunctional reagent. The method of Zukin and Kream is as follows (27):

Protocol 13. Covalent labelling of receptors using a bifunctional cross-linking reagent

1. Incubate rat brain membranes with [^{3}H]enkephalin analogues (2$-$8 nM) in the presence and absence of competing ligands (10 μM) in 50 mM Tris-HCl (pH 7.4) (45 min, 4°C).

2. Pellet the membranes by centrifugation (30 000 g, 15 min), resuspend in one-tenth the original volume of an ice cold 1% Brij 36T (Sigma) solution in phosphate buffer (pH 7.4), vortex briefly, and centrifuge (90 000 g, 90 min).

3. Remove unbound [^{3}H]enkephalin on a Sephadex G-25 column (1 $\times$ 20 cm) in 1% Brij 36T phosphate buffer pH 8.0.

4. Cross-link the protein-containing fractions (2 ml) for 30 min, at room temperature with dimethyl suberimidate (DMS, 2 mg/ml; Sigma) in 100 mM triethanolamine (pH 8.8).

5. Fractionate the radiolabelled proteins on a Sephadex G150 and then on a Sephadex G200 column (1.5 $\times$ 30 cm) in 50 mM Tris-HCl (pH 7.4).

A protein with M_r of 35 000 was labelled specifically by this method.
The method of Howard *et al.* (50) proceeds as follows:

Protocol 14. Covalent labelling of receptors with [^{125}I]-β-endorphin

1. Incubate crude membranes (1.6 mg protein/ml) in 50 mM K$_2$HPO$_4$ (pH 7.4)

containing 50 μg/ml bacitracin and 2 nM ^{125}I-β-endorphin (human, Amersham, NEN), plus or minus competing ligands, for 1 h at 25°C.

2. Dilute the membranes two-fold with ice-cold buffer, centrifuge (20 000 *g*, 15 min, 4°C), and resuspend to the original volume.

3. Cross-link with 1 mM bis {2-[(succinimidooxyl)carbonyl]oxy}-ethyl sulphone (BISCOES; Pierce) freshly prepared in dimethyl sulphoxide. Terminate the cross-linking (15 min, 4°C) by the addition of ten-fold excess of 50 mM Tris-HCl (pH 7.4), 1 mM EDTA and centrifugation of the membranes.

Proteins with M_rs of 65 000 (μ), 53 000 (δ), 41 000, and 38 000 were identified by this method, the two smaller species showed protection by both μ and δ ligands.

Using a similar method, Helmeste *et al.* (51) cross-linked specifically ^{125}I-[Tyr^{2+},Leu^5]β-endorphin with disuccinimidyl suberate (1 mM) in 50 mM K$_2$HPO$_4$ (pH 7.4) to peptides with M_rs of 40 000, 30 000, and 25 000 in membranes of human caudate and putamen, and of 92 000, 56 000, 38 000, and 23 000 in membranes of NG108-15 cells.

3.5 Affinity chromatography

The purification of opiate-binding proteins by affinity chromatography using opiate or enkephalin peptide analogues covalently coupled to a solid support has been reported by several groups (*Table 4*). Generally, detergent extracts are either pretreated with Bio-beads or diluted to reduce the detergent concentration and either incubated batchwise with the pre-equilibrated affinity gel or applied to the gel prepacked onto a column at a flow rate of 10−20 ml/h. In batchwise procedures the gel can be subsequently packed onto a column for the washing and elution steps. Usually 10−15 ml of soluble extract per ml of packed gel are used.

Variations in the treatment of the solubilized membrane fractions before application to the affinity gel include the pre-incubation of the extracts at 37°C for 30 min (i.e. Triton extracts in 50 mM Tris-HCl, pH 7.4; references 14), or the passage of the extracts first through the gel support that does not contain the ligand (37).

Various buffer systems and conditions have also been used for the washing of the affinity gel: Tris-HCl or imidazole buffers (50 mM; (pH 7.4)) have been used (10−20 column volumes) to wash off non-specifically bound proteins present in Triton brain extracts (14,37). In the purification of μ-receptors from digitonin frog-brain extracts, A. Borsodi and co-workers (57) used a linear salt gradient (0−0.5 M NaCl) in 50 mM Tris-HCl (pH 7.4)/1 mM EDTA/40 KIU/ml trasylol/20 μg/ml bacitracin/1 mM PMSF/0.05% digitonin. Gioannini *et al.* (58) used 46−50 column volumes of 50 mM Tris-HCl (pH 7.4)/1 mM EDTA-K$_2$/0.5 M NaCl/0.05% digitonin followed by one column volume of the same buffer containing 0.25% digitonin and one column volume containing 0.1% digitonin.

Opiate-binding proteins are specifically eluted from the affinity gel within 2−3 column volumes of buffer containing an agonist or an antagonist in μM concentra-

Table 4. Affinity chromatography purification of opiate-binding proteins from brain tissue.

Species	Detergent	Ligand	M_r (kDa)	% Yield[a]	Purification[a] (fold, approx.)	Reference
rat	Triton X-100	14-β-bromoacetamidomorphine	23; 35; 45		2000	60
rat	Triton X-100	6-succinylmorphine	58	6	68 000	59
rat	Triton X-100	6-succinylmorphine	58		1500–5000	37
rat	CHAPS	Hydromet	35; 44; 84	0.6	506	14
rat	digitonin/Mg^{2+}	DALE	62		450	72
rat	digitonin/Mg^{2+}	DALECK	62; 54	10	20 000–30 000	73
rat	digitonin/Mg^{2+}	GANC	62	8	60 000–70 000	73
bovine	digitonin/NaCl	β-naltrexethylene-diamine	65	5.8	65 000–75 000	58
guinea-pig	digitonin	dynorphin (1–10)	62	6	10 000–20 000	74
frog		DALE	65; 58	0.07	4300	57
frog		dynorphin (1–10)	65	0.3	6000	75

[a] Fold purification and yield are values obtained after removal of ligand by additional step(s).
DALE, D-Ala^2-Leu^{5}enkephalin; DALECK, D-ala^2-Leu^5-chloromethyl ketone; GANC, 14-β-(C-glycyl)-amido-(N-cyclopropylmethyl)norcodeinone; Hydromet, 7α-(1R)-hydroxy-1-methyl-3-p[4-(3'Bromomercuri-2'-methoxy-propoxy)phenyl]-propyl-6,14-endoethenotetrahydrothebaine.

tions. Cho *et al.* (59) have used, instead, a linear salt gradient (0−1 M NaCl) in 50 mM Tris-HCl (pH 7.4) to elute opiate-binding proteins from an agonist column. However, this results in the co-elution of non-specifically bound proteins and further purification is required. The activity of the purified opioid receptor protein(s) is assayed after removal of the ligand by either dialysis, lectin affinity chromatography, gel-filtration, hydroxyapatite chromatography, or a combination of those methods.

As an example of the use of the affinity gels specified in *Table 4*, in our laboratory we have used a norcodeinone analogue (77) coupled to Affi-Gel-10 (Bio-Rad) to purify the μ-opiate-binding protein from rat brain membranes solubilized with Mg^{2+}-digitonin (see Section 2.4.2).

Protocol 15. Receptor purification by ligand-affinity chromatography

1. Dilute the crude digitonin extracts 20-fold with Mg^{2+}-buffer to a final digitonin concentration of 0.1%.

2. Incubated batchwise with the affinity gel (25 ml), at room temperature for 2 h.

3. Remove non-bound proteins by filtration on a sintered-glass funnel and pack the gel into a column.

4. Wash the resin (20 ml/h) with 10 volumes of Mg-buffer containing 0.1% digitonin, followed by 10 volumes of the same buffer from which the protease inhibitors soya bean trypsin and bacitracin are omitted. Bound receptor proteins are specifically eluted with 2−3 column volumes of the latter buffer containing 0.5 mg/ml naloxone.

5. Apply the receptor−ligand eluate directly onto a WGA column (see Section 3.2) and elute with 0.2 M *N*-acetyl-glucosamine. Assay the activity of the purified protein after removal of the sugar by dialysis against 2 l Mg^{2+}-buffer (2 h, 4°C). The dialysis tubing should be pretreated as described in Chapter 10, Section 6, with substitution of digitonin for SDS.

---

4. The status of the opioid receptors at present

4.1 Molecular studies: outstanding issues

The goal of studies such as those outlined above can be considered to be to find answers to such questions as:

(a) Do the μ, δ, and $\varkappa$ pharmacological activities each reside on a separate polypeptide of different primary structure, i.e. μ, δ, and $\varkappa$ subunits? Does this exhaust the categories of opioid-receptor subunits? Within these categories, are there further subunits specific for sub-subtypes, e.g. for μ_1 and μ_2 (61) or $\varkappa_1$ and $\varkappa_2$ (62,63) subunits?

(b) Assuming that there are subtype-specific subunits, do these occur in the neuronal membrane as monomers or oligomers? Assuming the latter, do the oligomers contain one or more than one subtype?

(c) Are there different oligomeric compositions for presynaptic *versus* postsynaptic receptors, and again for extrasynaptic receptors? Do opioid receptors occur at all three locations?

(d) Is there a common transduction intermediary system (e.g. cyclase inhibition) for all subtypes? Are there opioid-specific intermediary elements, e.g. specific G-proteins for each receptor type?

(e) Is opioid-receptor activation *in vivo* always linked to the opening or closing of an ion channel? Is a specific type of ion channel always associated with specific opioid receptor subtypes?

(f) Do opioid receptor molecules occur at some sites associated in a native complex with another receptor, e.g. an α-adrenergic receptor?

(g) Do covalent protein modifications, such as phosphorylation, regulate the activity of the receptors? Are these events associated with down-regulation induced by chronic exposure to opioid agonists?

(h) Do the peripheral opioid receptors (as in the myenteric plexus) have subunits identical with some of those in the brain?

To none of these questions can a definite answer be given at present. Answers to all of them are required for an understanding of the opioid system at the molecular level.

4.2 Opioid-receptor heterogeneity

The question as to whether the μ, δ, and κ-binding activities arise from different polypeptides, each the product of a different gene, is all-important for guiding the purification and DNA-cloning studies but remains unresolved at present. Binding studies on brain membranes, when performed under optimal conditions, indicate that these three subtypes are indeed distinct entities (10) as do experiments in which soluble extracts on the membranes have been fractionated to give partial separations of the three activities (see *Table 3*). Allosteric interactions between these three classes of binding site can also be inferred from detailed analysis of the binding behaviour (10). None of the observations made on binding properties or on regional distributions in the brain are competent to establish that the subtypes arise from different amino-acid sequences and not from post-translational changes or interactions with other components in the membranes. Nor do they show definitively whether or not there exist hetero-oligomers (containing μ, δ, and κ sites together, e.g. in pairs, in one multi-subunit receptor). Studies in detergent solution are particularly suspect in this regard, since re-equilibration of the monomers among the complexes can be shown experimentally to be capable of occurring there (54).

The best approach at present, therefore, is to employ for the receptor purification tissue sources in which one subtype predominates, where this is possible, and to use a subtype-relative ligand in the affinity chromatography. For example, the NG108-15 neuroblastoma $\times$ glioma hybrid cell line shows only the δ binding activity; this does not prove that δ does not occur in a mixed oligomer with another subtype in some brain regions, nor does it exclude the possibility that some modification or interaction of the δ-chain could change it to the μ binding type, although there is no clear evidence in favour of this. It does exemplify the possibility of isolating a receptor with a single type of binding activity, from a suitable source. This has been used, for example, by Klee *et al.* (45) to identify a δ-type polypeptide of 58 000 M_r. For μ, generally the whole rat brain has been used as the source since this is the predominant subtype there, but δ and $\varkappa$ are also present and selective affinity chromatography has been relied upon in the studies to date, as shown in *Table 4*.

For $\varkappa$ receptors, several sources in which $\varkappa$ greatly predominates have been identified. The human placenta has >95% of its opioid binding in the $\varkappa$ category (64−66). In our hands, all of the digitonin-solubilized [^{3}H]bremazocine (5 nM; NEN) binding there is displaced by the ligands U50488H (K_i 20 nM in solution) or CAM-20 (K_i 2 nM) and none by DAGO or DADLE up to 1000 nM levels. However, while average content in the starting placental membranes is 170 fmol/mg (B_{max} for bremazocine), this varies greatly between individuals, and the quantities of tissue required for a preparation are very large, so that in practice this is not the optimal source for $\varkappa$-receptor isolation.

The other suitable sources for $\varkappa$ are the guinea-pig cerebellum and the frog brain. The latter has been studied extensively by A. Borsodi and co-workers, who have used the digitonin solubilization method and affinity chromatography to isolate a frog brain $\varkappa$ receptor (*Table 4*). A more abundant source of the $\varkappa$ subtype is the guinea-pig cerebellum, where >80% of the naloxone-binding sites are in this class (67). Employing solubilization (in digitonin) and purification of this receptor on an affinity column containing an immobilized derivative of dynorphin (1−10) (*Table 4*), we have found that, indeed, this is a suitable source for the purifiation of the $\varkappa$ subtype as a single polypeptide of M_r ∼62 000.

4.3 Approaches to the molecular biology of the opioid receptors

Clearly the most effective route to answering some of the questions listed in Section 4.1 would be to clone the cDNAs encoding the μ, δ, and $\varkappa$-receptors, since many of the other studies needed would flow naturally from this. At the time of writing, no cloning has been reported of a DNA encoding any opioid receptor; some leads described in two cases by methods other than via protein purification did not yield a cDNA with a recognizable receptor sequence.

The opioid receptors have proven particularly difficult to subject to molecular cloning. Some of the problems involved are as follows.

(a) The *Xenopus* oocyte receptor translation system (68) does not function well for the opioid receptors. After injection therein of poly(A)$^+$ RNA from neonatal rat or chick brain, when many other receptors mRNAs become translated to yield the functional receptor, little or no response to opioids can be detected. This precludes a route to cloning via oocyte expression such as that taken for the only other neuropeptide receptor to be cloned so far, the substance K-receptor (70). This inability to express may reflect the difficulty of finding in the oocyte a suitable coupled channel or intermediary system for opioid receptors. It may also reflect a very low mRNA level for this receptor, due to a low receptor abundance and/or a slow turover rate.

(b) The more conventional route via protein purification has encountered problems due to the very low level of opioid receptor that can be isolated from nervous tissue, its lability in extracts, and the co-extraction of the multiple receptor subtypes.

(c) Extraction and retention of activities of the opioid receptors are particularly sensitive to the detergent and other conditions used. The detergent which best retains their activity and maximizes the yield in extraction is digitonin (13) but this detergent is not a pure product and its composition varies between sources and batches. The μ-opioid-receptor isolation is very sensitive to this composition, whereas several other receptors purified in digitonin are not.

(d) Binding activity after purification has hitherto had to be tested with radiolabelled agonists. Maximal binding is dependent, for this receptor type, upon the presence of specific G-proteins and phospholipids (69). This makes the study of the completely pure receptor more difficult. Labelled antagonists have been used in assaying G-protein-linked receptors when purified, when those problems disappear since G-protein coupling is not then important. However, until now no subtype-selective radiolabelled antagonist for an opioid receptor has been available. This situation is expected to change soon, when the prospect should be much improved.

4.4 Conclusions

Since the opioid receptors are linked to G-proteins in all cases where there is evidence on their transduction route, on all present knowledge one could predict that the structure to be found for any of its subtypes would be a single-chain subunit (which may occur in oligomers) with seven hydrophobic membrane-spanning α-helices. It appears that such a structure for opioid receptor subtypes is not very homologous to other receptors involved in adenylate cyclase inhibition, such as that for substance K (70) or the muscarinic receptor subtypes (71), since cross-hybridization based on conservative regions of their cDNAs has not recognized opioid receptors in trials made to date. Despite the difficulties of the opioid receptor system noted above, there is no reason why these should not be overcome soon, and pure receptor preparations from several laboratories have recently been achieved. It can be expected with confidence, therefore, that before long one or other of the routes noted will yield at least one of the subtypes in cloned form. Cross-hybridization between subtypes would then be pursued at low stringency, hopefully to yield clones for the others. Evaluation of the molecular features of the opioid receptor family must now await that conclusion.

References

1. Paterson, S. J., Robson, L. E., and Kosterlitz, H. W. (1983). *Br. Med. Bull.*, **39**, 31.
2. Simon, E. J. (1987). *J. Receptor Res.*, **7**, 105.
3. Glazel, J. A., Venn, R. F., and Barnard, E. A. (1980). *Biochem. Biophys. Res. Comm.*, **95**, 263.
4. Fischel, S. V. and Medzihradsky, F. (1981). *Mol. Pharmacol.*, **20**, 269.
5. Zukin, R. S. and Zukin, S. R. (1981). *Mol. Pharmacol.*, **20**, 246.
6. Cooper, D. M. F., Londos, C., Gill, D. L., and Rodbell, M. (1982). *J. Neurochem.*, **38**, 1164.
7. Kurose, H., Katada, T., Amano, T., and Ui, M. (1983). *J. Biol. Chem.*, **258**, 4870.
8. Wong, Y. H., Demoliou-Mason, C. D., and Barnard, E. A. (1988). *J. Neurochem.*, **51**, 114.
9. Sadee, W., Pfeifer, A., and Hertz, A. (1982). *J. Neurochem.*, **39**, 659.
10. Demoliou-Mason, C. D. and Barnard, E. A. (1985). *J. Neurochem.*, **46**, 1118.
11. Larsen, N. E., Mullikin-Kilpatrick, D., and Blume, A. J. (1981). *Mol. Pharmacol.*, **20**, 255.
12. Bowen, W. D. and Pert, C. B. (1982). *Cell. Mol. Neurobiol.*, **2**, 115.
13. Demoliou-Mason, C. D. and Barnard, E. A. (1984). *FEBS Lett.*, **170**, 378.
14. Maneckjee, R., Zukin, R. S., Archer, S., Mickael, J., and Osei-Gyimah, P. (1985). *Proc. Natl. Acad. Sci. USA*, **82**, 594.
15. Abood, L. G., Salem, N., MacNeil, and M., Butler, M. (1978). *Biochim. Biophys. Acta.*, **530**, 35.
16. Hasegawa, J. -I., Loh, H. H., and Lee, N. M. (1987). *J. Neurochem.*, **49**, 1007.
17. Puget, A., Jauzac, P., and Meunier, J. C. (1980). *FEBS Lett.*, **122**, 199.
18. Gioannini, T. L., Howells, R. D., Hiller, J. M., and Simon, E. S. (1982). *Life Sci.*, **31**, 1315.
19. Simonds, W. P., Koski, G., Streaty, R. A., Hjeimeland, L. M., and Klee, W. A. (1980). *Proc. Natl. Acad. Sci. USA*, **77**, 4623.
20. Quirion, R., Bowen, W. D., Herkenham, M., and Pert, C. B. (1982). *Cell. Mol. Neurobiol.*, **2**, 333.
21. Chow, T. and Zukin, S. (1983). *Mol. Pharmacol.*, **24**, 203.
22. Koski, G., Simonds, W. F., and Klee, W. A. (1981) *J. Biol. Chem.*, **256**, 1536.
23. Zukin, R. S. and Maneckjee, R. (1985). *Meth. Enzymol.*, **124**, 72.
24. Simon, J., Benyhe, S., Abutidze, K., Borsodi, A., Szücs, M., Toth, G., and Wollemann, M. (1986). *J. Neurochem.*, **46**, 695.
25. Simon, E. J., Hiller, J. M., and Edelman, I. (1975). *Science*, **190**, 389.
26. Smith, A. P. and Loh, H. H. (1979). *Mol. Pharmacol.*, **16**, 757.
27. Zukin, R. S. and Kream, R. M. (1979). *Proc. Natl. Acad. Sci. USA*, **76**, 1593.
28. Dornay, M. and Simantov, R. (1982). *J. Neurochem.*, **38**, 1524.
29. Ruegg, U. T., Guenod, S., Fulpians, B. W., and Simon, E. A. (1982). *Biochim. Biophys. Acta*, **685**, 241.
30. Ogawa, N., Yamawaki, Y., Kuroda, H., Nukina, I., and Ofuji, T. (1981). *Neurosci. Lett.*, **27**, 205.
31. Bidlack, J. M. and Abood, L. G. (1980). *Life Sci.*, **27**, 331.
32. Ruegg, U. T., Guenod, S., Hiller, J. M., Gioannini, T., Howells, R. B., and Simon, E. A. (1981). *Proc. Natl. Acad. Sci. USA*, **78**, 4635.

33. Gioannini, T., Foucaud, B., Hiller, J. M., Hatten, M. E., and Simon, E. J. (1982). *Biochem. Biophys. Res. Comm.*, **105**, 1128.

34. Simon, J., Szücs, M., Benyhe, S., Borsodi, A., Zeman, P., and Wollemann, M. (1984). *J. Neurochem.*, **43**, 395.

35. Cho, T. M., Ge, B. -L., Yamato, C., Smith, A. P., and Loh, H. H. (1983). *Proc. Natl. Acad. Sci. USA*, **80**, 5176.

36. Cho, T. M., Ge, B. -L., and Loh, H. H. (1985). *Life Sci.*, **36**, 1075.

37. Ueda, H., Harada, H., Misawa, H., Nozaki, M., and Takagi, H. (1987). *Neurosci. Lett.*, **75**, 339.

38. Itzhak, Y., Hiller, J. M., Gioannini, T. L., and Simon, E. J. (1983). *Life Sci. Suppl. I*, **33**, 191.

39. Itzhak, Y., Hiller, J. M., and Simon, E. J. (1984). *Proc. Natl. Acad. Sci. USA*, **81**, 4217.

40. Wong, Y. H., Demoliou-Mason, C. D., and Barnard, E. A. (1989). *J. Neurochem.*, **52**, 999.

41. Borsodi, A., Khan, A., Simon, J., Benyhe, S., Hepp, J., Wollemann, M., and Medzihradsky, K. (1986). In *NIDA Res. Monograph Ser.*, (eds S. W. Holiday, P. Y. Law, and A. Hertz), Vol. 75, pp. 1−4.

42. Casy, A. F. and Parfitt, R. T. (eds) *Opioid analgesics, chemistry and receptors*. Plenum Press, New York.

43. Caruso, T. P., Larson, D. L., Portoghese, P. S., and Takemori, A. E. (1980). *Life Sci.*, **27**, 2063.

44. Bidlack, J. M., Abood, L. G., Munemitsu, S. M., Archer, S., Gala, D., and Krielick, R. W. (1982). *Adv. Biochem. Pharmacol.*, **33**, 301.

45. Klee, W. A., Simonds, W. F., Sweat, F. W., Burke, T. R., Jr., Jacobson, A. E., and Rice, K. C. (1982). *FEBS Lett.*, **150**, 125.

46. Newman, L. and Barnard, E. A. (1984). *Biochemistry*, **23**, 5385.

47. Newman, E. L., Borsodi, A., Toth, G., Hepp, F., and Barnard, E. A. (1986). *Neuropeptides*, **8**, 305.

48. Glasel, J. A. and Venn, R. F. (1981). *Life Sci.*, **29**, 221.

49. Zajac, J. M., Rotene, W., and Roques, B. P. (1987). *Neuropeptides*, **9**, 295.

50. Howard, A. D., Sarne, Y., Gioannini, T. L., Hiller, J. M., and Simon, E. J. (1986). *Biochemistry*, **25**, 357.

51. Helmesta, D. M., Hammonds, R. G., and Li, C. H. (1986). *Proc. Natl. Acad. Sci. USA*, **83**, 4622.

52. Hiller, J. M., Edelman, I., and Simon, E. J. (1975). *Science*, **190**, 389.

53. Hammonds, R. G., Nicolas, P., Jr., and Li, C. H. (1982). *Proc. Natl. Acad. Sci. USA*, **79**, 6494.

54. Demoliou-Mason, C. D. and Barnard, E. A. (1986). *J. Neurochem.*, **46**, 1129.

55. Puget, A., Jauzac, P., and Meunier, J. C. (1983). *Life Sci. Suppl. I*, **33**, 199.

56. Simon, J., Benyhe, S., Borsodi, A. and Wollemann, M. (1986). *Neuropeptides*, **7**, 23.

57. Simon, J., Benyhe, S., Hepp, J., Khan, A., Borsodi, A., Szücs, M., Medzihradszky, K., and Wollemann, M. (1987). *Neuropeptides*, **10**, 19.

58. Gioannini, T. L., Howard, A. D., Hiller, J. M., and Simon, E. J. (1985). *J. Biol. Chem.*, **260**, 15117.

59. Cho, T. M., Hasegawa, J. -I., Ge, B. -L., and Loh, H. H. (1986). *Proc. Natl. Acad. Sci. USA*, **83**, 4138.

60. Bidlack, J. M., Abood, L. G., Osei-Guimah, P., and Archer, S. (1981). *Proc. Natl. Acad. Sci. USA*, **78**, 636.

61. Pasternak, G. W. (1986). *Biochem. Pharmacol.*, **35**, 361.
62. Zukin, R. S., Eghbali, M., Olive, D., Underwald, E. M., and Tempel, A. (1988). *Proc. Natl. Acad. Sci. USA*, **85**, 4061.
63. Gouardere, C. and Cros, J. (1984) *Neuropeptides*, **5**, 113.
64. Ahmed, S. M. (1983). *Membrane Biochem.*, **5**, 35.
65. Agbas, A., Simon, J., Oktem, H. A., Varga, E., and Borsodi, A. (1988). *Neuropeptides*, **12**, 171.
66. Richardson, A., Brugger, F., Demoliou-Mason, C., and Barnard, E. A. (1989). *Proc. INRC. Advances in the Biosciences*, **75**, 13.
67. Robson, L. E., Foote, R. W., Maurer, R., and Kosterlitz, H. W. (1984). *Neuroscience*, **12**, 621.
68. Barnard, E. A. and Bilbe, G. (1987). In *Neurochemistry: a practical approach.* (eds A. J. Turner and H. C. Bachelard). IRL Press, Oxford.
69. Ueda, H., Harada, H., Nozaki, M., Katada, T., Ui, M., Satoh, M., and Takagi, H. (1988). *Proc. Natl. Acad. Sci. USA*, **85**, 7013.
70. Masu, Y., Nakayama, K., Tamaki, H., Harada, Y., Kuno, M., and Nakanishi, S. (1987). *Nature*, **329**, 836.
71. Barnard, E. A. (1988). *Nature*, **335**, 301.
72. Fujioka, T., Inoue, F., and Kuriyama, M. (1985). *Biochem. Biophys. Res. Comm.*, **131**, 640.
73. Barnard, E. A, Demoliou-Mason, C. D., and Wong, Y. (1986). *Proc. INRC*, San Francisco.
74. Simon, J., Demoliou-Mason, C. D., Richardson, A., Hepp, J., and Barnard, E. A. (1990). Manuscript in preparation.
75. Simon, J., Benyhe, S., Hepp, J., Varga, E., Medzihradszky, K., Borsodi, A., and Wollemann, M. (1990). *J. Neurosc. Res.*, in press.
76. Hulme, E. C. (ed.) *Receptor−ligand interactions, a practical approach*, IRL Press, Oxford, in press.
77. Peers, E. M., Rance, M. J., Barnard, E. A., Haynes, A. S., and Smith, C. F. (1983) *Life Sci. Suppl. 1*, **33**, 439.

5

Photoaffinity labelling and purification of the β-adrenergic receptor

JEFFREY L. BENOVIC

1. Introduction

Catecholamines regulate a wide variety of physiological responses via their inter-action with specific cellular receptors (1). The catecholamine receptors have been divided into two major types, α and β, each of which have been additionally divided into two subtypes. The β_1- and β_2-adrenergic receptors are coupled to stimulation of the enzyme adenylate cyclase which regulates the intracellular level of the second messenger, cyclic AMP. Although very similar in function, the β_1- and β_2-receptors are distinct in structure and pharmacological characteristics. The α_1- and α_2-adrenergic receptors are even more distinct, being coupled to different second messenger systems. The α_1-receptor is coupled to stimulation of phosphatidyl-inositol turnover, which regulates the level of the second messengers diacylglycerol and inositol trisphosphate, while the α_2-receptor is coupled to inhibition of adenylate cyclase.

This chapter describes methods for the identification, photoaffinity labelling, solubilization, and purification of the mammalian β_2-adrenergic receptor. Most of the methods described are also applicable to the study of the β_1-adrenergic receptor.

2. Tissue source

One of the most important considerations in the purification of a low-abundance membrane protein is the source of the protein. In the purification of the β-adrenergic receptor there are several important points to consider. First, what is the ratio of β_1- to β_2-receptor in the tissue of interest? Obviously, if one wishes to purify the β_2-receptor, a tissue which contains a high percentage of β_2 should be chosen. Another obvious issue is the total amount of β-receptor present in a particular tissue. Again, a source with a high level of the receptor of interest is preferred. Finally, what is the stability of the receptor, i.e. is it readily proteolyzed or degraded? This can also be a tissue- and/or species-dependent phenomenon, as is apparent in tissues that contain a high endogenous level of protease activity, such as liver.

Table 1. β_1- and β_2-adrenergic receptor subtypes and density in various tissues[a]

Species and tissue	Ratio of β_1- to β_2-adrenergic receptor	Total receptor density (pmol/mg protein)	Reference
Frog			
erythrocyte	0:100	1.6[b]	3
Turkey			
erythrocyte	100:0	0.6 – 0.8	4
Rat			
lung	20.80	0.4 – 0.6	5,6
adipose	100:0	0.24	7
erythrocytes	0:100	0.20	8
reticulocytes	0:100	0.76	8
cerebral cortex	81:19	0.07	9
cerebellum	15:85	0.02	9
Hamster			
lung	2:98	1.4 – 1.7	2,6
Guinea pig			
lung	20.80	1.2	2
Rabbit			
lung	70:30	0.23 – 0.39	5,10
Canine			
lung	5:95	0.23	11,12
Human			
lung	30:70	0.10	13,14
placenta (basal membrane)	65:35	1.2 – 2.5	15,16
heart (right atrium)	74:26	0.05	14,17

[a] This table is not meant to be a totally comprehensive review of the literature on β-adrenergic receptor subtypes and density.
[b] Plasma membrane preparation was purified on a sucrose density gradient.

As shown in *Table 1*, tissues vary significantly in their $\beta_1:\beta_2$ receptor ratios and in their receptor levels. We routinely purify the β_2-adrenergic receptor from hamster lung (2). This tissue contains predominantly the β_2 subtype ($\sim 98\%$) at a relatively high concentration (~ 1.7 pmol β-adrenergic receptor binding sites/mg of protein) in partially purified plasma membrane preparations. The tissue is readily available in large quantity from Pel Freeze Biologicals. Guinea-pig lung also serves as a very good source for purification of the β_2-adrenergic receptor (2). The avian β-adrenergic receptor, which shares some pharmacological characteristics with the mammalian β_1- receptor, has been successfully purified from turkey erythrocytes (4). A potentially good source for purification of the mammalian β_1-adrenergic receptor is the human placenta which contains $\sim 65\%$ β_1-receptors at a concentration of $1-2$ pmol β_1-receptor/mg protein in basal membranes (15,16).

3. Radioligand binding

Ligand binding represents a valuable and essential tool for the study of receptors. There is a wide variety of methods for assessing the level of a receptor in either

Table 2. Radioligands for β-adrenergic receptors

Radioligand	Maximum specific activity (Ci/mmol)	Source
[^{3}H]Dihydroalprenolol ([^{3}H]DHA)	120	NEN, Amersham
[^{3}H]Propranolol ([^{3}H]PROP)	30	NEN, Amersham
[^{3}H]CGP 12177 ([^{3}H]CGP)	50	Amersham
[^{3}H]Hydroxybenzylisoproterenol ([^{3}H]HBI)	20	NEN
[^{125}I]Iodocyanopindolol ([^{125}I]CYP)	2200	NEN, Amersham
[^{125}I]Iodohydroxybenzylpindolol ([^{125}I]HYP)	2200	NEN, ICN
[^{125}I]Iodopindolol ([^{125}I]PIN)	2200	NEN

particulate or soluble form (see Chapter 1; see also reference 22). Several ligands that are commercially available for studying β-adrenergic receptors are listed in *Table 2*. A number of additional radioligands, including [^{3}H]norepinephrine and [^{3}H]isoproterenol, are also available but seldom used. The most commonly used radioligands are [^{3}H]dihydroalprenolol, which binds to $β_1$- and $β_2$-receptors with comparable affinity (K_d ~1 nM), and [^{125}I]cyanopindolol, which is also non-subtype selective with a K_d ~20–50 pM. The procedures used for [^{3}H]DHA binding are shown below. When [^{125}I]CYP is used as the ligand the receptor level is reduced 10–20-fold while the ligand concentration used is typically ~200 pM.

3.1 Radioligand binding to particulate β-adrenergic receptors

Protocol 1. Assay of membrane-bound βARS

1. Prepare hamster lung membranes as described in Section 4. The membranes from 1 g of tissue should contain 20–25 pmol of β-receptor ligand-binding sites.

2. Set up duplicate 0.5 ml incubations containing:

   0.1–0.5 pmol β-receptor
   10–20 nM [^{3}H]DHA
   25 mM Tris-HCl, pH 7.2, 2 mM MgCl$_2$ (buffer A) with or without 10 μM (−)alprenolol or (−)isoproterenol (Sigma) (to define non-specific binding).

3. Shake for 1 h at room temperature.

4. Stop the incubations with 3–5 ml of ice-cold buffer A and immediately filter the samples through GF/C glass-fibre filters (Whatman). Wash each filter 3–4 times with 3–5 ml of ice-cold buffer A. The washing procedure is usually complete within 15–30 seconds.

5. Remove the filters and count in a scintillation counter with 10 ml of aqueous

scintillation fluid. It takes a few hours for the filters to solubilize before the maximum counts are obtained.

6. The counts obtained from the samples without the cold ligand are the total counts, while the samples in the presence of unlabelled ligands (with alprenolol or isoproterenol) are the non-specific counts. The difference between the two sets of counts represents specific radioligand binding and this number can be used to determine the amount of β-adrenergic receptors in the sample. For example, if a 0.1 ml sample yields 10 000 specific counts using [^{3}H]DHA with a specific activity of 50 Ci/mmol (50 000 cpm/pmol assuming a 45% efficiency of counting) then the sample contains 2 pmol β-receptors/ml (10 000 cpm divided by 50 000 cpm/pmol divided by 0.1 ml).

3.2 Radioligand binding to soluble β-adrenergic receptors

Protocol 2. Assay of soluble βARS

1. Prepare a hamster lung soluble preparation as described in Sections 6 or 7.
2. Set up duplicate 0.5 ml incubations containing:

   0.1−0.5 pmol β-receptor
   10−20 nM [^{3}H]DHA
   100 mM NaCl, 10 mM Tris-HCl, pH 7.2, 0.1% digitonin (buffer B) with or without 10 μM (−) alprenolol or (−)-isoproterenol.

   Note: Digitonin (Gallard Schlesinger or Wako Chemical Co.) is prepared as a 5% solution in H$_2$O by heating to ~90°C. After cooling, the solution is filtered and can be stored at 4°C as a stock solution. Before use the digitonin is diluted into the appropriate buffer.
3. Incubate for 2−3 h at 25°C followed by 10 min at 4°C (for most soluble receptor preparations assays can also be incubated at 4°C for 12−18 h).
4. The samples are then desalted on 3.5 ml−Sephadex G50 (fine) columns (12 × 0.6 cm) equilibrated with buffer B at 4°C.
5. Typically, the receptor/ligand complex elutes between 1.1 and 2.1 ml (this should be determined using blue dextran). The samples can be collected directly into scintillation vials before the addition of 10 ml of scintillation fluid. The time required for the separation procedure is usually less than 15 min. The data are analysed as described in *Protocol 1, step 6*. The Sephadex columns may be reused after washing with 15 ml of buffer A. The use of a column rack customized to fit the scintillation vial box allows up to 100 samples to be run simultaneously (see Chapter 6).

4. Membrane preparation

Once the choice of tissue is made the next step in the purification of the β-receptor is the membrane preparation. This is illustrated below for a 150 g preparation of hamster lung, and may be scaled up or down accordingly.

Protocol 3. Membrane preparation

1. 150 g of frozen hamster lungs (from Pel-Freeze Biologicals) are partially thawed in 1500 ml of ice-cold buffer C (50 mM Tris-HCl, pH 7.2, 5 mM EDTA, 2 μg/ml leupeptin, 5 μg/ml soybean trypsin inhibitor, 15 μg/ml benzamidine).

2. Homogenize with three 20 s bursts of a tissue disrupter (Brinkmann PT 10/35 homogenizer with a PTA 20TS probe) at a setting of 10. The suspension should be kept on ice at all times. It is important that the sample does not warm up appreciably as this is the major step where proteolysis occurs.

3. Centrifuge at 200 g (1000 r.p.m.; Sorvall RT 6000 centrifuge) in 250 ml bottles (Corning) for 10 min.

4. The pellet is rehomogenized in 1000 ml of buffer C with two 20 s bursts of the tissue disrupter. Recentrifuge at 200 g for 10 min.

5. The supernatants from the two spins are pooled and centrifuged at 40 000 g (20 000 r.p.m.; Sorvall RC-5B centrifuge with a TZ-28 rotor) for 20 min. This rotor holds ~ 1350 ml of liquid which, following centrifugation, can be rapidly removed with suction. The resulting membrane pellet, which forms on the outside wall of the rotor, can be removed with a Teflon spatula. Alternatively, membranes may be pelleted in an SS34 rotor at 19 000 r.p.m. for 20 min.

6. The pellet is resuspended in ~ 1250 ml of buffer C by use of the tissue disrupter on low speed. Recentrifuge at 40 000 g for 20 min.

7. Repeat the washing procedure (*step 6*) 2−3 times.

8. The final washed membrane pellet can be stored at −80°C as a paste, or resuspended in buffer C. Alternatively, the membranes can be immediately resuspended in a detergent-containing buffer for solubilization (see Section 6).

5. Photoaffinity labelling of the β-adrenergic receptor

The technique of affinity or photoaffinity labelling has proven to be an invaluable tool in the biochemical characterization of the β-adrenergic receptor. Recent work has involved the use of radioiodinated photoaffinity probes, which possess high specific activity and high affinity ($K_d < 10^{-9}$ M) for the β-adrenergic receptor. Some of these ligands are listed in *Table 3*. These probes have been extremely useful in studying the molecular size of the receptor in cells and membrane preparations, and in characterizing the receptor at various stages of purification. In several systems, such as the frog (3) and turkey erythrocyte (4) and the mammalian lung (2), it has been shown, using photoaffinity labelling and purification techniques, that the purified β-adrenergic receptors are in fact identical to those labelled in the membrane. In a number of studies, significant heterogeneity exists in the photoaffinity labelling patterns of the β-receptor in several systems. This heterogeneity appears to result from endogenous proteolytic activity (6) which varies among species and tissues.

Table 3. Photoaffinity ligands for β-adrenergic receptors

Ligand	Maximum specific activity (Ci/mmol)	Source
[125I]15-(4′-Azido-3′-iodobenzyl)-carazolol ([125I]pABC)	2200	NEN
(±)3-[125I]Iodocyanopindolol diazirine ([125I]diazirine CYP)	2200	Amersham
[125I]Iodoazidobenzylpindolol ([125I]IABP)	2200	NA[a]

[a] NA, not commercially available.

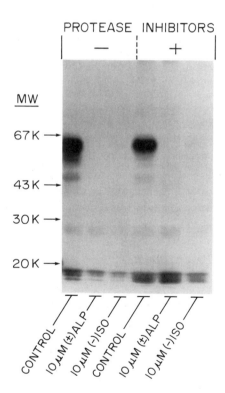

Figure 1. Effect of protease inhibitors on [125I]pABC photoaffinity labelling patterns of hamster lung membranes. Hamster lung membranes (50 pM β-receptor) were prepared in 50 mM Tris-HCl, pH 7.2 buffer containing either no protease inhibitors or 5 µg/ml soybean trypsin inhibitor, 100 µM benzamidine, 5 µg/ml leupeptin, and 5 mM EDTA. Membranes were incubated with [125I]pABC (100 pM) alone or in the presence of (±)alprenolol (10 µM) or (−)isoproterenol (10 µM). Photolysis and electrophoresis conditions are described in the photoaffinity labelling procedures.

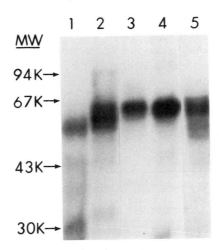

Figure 2. Photoaffinity labelling of dog, bovine, guinea-pig, and hamster lung and rat erythrocyte membranes with [^{125}I]pABC. Membranes (20 – 50 pM receptor), prepared in the presence of 50 mM Tris-HCl, pH 7.2, 5 mM EDTA, were incubated with [^{125}I]pABC (100 pM). Lane 1, dog lung; lane 2, bovine lung; lane 3, guinea-pig lung; lane 4, hamster lung; lane 5, rat erythrocyte. Photolysis and electrophoresis conditions are as described in the photoaffinity labelling procedure.

Figure 1 presents results from photoaffinity labelling studies using hamster lung membranes and the photoaffinity probe [^{125}I]pABC. These membranes contain β-adrenergic receptors which are predominately of the β_2 subtype ($\sim 98\%$) as determined by radioligand-binding studies with [^{125}I]CYP and subtype selective antagonists. It is apparent that only one peptide, at $M_r \sim 64\ 000$, is observed when protease inhibitors are included, while additional lower molecular weight peptides are present in the absence of inhibitors.

Photoaffinity labelling of several other mammalian β-adrenergic receptors is shown in *Figure 2*. Labelling of dog lung membranes (lane 1) reveals a major peptide at $M_r \sim 56\ 000$ even in the presence of protease inhibitors. Some experiments, however, reveal a minor ($\sim 7\%$) specifically labelled peptide at $M_r\ 64\ 000$, suggesting that the smaller peptide is a proteolytic product (data not shown). Bovine lung membranes (lane 2) also appear heterogeneous with peptides at $M_r \sim 64\ 000$, 60 000, and 56 000. Conversely, guinea-pig lung (lane 3), hamster lung (lane 4), and rat erythrocytes (lane 5) share similar labelling patterns, with major peptides at $M_r \sim 64\ 000$. Sheep and pig lung also contain a major $M_r \sim 64\ 000$ specifically labelled peptide (data not shown). In addition, work by Stiles *et al.* (18) has demonstrated that the heart β_1-adrenergic receptor from several species also resides on a $M_r \sim 64\ 000$ peptide which is susceptible to endogenous proteolysis.

The findings obtained suggest that the heterogeneity of β-adrenergic receptor peptides observed in many systems might be due to proteolysis. Since the various proteolyzed forms of the receptors are all specifically labelled by [^{125}I]pABC, they each contain the intact ligand-binding site of the β-adrenergic receptor. These results suggest that the mammalian β-adrenergic receptor is contained on a peptide of apparent M_r = 62 000 − 65 000. Moreover, they demonstrate that photoaffinity labelling is a viable means for assessing the structural integrity of β-receptor preparations.

5.1 Photoaffinity labelling of particulate β-adrenergic receptors

---

Protocol 4. Photoaffinity labelling of membrane-bound βARS

1. Prepare hamster lung membranes as described in Protocol 3.

2. Dilute membranes (containing 1 − 2 pmol of β-receptor) with 50 mM Tris-HCl, pH 7.2, 5 mM EDTA, 2 μg/ml leupeptin, 5 μg/ml soybean trypsin inhibitor, 15 μg/ml benzamidine (buffer C) to yield a β-receptor concentration of 30 − 50 pM.

3. Dilute [^{125}I]pABC in 10% ethanol, 5 mM HCl before addition to the membranes, yielding a final [^{125}I]pABC concentration of 100 pM and an ethanol concentration < 0.1%. This step should be performed in the dark or under dim red light. [^{125}I]diazirine CYP is utilized with a procedure comparable to that for [^{125}I]pABC.

4. Include a sample containing 10 μM (−)alprenolol to determine non-specific binding.

5. Incubate membrane suspensions for 90 − 120 min in the dark at 25°C. The incubations are done directly in centrifuge tubes (for SS34 rotor).

6. Following the incubation, pellet the membranes (19 000 r.p.m., 10 min) and wash them twice by centrifugation in buffer C containing 0.5% bovine serum albumin and once in buffer C. The albumin wash helps to reduce non-specifically bound ligand.

7. Resuspend the final pellet in 15 ml buffer C, transfer it to a Petri dish on ice, and photolyse for 60 − 90 s, 12 cm from a Hanovia 450 W medium pressure mercury arc lamp filtered with 5 mm of Pyrex glass. Alternatively, the sample may be photolysed for 5 − 10 min with a short-wave UV hand lamp (6 watt).

8. Centrifuge the sample (19 000 r.p.m., 10 min) and resuspend the pellet (by sonication) in 500 μl of SDS sample buffer (8% SDS, 10% glycerol, 5% β-mercaptoethanol, 25 mM Tris-HCl, pH 6.5, 0.003% bromphenol blue).

9. Electrophorese the samples (∼ 150 μl) on a 10% polyacrylamide slab gel followed by drying (Bio-Rad Model 224 gel dryer) and autoradiography at −90°C with Kodak XAR-5 film and an intensifying screen (Cronex Plus; Dupont). See Chapter 10, Section 3.3 for further details.

5.2 Photoaffinity labelling of soluble β-adrenergic receptors

---

Protocol 5. Photoaffinity labelling of soluble βARS

1. Partially purified or purified β-adrenergic receptor in 100 mM NaCl, 10 mM Tris-HCl, pH 7.2, 0.05% digitonin is used. Crude soluble β-receptor preparations are not effectively labelled with [^{125}I]pABC because of the hydrophobic nature of the ligand, however, [^{125}I]diazirine CYP may be usable for this purpose.

2. To 1−2 pmol of β-receptor in a total volume of 0.5−1 ml (2 nM receptor) add a comparable concentration of [^{125}I]pABC. 10 μM (−)alprenolol can be included in a separate incubation to determine non-specific labelling. This step should be performed in the dark or under dim red light.

3. Incubate the sample for 2−3 h at 25°C or overnight at 4°C and then desalt the sample on a small (3.5 ml) Sephadex G50 (fine) column equilibrated with 5 mM Tris-HCl, pH 7.2, 0.05% digitonin at 4°C.

4. The sample is then photolysed for 60−90 s using a Hanovia 450 W lamp.

5. The sample is lyophilized, resuspended in SDS sample buffer and electrophoresed.

---

6. Solubilization procedures

A critical step in the purification of a membrane-bound protein is its extraction or solubilization from the phospholipid bilayer. There are a wide variety of detergents available for solubilizing membrane proteins. These include non-ionic detergents, such as Lubrol PX, Triton X-100, octyl glucoside, and digitonin, and ionic detergents, such as sodium cholate and CHAPS, to name a few (for review see reference 19. See also Chapter 1, Section 2). A large number of different detergents have been used to solubilize the β-adrenergic receptor. However, digitonin is the only commercially available detergent that solubilizes the β-receptor in an active state, i.e. a state which still binds adrenergic ligands (3). The procedure described below has been used successfully to solubilize the β$_2$-adrenergic receptor from hamster lung membranes. This involves an initial presolubilization step at a low digitonin concentration, which removes significant amounts of protein without solubilizing the receptor. The β$_2$-receptor is then solubilized with a higher concentration of digitonin, resulting in a recovery of *c.* 50% of the binding sites and, approximately a twofold purification.

Protocol 6. Solubilization of active βARS

1. Prepare membranes from 150g of hamster lung as described in *Protocol 3*. Pellet the membranes by centrifugation and resuspend the pellet in 600 ml of low-digitonin buffer (100 mM NaCl, 10 mM Tris-HCl, pH 7.2, 5 mM EDTA,

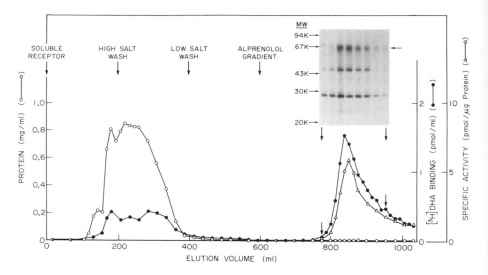

Figure 3. Sepharose – alprenolol chromatography of digitonin-solubilized hamster lung β-adrenergic receptor activity. Approximately 340 pmol of digitonin-solubilized hamster lung receptor (200 ml) was applied to a 200 ml column of Sepharose – alprenolol previously equilibrated with 100 mM NaCl, 10 mM Tris-HCl, 0.05% digitonin, 2 mM EDTA, pH 7.2. The column was then washed and eluted as described in Protocol 8. Individual fractions (12.9 ml) were assayed for receptor (after desalting a 0.2 ml aliquot) by [³H]DHA binding (●) and for protein by the Amido Schwarz assay (○, see Chapter 6). These results were used to calculate specific activities for the eluted receptor (△). Additionally, 300 μl aliquots of several alprenolol-eluted fractions were iodinated by the choramine-T method (Chapters 2, 8 and 10) and electro-phoresed on a 10% SDS-polyacrylamide gel. The autoradiogram of the dried gel is shown in the inset at upper right. The arrow indicates where [¹²⁵I]pABC-labelled hamster lung β-receptor electrophoresed in this experiment. The molecular weight standards (MW) are shown × 1000 (K). The alprenolol-eluted receptor activity in the experiment shown (160 pmol) represents a 47% recovery of the applied digitonin-solubilized receptor activity.

2 μg/ml leupeptin, 5 μg/ml soybean trypsin inhibitor, 15 μg/ml benzamidine, 0.3% digitonin) by use of a Brinkmann tissue disrupter at low speed.

2. Stir the suspension for 20 min at 4°C before centrifugation (19 000 r.p.m., 20 min).

3. Resuspend the pellet in 600 ml of high-digitonin buffer (as above except the concentration of digitonin is 1.5%) again by use of a Brinkman tissue disrupter.

4. Stir suspension for 60 min at 4°C before centrifugation (19 000 r.p.m., 20 min). The supernatant contains the solubilized receptor. High-speed centri-fugation (48 000 r.p.m.) improves the specific activity of the receptor at this step only minimally and thus is not necessary.

5. Typically, ~50% of the initial particulate β-receptor-binding activity is recovered in the soluble form. This material can be frozen and stored at −80°C

or it can be immediately loaded on a Sepharose – alprenolol affinity column (see below).

7. Purification of the β-adrenergic receptor

Most hormone receptors exist in cells in extremely low concentrations and thus require extensive purification after their solubilization from the plasma membrane. This task is greatly facilitated by utilizing the ability of receptors to selectively bind specific ligands. Thus receptors are ideal candidates for the use of affinity chromatography. The use of affinity chromatography for purification of the β-adrenergic receptor is an essential step which provides a *c.* 1000-fold purification of the receptor with a recovery of *c.* 50% (*Figure 3, Table 4*). The receptor can then be purified to homogeneity by chromatography on gel-permeation HPLC (*Figure 4*). Obviously, there are a wide variety of additional procedures that could be utilized in the purification of the β-receptor (e.g. ion exchange, hydrophobic interaction, and lectin affinity chromatography). However, the use of affinity chromatography followed by gel-permeation HPLC results in a highly purified β_2-adrenergic receptor preparation (*Figure 5*). Moreover, the receptor purified using these techniques is functionally intact as:

(a) it binds ligands with a β-adrenergic specificity (2);

(b) it interacts with the stimulatory guanine nucleotide regulatory protein (G_s) in a reconstituted system (20); and

(c) it stimulates adenylate cyclase in a hybrid cell system (21).

Table 4. Purification of the β-Adrenergic Receptor of Hamster Lung. Typically, 40 – 60 g of tissue was homogenized, yielding 26 ± 1 pmol [³H]DHA binding sites per g of hamster lung. Purified membranes were solubilized with digitonin before undergoing successive Sepharose – alprenolol affinity and high-performance steric exclusion chromatography, as described in Sections 7.2 and 7.3. The results are for 50 g hamster lung preparations and are expressed as the mean ± standard error as determined from a minimum of four separate experiments. This purification scheme yields ~ 150 pmol (10 μg) of pure β-adrenergic receptor.

Step	Overall yield (%)	Specific activity[a] (pmol/mg protein)		Overall purification (fold)
Crude homogenate	100	0.44 ±	0.01	1
Purified membranes	82	1.7 ±	0.1	3.8
Digitonin extract	41	3.2 ±	0.2	7.3
Sepharose – alprenolol eluate[b]	13	3630 ±	710	8250
HPLC pass	11	14 660 ±	1170	33 320

[a] As measured by [³H]DHA binding and Amido Schwarz protein assay.
[b] Ultrafiltration using an Amicon concentrator with a YM-30 membrane.

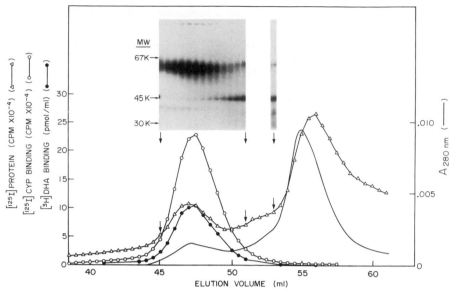

Figure 4. High-performance steric exclusion chromatography of affinity-purified hamster lung receptor activity. Receptor-containing fractions after Sepharose – alprenolol chromatography were pooled and concentrated to 2 ml by ultrafiltration using an Amicon concentration cell and YM-30 membrane. The concentrated receptor was chromatographed on two TSK-4000 and one TSK-3000 columns tandem-linked (total volume = 72 ml). Receptor activity was located by [³H]DHA binding (●) or by chromatography of an aliquot of the affinity column eluate which had been desalted and incubated with [¹²⁵I]CYP prior to chromatography (○). The absorbance at 280 nm (——) is also shown in this profile. Additionally, an aliquot of the affinity column eluate concentrate was iodinated prior to chromatography. After chromatography, fractions were counted for [¹²⁵I] (△) and 10 μl aliquots were electrophoresed on an 8% SDS-polyacrylamide gel. The resulting autoradiogram (top inset) is shown for the corresponding fractions. The molecular weight standards (MW) are shown × 1000 (K). Mobile phase conditions were 0.1% digitonin, 100 mM Tris-SO₄, pH 7.5 at a flow rate of 1 ml/min, 0.5 ml fractions were collected 36 min after sample injection.

7.1 Synthesis of the Sepharose – alprenolol affinity resin

Protocol 7. Synthesis of Sepharose – alprenolol affinity resin

1. Wash 150 ml of Sepahrose 4B or 6B (Pharmacia) with 8 litres of distilled H_2O using a 2 litre Kimax fritted-disc funnel (medium).

2. Add 315 ml of 0.3 N NaOH to the moist gel in a 1 litre beaker. Add 46.5 g of 1,4-butanediol diglycidyl ether (Aldrich, stored under N_2 at $-20°C$) to the mixture. Gently stir or rotate for ~ 15 h at room temperature. After the reaction, wash the gel with 8 – 10 litres of H_2O (until the effluent is neutral). Note: the 1,4-butanediol diglycidyl ether is not completely miscible with aqueous solution, and after 15 h there will be a few drops left at the bottom. The left over butanediol droplets should not be decanted with the gel when washing.

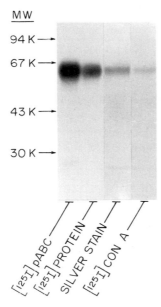

Figure 5. SDS-PAGE of HPLC-purified hamster lung receptor activity. An aliquot (3 pmol) of HPLC-purified hamster lung receptor was incubated with [^{125}I]pABC for 12 h at 4°C and photolysed as outlined in Section 5.2. 20 μl of the labelled receptor was combined with 40 μl of SDS sample buffer and electrophoresed on a 10% SDS-polyacrylamide gel. The resulting autoradiogram is shown in lane 1. Iodinated HPLC-purified hamster lung receptor was also electrophoresed on a 10% gel, and is shown in lane 2. Additionally, a 10 pmol aliquot of HPLC-purified hamster lung receptor was lyophilized and dissolved in 100 μl of SDS sample buffer. After electrophoresis, the gel was silver stained and photographed (lane 3). After destaining with Farmer's reducer (Kodak), the gel was overlayed with 4 ml of [^{125}I]Con A (Sigma; see also Chapter 8, Section 3 and Chapter 10, Section 3), and after a 12 h incubation was washed for 24 h with 50 mM Tris-HCl, pH 7.5, 0.1 mM CaCl$_2$, 0.1 mM MgCl$_2$, 600 mM NaCl, dried and exposed to Kodak XAR-5 film. The resulting autoradiogram is shown in lane 4. The molecular weight standards (MW) are shown × 1000 (K).

3. To the moist gel add 190 ml of 0.1 M potassium phosphate, pH 6.3, and 190 ml of 2 M sodium thiosulphate. Stir slowly for 3 h at room temperature. Wash with 16−18 litres of H$_2$O.

4. Equilibrate the gel with 0.1 M NaHCO$_3$ (containing 0.4 g disodium EDTA/litre). Add 230 ml 0.1 M NaHCO$_3$ with EDTA to the moist gel. Add 10 g of solid dithiothreitol and react for 3 h with gentle stirring. Wash with 6−8 litres of cold N$_2$-bubbled H$_2$O as rapidly as possible.

5. To the moist gel add 2.3 g of alprenolol hydrochloride dissolved in 40 ml of N$_2$-saturated H$_2$O. Warm the mixture to ~40°C and then add 2 ml of potassium persulphate (46 mg/ml) every 12 min for 2 h while stirring. Wash the gel with 8 litres of H$_2$O. The gel can be stored after this step.

6. Add 200 ml of 0.1 M NaHCO$_3$/Na$_2$CO$_3$, pH 9.3 and 4 g of sodium borohydride. Stir for 2 h at 4°C (this reaction can go overnight). Wash with 4−6 litres of H$_2$O. The gel can be stored after this step.

7. Resuspend the gel in 200 ml of 0.2 M NaHCO$_3$ containing 10 mM iodoacetamide. React for 2 h at room temperature. Wash with 6−8 litres of H$_2$O.

8. Before use the gel should be equilibrated with protein to reduce non-specific binding. Generally an affinity column pass-through fraction from a previous β-receptor purification is recycled over the affinity matrix. The gel should then be washed with several litres of 1 M NaCl and 6−8 litres of H$_2$O before use.

7.2 Chromatography of solubilized β-adrenergic receptor on Sepharose – alprenolol

Protocol 8. Affinity chromatography of solubilized βARS

1. Prepare a 200 ml column of Sepharose−alprenolol (LKB 2.6 × 40 cm column) and equilibrate with 400 ml of 100 mM NaCl, 10 mM Tris-HCl, pH 7.2, 2 mM EDTA, 0.05% digitonin (buffer D).

2. Load the solubilized β-adrenergic receptor preparation onto the Sepharose−alprenolol column at a flow rate of ∼100 ml/h at 25°C or ∼50 ml/h at 4°C.

3. Wash the column at 4°C with 300 ml of 500 mM NaCl, 50 mM Tris-HCl, pH 7.2, 2 mM EDTA, 0.5% digitonin followed by 300 ml of buffer D.

4. Return the column temperature to 25°C and elute the receptor activity with a 400 ml linear gradient of 0−40 μM (±)alprenolol in buffer D.

5. Typically the first 170−180 ml of the eluate contains no receptor activity and is discarded, while the next 180−220 ml is pooled. The β-receptor activity can be assayed by desalting an aliquot (0.1−0.5 ml) of the eluted material on a Sephadex G50 (fine) column equilibrated with buffer D. This effectively removes the free alprenolol. The sample can then be assayed as described in Protocol 2.

6. The sample can be concentrated to 2−3 ml using a 50 ml Amicon concentrator with a YM-30 membrane. The concentrated receptor can be stored at −80°C.

7.3 Chromatography of affinity-purified β-adrenergic receptor on gel-permeation HPLC

Protocol 9. HPLC gel-filtration of purified βARS

1. 1 TSK-3000 and 2 TSK-4000 HPLC columns (0.75 × 60 cm) are linked in tandem and equilibrated with 50 mM Tris-SO$_4$, pH 7.5, 0.05% digitonin at

a flow rate of 1 ml/min. The buffer should be filtered just prior to use through a 0.2 μ Millipore membrane.

2. The concentrated affinity-purified receptor preparation, also filtered through a 0.2 μ Millipore membrane, is then manually injected onto the HPLC in 1−2 ml aliquots.

3. The β-receptor elutes from the column at about 45−50 ml following injection. The peak of receptor activity is determined by radioligand binding as described in Section 3.2.

4. The peak β-receptor fractions are pooled and concentrated (if desired) using a 10 ml Amicon concentrator with a YM-30 membrane.

5. The receptor can be stored at −80°C for a few weeks without significant loss of activity. The inclusion of 1 μM (±)alprenolol to the preparation allows longer storage at −80°C.

Acknowledgements

I would like to thank Drs Robert Lefkowitz and Marc Caron for their continuous support, Drs Michel Bouvier and John Regan for careful reading of the manuscript, and Mary Holben for her help in preparing the manuscript.

References

1. Lefkowitz, R. J., Stadel, J. M., and Caron, M. G. (1983). *Annu. Rev. Biochem.*, **52**, 159.
2. Benovic, J. L., Shorr, R. G. L., Caron, M. G., and Lefkowitz, R. J. (1984). *Biochemistry*, **23**, 4510.
3. Caron, M. G. and Lefkowitz, R. J. (1976). *J. Biol. Chem.*, **251**, 2374.
4. Shorr, R. G. L., Strohsacker, M. W., Lavin, T. N., Lefkowitz, R. J., and Caron, M. G. (1982). *J. Biol. Chem.*, **257**, 12341.
5. Dickinson, K. E. J. and Nahorski, S. R. (1981). *Life Sci.*, **29**, 2527.
6. Benovic, J. L., Stiles, G. L., Lefkowitz, R. J., and Caron, M. G. (1983). *Biochem. Biophys. Res. Commun.*, **110**, 504.
7. Williams, L. T., Jarett, L., and Lefkowitz, R. J. (1976) *J. Biol. Chem.*, **251**, 3096
8. Dickinson, K., Richardson, A., and Nahorski, S. R. (1978). *Mol. Pharmacol.*, **19**, 194.
9. Rugg, E. L., Barnett, D. B., and Nahorski, S. R. (1978). *Mol. Pharmacol.*, **14**, 996.
10. Minneman, K. P., Hegstrand, L. R., and Molinoff, P. B. (1979). *Mol. Pharmacol.*, **16**, 34.
11. Homcy, C. J., Rockson, S. G., Countaway, J. and Egan, D. A. (1983). *Biochemistry*, **22**, 660.
12. Manalan, A. S., Besch, H. R., Jr, and Watanabe, A. M. (1981). *Circ. Res.*, **49**, 326.
13. Engel, G. (1981). *Postgrad. Med. J.*, **57** (Suppl. I), 77.
14. Brodde, O.-E., Karad, K., Zerkowski, H.-R., Rohm, N., and Reidemeister, J. C. (1983). *Br. J. Pharmacol.*, **78**, 72P.
15. Kelley, L. K., Smith, G. H., and King, B. F. (1983). *Biochim. Biophys. Acta*, **734**, 91.

16. Bahouth, S. W. and Malbon, C. C. (1987). *Biochem. J.,* **248**, 557.
17. Stiles, G. L., Taylor, S., and Lefkowitz, R. J. (1983). *Life Sci.,* **33**, 467.
18. Stiles, G. L., Strasser, R. H., Lavin, T. N., Jones, L. R., Caron, M. G., and Lefkowitz, R. J. (1983). *J. Biol. Chem.,* **258**, 8443.
19. Hjelmeland, L. M. and Chrambach, A. (1984). *Methods Enzymol.,* **104**, 305.
20. Cerione, R. A., Codina, J., Benovic, J. L., Lefkowitz, R. J., Birnbaumer, L., and Caron, M. G. (1984). *Biochemistry,* **23**, 4519.
21. Cerione, R. A., Strulovici, B., Benovic, J. L., Lefkowitz, R. J., and Caron, M. G. (1983). *Nature,* **306**, 562.
22. Hulme, E. C. (ed.) *Receptor−Ligand Interactions, a Practical Approach*, IRL Press, Oxford, in press.

α_2-Adrenergic receptor purification

JOHN W. REGAN and HIROAKI MATSUI

1. Introduction

Despite recent progress in the purification and cloning of α_2-adrenergic receptors (1,2), the purification of sufficient quantities of receptor for structural character-ization remains a formidable challenge. This is due to the low natural abundance of α_2-adrenergic receptors which necessitates a substantial purification (100 000-fold) to obtain a pure preparation of the receptor. With the advent of the molecular cloning of the gene for the human platelet α_2-adrenergic receptor and the possibility of designing high-yield expression systems, this purification problem will be reduced but probably not eliminated. A useful and presently essential tool for the purification of α_2-adrenergic receptors is affinity chromatography.

Two affinity adsorbants have been developed for the purification of α_2-adrenergic receptors. Both adsorbants take advantage of α_2-selective antagonists that are covalently coupled to agarose and thereby immobilized. One employs SKF 101253, a benzazepine derivative, and the other uses yohimbinic acid, a congener of the α_2-adrenergic antagonist yohimbine. SKF 101253 was one of a series of compounds that were designed expressly for the purpose of being immobilized to agarose for use in affinity chromatography (3). The parent compound of this series, SKF 86466, is a selective α_2-adrenergic antagonist with high affinity for α_2-adrenergic receptors. SKF 101253 differs in having an allyl substituent which serves as the attachment site to the agarose support. Covalent coupling is achieved by means of a free-radical reaction between the allyl group and sulphydryl-activated agarose. The expected product of this reaction is a stable thio-ether (sulphide) bond. This type of chemical immobilization was used with success previously for the β-adrenergic receptor affinity adsorbent, Sepharose−alprenolol (see Chapter 5). As shown in *Figure 1*, however, a number of other linkages could be expected and, depending upon the reaction conditions, the relative yield of these products could vary. This might explain differences which occur between labs using what is ostensibly the same affinity adsorbent.

The basic reaction shown in *Figure 1a* is between a thiyl radical (**1**) and the allyl group (**2**) leading to the formation of an intermediate β-alkyl radical (**3**) and then the sulphide adduct (**4**). The thiyl radical is generated by adding potassium persulphate, which can breakdown or react with water to generate the initiaiting

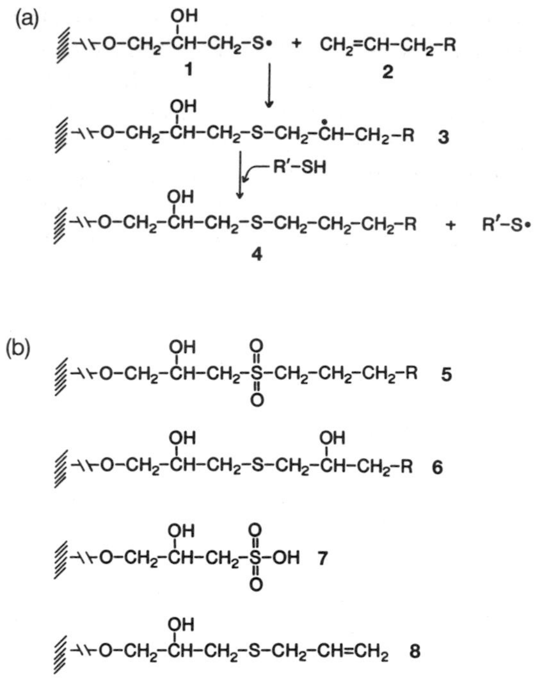

(a)

(b)

Figure 1. Likely products of the coupling reaction between thiol-activated agarose and an allyl containing ligand: (a) initial reactants and expected product; (b) some additional products favoured under oxidative conditions. (1) Thiyl-activated agarose; (2) allyl-containing ligand; (3) β-alkyl radical intermediate; (4) sulphide adduct; (5) sulphone adduct; (6) β-hydroxy sulphide adduct; (7) sulphonic acid agarose; (8) allyl agarose. R represents the remainder of the ligand molecule. R′–SH equals thiol-activated agarose.

sulphate ion radicals or hydroxyl radicals and oxygen, respectively. The reaction is propagated when the intermediate β-alkyl radical reacts with another free thiol to yield the sulphide and a new thiyl radical. Under the actual reaction conditions used for the synthesis of Sepharose–alprenolol (Chapter 5) and SKF 101253–Sepharose (4) the amount of potassium persulphate used is probably in excess. This could lead to oxidative conditions where additional products could be expected (*Figure 1b*). Thus, the corresponding sulphoxide (not shown) and sulphone (5) may be generated. During a subsequent reduction with sodium borohydride, however, the sulphoxide would be converted back to the expected sulphide whereas the sulphone would remain. Another product which might be expected is the β-hydroxy sulphide adduct (6). This would be generated when the intermediate β-alkyl radical reacts with oxygen to give a β-peroxyl radical (not shown). The β-peroxyl radical would then be expected to undergo further reactions to yield the β-hydroxy sulphide adduct and its corresponding sulphoxide and sulphone (not shown). Under the oxidative conditions, one would also expect that the relative excess of thiol groups from the agarose would lead to the formation of disulphides and a variety of oxidation products, including sulphonic acid residues (**7**). The sodium borohydride reduction step would convert many of these products back to the free thiols but not the sulphonic acids. A final product that may be present is an allyl sulphide linked to the agarose (**8**). This would be the product of a β-elimination reaction from the β-alkyl radical intermediate. In some procedures, after the reduction step there is a reaction with iodoacetamide to convert the free thiol groups to the corresponding amides. There is also the distinct possibility that the iodoacetamide could react with the sulphide adduct or with amino groups, if they are present in the ligand, to form sulphonium or quaternary ammonium salts, respectively. The sulphonium salt is relatively unstable, which could lead to problems associated with the slow release of the ligand from the gel. Nevertheless, it is very important to note, especially with respect to the synthesis of Sepharose-alprenolol, that the final product works well.

With regard to the synthesis of SKF 101253−Sepharose, the product shows biospecific adsorption and elution of receptor activity and can be used in combination with other chromatographic steps for the complete purification of human platelet α_2-adrenergic receptors (**4**). There are, however, shortcomings to the use of SKF 101253−Sepharose. One is that the fold purification is relatively low (25−50) by affinity chromatographic standards. A second is that SKF 101253 is not commercially available, and a third is that there is significant variability between different synthetic batches of SKF 101253−Sepharose. The latter characteristic, perhaps a result of the chemistry of immobilization, means that each batch must be characterized in terms of the conditions needed for the adsorption and elution of receptor activity.

Yohimbinic acid, coupled to amino-activated Sepharose by means of a carbodiimide reaction, has been used for the purification of α_2-adrenergic receptors from porcine brain (**5**). This support has the advantages that a better fold purification of the receptor can be obtained and that yohimbinic acid is commercially available. An interesting aspect of this gel is that the product appears to have significantly higher affinity than the starting material. Thus, yohimbinic acid itself has an affinity in the μM range, whereas, yohimbinic acid−agarose appears to have an affinity in the nM range. This is supported by studies which show that replacement of the original methyl ester of rauwolscine (a diastereomer of yohimbine) with a variety of amide substituents yields compounds with higher affinity for α_2-adrenergic receptors (**6**).

We have synthesized a number of yohimbinic acid−agarose affinity gels including the original (**5**). Although the coupling chemistry of the carbodiimide reaction is more predictable than the free-radical reaction, there is still significant variability between different synthetic batches of yohimbinic acid−agarose. The reason for this variability is unknown and therefore it is important for the reader to follow the protocol described here, or by Repaske *et al.* (**5**), closely. It is perhaps more important to note any changes that are made and to record the conditions of variables that are not explicitly stated. We do the synthesis of the ethylenediamine-activated gel on a relatively large scale for two reasons:

(a) this appears to be one of the critical steps in the procedure, and

(b) it is desirable to have a supply of the activated gel on hand if affinity gels with particularly good characteristics are obtained.

The coupling reaction of the ligand with the gel is done on a smaller scale and then the gel is tested for its adsorption and elution characteristics. If the characteristics are good, the coupling reaction is repeated several times under the same conditions (i.e. it is not scaled up). If the characteristics are poor, the conditions of the coupling reaction can be varied; however, it may be more expedient to repeat the gel activation step using new reagents (different lot numbers). For a given set of reagents, gels that are made over a period of 2−3 weeks usually have similar properties with respect to the adsorption and elution of receptor activity. The ligand-coupled gels are stable for up to a year when stored in the dark at 4−6°C.

The procedures described here were used successfully for the purification of 3 nmoles of human platelet α_2-adrenergic receptors for the purpose of obtaining amino-acid sequence and cloning (1). Yohimbinic acid−Separose was used as the affinity support and it differed from the original gel (5) in the type of spacer arm employed and in the conditions of the coupling reaction. The present gel has a longer spacer and it employs a water-soluble carbodiimide for the coupling reaction. *Figure 2* shows this spacer and the steps involved in the synthesis of yohimbinic acid−Sepharose as described here. With respect to the use of this affinity gel, the reader may note that the procedures we describe take a distinctly low-tech. approach to liquid chromatography. Thus, instead of using a single large column with a pump and detectors, etc., a large number of small columns and gravity are employed. This technique gave us better performance than the use of large columns and should be considered when difficulties are encountered during scale-up.

The use of a large number of small columns was chosen to alleviate two problems. The first was that of the batch-to-batch variability of the affinity gels, and the second concerned difficulties associated with scale-up. Because of the variability problem it was necessary to characterize the adsorption and elution conditions for each new batch of gel. Doing this on a preparative scale can be very time-consuming and costly. On the other hand, these conditions can be rapidly established on an analytical scale. Unfortunately the adsorption and elution conditions that work on an analytical scale (1 ml columns) do not necessarily work on a preparative scale (50−100 ml columns); however, preparative-scale work can also be achieved by simply increasing the number of analytical columns. The most significant aspect of this approach is that it guarantees a true linear scale-up. Thus, the yield and fold purifications obtained on the analytical scale can be obtained predictably on the preparative scale. This approach also provides flexibility. Therefore, depending upon the volume of the soluble preparation, one uses the appropriate number of columns so that the ratio of the soluble prep to gel volume is preserved. This approach to scale-up, as others, assumes a certain amount of consistency between successive runs using the same affinity gel. For a given batch of yohimbinic acid−Sepharose, approximately 20 successive runs could be obtained before yields began to deteriorate.

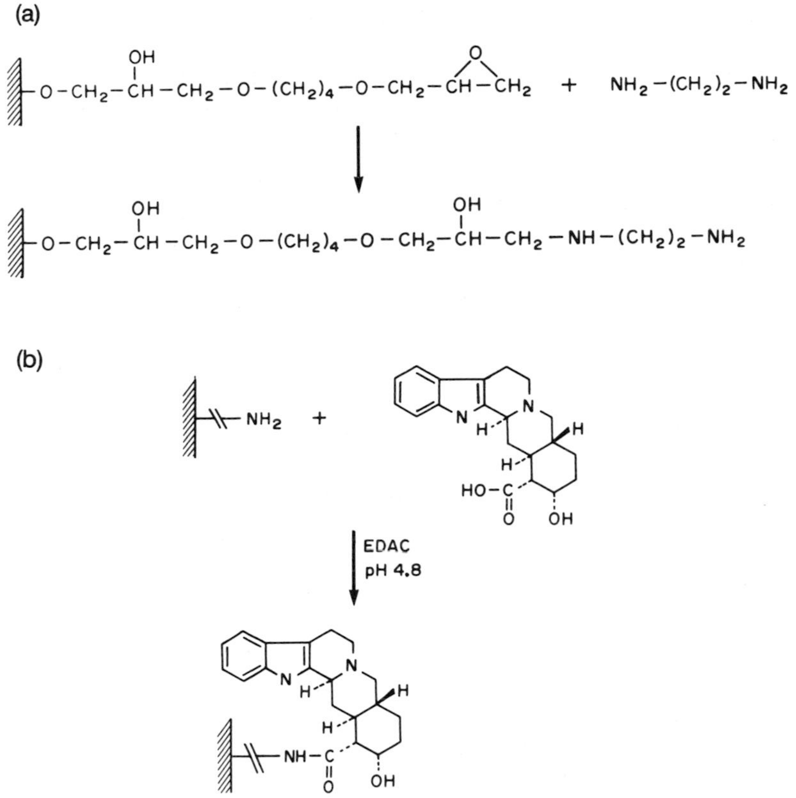

(a)

(b)

Figure 2. Synthesis of yohimbinic acid – Sepharose: (a) reaction of epoxy-activated Sepharose with ethylenediamine; (b) carbodiimide coupling between amino-activated Sepharose and yohimbinic acid. EDAC is 1-ethyl-3-(3-dimethylaminopyropyl)carbodiimide.

2. Synthesis of yohimbinic acid – Sepharose

2.1 Ethylenediamine-activated Sepharose

Protocol 1. Ethylenediamine-activated Sepharose

1. Wash 500 g of Sepharose 4B (Pharmacia) with 20 l of deionized water in a 2 l Buchner funnel (medium porosity, $10-20$ μm) as described in *Figure 3*. At the end of the washing procedure filter until there is no free liquid remaining. Do not dry out the gel. The gel should have the consistency of a moist cake as opposed to a slurry or a crumbly dry cake. One ml of the filtered gel will weigh approximately one 1 g.

145

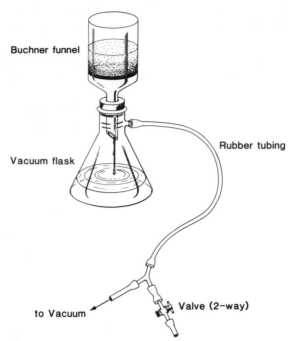

Buchner funnel

Vacuum flask

Rubber tubing

to Vacuum

Valve (2–way)

Figure 3. Washing agarose gels. The vacuum source that we use is a house vacuum providing between 18 – 32″ Hg. Best results are obtained when the packed gel volume is not more than 30% of the total volume of the Buchner. After initially resuspending the gel with a rubber spatula, the suspension should be stirred with the spatula until about half, or three-quarters, of the total volume is filtered. At this point the gel may be allowed to settle. As the surface of the water approaches the surface of the packed gel, open the valve to release the vacuum. With the vacuum still on, place a finger over the open valve, to control the rate of filtration. New funnels usually need to be treated with a solution of chromic acid to improve the flow rates. Add a volume of chromic acid equal to approximately 25% of the volume of the funnel and let it drip through the funnel (without vacuum) overnight. Remember that chromic acid is very corrosive and take appropriate precautions: wear gloves and eye and skin protection. The chromic acid treatment may need to be repeated, especially after washing gels. Always rinse the funnel thoroughly with water afterwards.

2. Transfer the gel to a 3 l suspension culture spinner flask (Wheaton, Corning, Bellco, etc.) and resuspend it in 1.1 l of 350 mM NaOH containing 2 mg/ml of NaBH$_4$. Place the flask on a magnetic stirrer and set the speed so that the gel does not settle. (The use of the suspension culture flask is desirable because the stirrer does not touch the bottom of the flask and the agarose will experience less physical abuse.)

3. Over the course of a few minutes, add 300 ml of 1,4-butanediol diglycidyl ether (Aldrich) and stir overnight at room temperature (22°C). Transfer the suspension to the Buchner funnel and wash as before with approximately 20 l of deionized water. Occasionally (2–3 times) during the wash resuspend the

gel and let it settle without applying the vacuum. Then aspirate the supernatant to remove the agarose fines and an oily residue that is present on the surface of the wash.

4. Transfer the moist gel back to the spinner flask. Mix 16.7 ml of ethylenediamine (Aldrich) with 500 ml of water and add it to the flask. Stir overnight at room temperature and then wash the gel as before with 20 l of deionized water. {This reaction can be monitored by adding 25 μCi of [^{14}C]ethylenediamine (specific activity 44 Ci/mol; Amersham) to the ethylenediamine solution. After the reaction there should be approximately 3000 cpm per gram of the washed moist gel (~ 16 μmol ethylenediamine/g). Most of the radioactivity will not be incorporated and will be removed in the initial wash.}

2.2 Coupling of yohimbinic acid to the activated Sepharose

Protocol 2. Coupling of yohimbinic acid to the activated Sepharose

1. Weigh out 50 g of the ethylenediamine-activated Sepharose and equilibrate it with 200 ml of 50% dimethylformamide (DMF) and water for 1−2 hours. Filter in a 300 ml Buchner funnel and transfer the moist gel to a 250 ml beaker.

2. Add a solution of 2.5 g yohimbinic acid (Aldrich) in 25 ml of DMF and resuspend the gel using an overhead stirrer or a magnetic stir bar that is not in direct contact with the bottom of the beaker ('floating' stir bar; Nalgene). A suspension culture flask may also be used.

3. Adjust the pH to 4.8 with 1 M HCl and leave the electrode in place to monitor the next reaction.

4. To a 50 ml beaker, add 25 ml of water and stir rapidly with a magnetic stirrer. Add 1.7 g of 1-ethyl-3-(3-dimethylaminopropyl)-carbodiimide (EDAC, BioRad; EDC, Pierce) all at once and when dissolved immediately add it to the stirring gel suspension. Keep the pH adjusted to approximately 4.8 with 1 M HCl.

5. After 1.5 h repeat *step 4*. After approximately 1 h the gel can be left to stir overnight without the need to monitor the pH.

6. The next day, repeat *step 4* and stir for 4 h. Transfer the gel suspension to a 300 ml Buchner funnel and wash sequentially with 500 ml of 50% DMF, 5 l of water and then 500 ml of 400 mM sodium acetate, pH 4.8. During the initial washing step with 50% DMF it is very important not to dry the gel: keep the gel thoroughly resuspended with a spatula and do not let the volume of the gel suspension fall below approximately 100 ml. Continue washing in this fashion with the first 1 l of water; afterwards with the remaining 4 l, resuspend the gel after each addition of water and then filter down to the surface of the gel.

7. Transfer the moist gel to a 250 ml beaker and resuspend in 50 ml of 400 mM sodium acetate pH 4.8. With stirring as before, monitor the pH and then add

3.8 g of EDAC in 50 ml of H_2O. Keep the pH adjusted to 4.8 with 1 M acetic acid. Let the gel stir overnight and then wash it as before with 3 l of water, 500 ml of 5 mM NaOH, 3 l of water and finally with 1 l of the storage buffer (20 mM Tris-HCl, 1 mM EDTA, 0.05% NaN_3, pH 7.5). Store as a 50% suspension in the dark at $4-6°C$.

---

3. Membrane preparation and solubilization

3.1 Membranes

All operations with respect to the membrane preparation, solubilization, and purification are done on ice or in the cold $(4-6°C)$.

Protocol 3. Platelet membrane preparation

1. Pool 200 units of platelet-rich plasma (approximately 10 l) and centrifuge for 30 minutes at 4000 r.p.m. using a HG-4L rotor (r_{ave} 15.6 cm; Sorvall) and 1 l polycarbonate centrifuge bottles.

2. Discard the supernatant and resuspend the pellet in lysis buffer [7.5 mM NaCl, 2.5 mM Tris-HCl, 1 mM EDTA, 5 mM EGTA, 100 μM phenylmethyl-sulphonyl fluoride (PMSF), pH 7.2] keeping the total volume under 1 l. Homogenize for 1 minute at maximum speed with a Polytron (Brinkman, PTA-10 rotor).

3. Centrifuge for 45 minutes at 20 000 r.p.m. using a TZ-28 rotor (r_{ave} 6.6 cm; Sorvall). Discard the supernatant and resuspend the pellet in approximately 800 ml of lysis buffer.

4. Repeat the previous homogenization and centrifugation procedure three more times.

5. Resuspend the final pellet with lysis buffer so that the final volume is approximately 200 ml (1 ml per original unit of platelets). Aliquot into 50 ml polypropylene screw-top centrifuge tubes and freeze in liquid nitrogen. Store at $-80°C$.

3.2 Solubilization

For the following steps, centrifugations are done at 20 000 r.p.m. with an SS-34 rotor (r_{ave} 8.2 cm; Sorvall).

Protocol 4. Solubilization of α_2Rs

1. Thaw 200 ml of frozen membrane concentrate and add 600 ml of lysis buffer. Homogenize twice for 1 minute with a Polytron at maximum power. Centrifuge for 30 minutes.

2. Discard the supernatant and resuspend the pellet in 1 l of presolubilization buffer (20 mM Tris-HCl, 15 mM EGTA, 1 mM DTT, 100 μM PMSF, 0.2% digitonin, pH 7.2). Homogenize wih three up-and-down strokes of a Dounce homogenizer (pestle clearance 0.03 − 0.08 mm; Kontes) and stir vigorously but without creating a deep vortex for 15 min at 4°C.

3. Centrifuge for 40 min. Discard the supernatant and resuspend the pellet in 1 l of solubilization buffer (the same as the presolubilization buffer but with 1% digitonin). Homogenize with 10 strokes of a Dounce homogenizer and stir for 90 min at 4°C.

4. Centrifuge for 60 min and *save the supernatant*. (This soluble prep. can be stored overnight at 4−6°C; however, once the next step is started you will need to continue until the affinity chromatography is completed.)

5. Place the soluble prep. in an ice bath and over 10 min, with stirring, add 45 g of polyethylene glycol (M_{ave} 3350; Sigma). Continue stirring for an additional 30 min and then centrifuge for 60 min. Save the supernatant for application to the affinity columns.

4. Receptor purification

4.1 Affinity chromatography

Protocol 5. Yohimbinic acid − Sepharose affinity chromatography

1. Equilibrate 100, 1 ml yohimbinic acid − Sepharose affinity columns (see *Figure 4*) with 10 ml each of buffer A (50 mM Tris-HCl, 1 mM EDTA, 1 mM DTT, 0.1% digitonin, pH 7.2).

2. Apply five, 2 ml volumes of the soluble prep. to each column over a period of 2−3 hours. A repeating pipette or a similar dispenser is useful for this purpose.

3. Wash each column with 6 ml of buffer B (buffer A containing 50 mM KCl).

4. Apply 0.8 ml of the elution buffer (buffer B containing 200 μM phentolamine). After 1 h apply 1 ml of the elution buffer and collect the eluate. Thirty minutes later apply another 1 ml of the elution buffer and collect the eluate. Repeat the last step: the total volume of the eluate should be approximately 300 ml.

5. Concentrate the eluate to approximately 10 ml using two stirred cells (180 ml; Amicon) with YM-30 membranes. This should take 4−6 hours. Both the affinity eluate and the concentrate can be stored overnight at 4−6°C without loss of binding activity. In the case of the concentrate, a precipitate may be present after storage which should be removed by centrifugation prior to the next step.

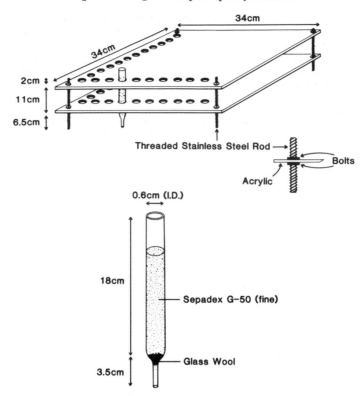

Figure 4. Sephadex G50 and affinity chromatography: rack and column design. The proportions of the rack and columns used for routine chromatographic applications are shown above. The rack is made from two pieces of clear acrylic that measure $34 \times 34 \times 0.5$ cm; four pieces of treated stainless-steel rod that are 0.6×20 cm; and 16 bolts that screw onto the rods. Ten rows of 10 holes (0.9 cm, diameter) are drilled through the top piece of acrylic and the same number of holes (0.6 cm, diameter) are drilled through the bottom piece. The rack, when assembled, can be placed over a box of 100 scintillation vials (20 ml) as supplied by Research Products International. The holes are approximately 3 cm apart from centre to centre but this can be varied depending upon the specific application. If reservoirs are used, the spacing between columns must be kept in mind. The diameter of the reservoirs that we use is 2.5 cm. The columns are glass with the dimensions listed above. They are plugged with a small piece of glass wool that is gently tapped into the bottom of the column with a wooden dowel. For size exclusion chromatography the columns are each filled with 3.4 ml (settled gel volume) of hydrated Sephadex G50 (fine); usually as a 50% aqueous suspension. The use of the 'fine' Sephadex keeps the columns from drying or cracking. In our experience, the columns can be stored, loosely covered, without adding water to them for up to a month at $4-6\,^{\circ}$C. For binding assays or buffer exchange (desalting), the following volumes are used. Each column is pre-equilibrated with $5-10$ ml of the appropriate buffer. The sample is applied in a final volume of 500 μl. It is chased with 600 μl of the pre-equilibration buffer and the flow-through is discarded. Buffer (1 ml) is added and the flow-through (void volume) containing the receptor-binding activity is collected. The columns should be checked with blue dextran (Sigma) or with other molecular weight markers to verify the volumes of the chase and of the void. The columns are routinely regenerated with $20-30$ ml of a low ionic strength buffer and are reused extensively. Depending upon what the columns have been exposed to, and what the next

intended application is, the columns can be regenerated with a variety of solutions, including ones that are basic, acidic, chaotropic, etc. (check the manufacturer's recommendations or other texts; for example, Kremmer and Boross (12). There may be some initial adsorption losses of binding activity but this decreases as the columns are reused. For affinity chromatography the columns are each filled with 1 ml of the affinity gel. Flow-throughs are collected with a large tray placed under the columns. The columns are stored, loosely covered, at 4 – 6°C and in the dark (yohimbine is light sensitive). If the gels are not used extensively a buffer (20 mM Tris – HCl, 1 mM EDTA, 0.05% NaN$_3$, pH 7.5) should be added every few weeks to inhibit microbial growth and to keep the gels from drying. To facilitate washing procedures, reservoirs may be attached to the columns. Disposable syringes (Becton Dickinson) or polypropylene funnels (Kontes) can be attached to the columns with pieces of rubber tubing. Kontes also makes disposable polypropylene columns and a 50-column rack similar to that described here.

4.2 Heparin – agarose chromatography

Protocol 6. Heparin – agarose chromatography

1. Equilibrate a 1 ml column of heparin – agarose (Bio-Rad) with 10 ml of buffer C (buffer A containing *0.05% digitionin* and 100 µM phentolamine).

2. Apply the concentrate in 1 ml aliquots over a period of approximately 2 h.

3. Wash the column in succession with: 10 ml of buffer C, 7 ml of buffer C containing 50 mM KCl, 3 ml of buffer C containing 100 mM KCl, and 0.7 ml of buffer C containing 500 mM KCl.

4. After 30 min apply 1.5 ml of buffer C containing 500 mM KCl and collect the eluate.

4.3 Inverse affinity chromatography

Protocol 7. Inverse affinity chromatography

1. Concentrate the eluate to approximately 300 µl using a stirred cell (10 ml; Amicon) with a YM-30 membrane.

2. Add 9 ml of buffer C to the stirred cell and concentrate again to 300 µl. Repeat this one more time.

3. Apply the concentrate to a 300µl column of yohimbinic acid-agarose that has been equilibrated with 3 ml of buffer C and *collect the flow through*. [To facilitate the transfer of these small volumes, 10 cm of Teflon tubing (0.8 mm, I.D.) is attached to a 1 ml Eppendorf pipette tip using a piece of silicon tubing. The pipette tip is attached to a pipettor and the tubing is inserted into the stirred cell. With the stirred cell tipped slightly it is possible to remove virtually all of the concentrate. The tubing is then inserted into the column, and with its tip near the surface of the gel, the concentrate is gently expelled.]

4. Briefly rinse the stirred cell with 500 μl of buffer C and immediately apply to the column while collecting the flow-through. The receptor activity will be in this flow-through since the column was equilibrated with a buffer containing the counterligand.

5. Repeat *step 4* and then apply 1 ml of buffer C to the column and collect the flow-through (total volume, 2.3 ml).

4.4 WGA – agarose chromatography

Protocol 8. WGA – agarose chromatography

1. Concentrate the flow-through to approximately 300 μl using a stirred cell (10 ml; Amicon) with a YM-30 membrane.

2. Equilibrate a 500 μl column of WGA–agarose (E.-Y. Labs Inc.) with 5 ml of buffer D (50 mM HEPES, 1 mM EDTA, 1 mM DTT, 0.05% digitonin, pH 7.4).

3. Apply the concentrate to the column and rinse the stirred cell three times with 1 ml of buffer D, applying each 1 ml rinse to the column.

4. Wash the column in succession with: 4 ml of buffer D, 4 ml of buffer D containing 300 mM NaCl, and 0.35 ml of buffer D containing 300 mM *N*-acetyl-D-glucosamine.

5. After 1 h apply 1 ml of buffer D containing 300 mM *N*-acetyl-D-glucosamine and collect the eluate. Repeat the last step.

6. After taking aliquots for the measurement of binding activity and for radio-iodination, add 10 μl of 10 mM phentolamine to the eluate. The eluate can be stored at 4−6°C for 2−3 weeks without appreciable loss of activity; for longer periods, freeze in liquid nitrogen and store at −80°C.

5. Comments

The digitonin:protein ratio has a marked effect on the solubilization of binding activity. During the presolubilization step the ratio is in the range of 0.4−0.5. This results in the solubilization of approximately 10% of the membrane protein and 5% of the binding activity. During the solubilization step the ratio is in the range of 2.3−2.9. This yields another 20−30% of the original membrane protein and 55−65% of the original binding activity. The result of using the two-step solubilization procedure is a three- to fourfold purification of the receptor. The concentration of digitonin can also have marked effects on binding activity. Following the initial solubilization, the amount of digitonin needed to maintain binding activity is less. In fact, within limits, as the digitonin concentration decreases, binding

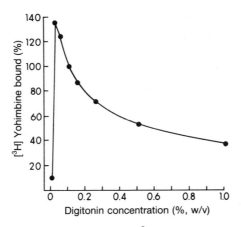

Figure 5. Effect of digitonin on the binding of [³H]yohimbine to partially purified human platelet α_2-adrenergic receptors. Partially purified receptors (specific activity 300 pmol/mg protein) were desalted (*Figure 4*) into a buffer containing 50 mM Tris-HCl, 1 mM EDTA, 1 mM DTT, and 0.2% digitonin (pH 7.2). Binding assays were started by adding an aliquot (40 μl) of the receptor preparation to a solution (final volume, 500 μl) containing 6 nM [³H]yohimbine and the final concentrations of digitonin listed above. After a 3 h incubation at 4 – 6°C, the assays were terminated by gel-filtration as described in *Figure 4*. 100% equals 27 fmol of receptor-binding activity. 1% digitonin is 8.1 mM.

increases. *Figure 5* shows the effects of the digitonin concentration on the binding of [³H]yohimbine to partially purified human platelet α_2-adrenergic receptors. As the concentration of digitonin is decreased from an initial value of 1%, the amount of specifically bound [³H]yohimbine increases until suddenly, between 0.03% and 0.01%, a nearly complete loss of binding activity occurs. This abrupt transition is probably due to protein aggregation resulting from the decrease of the digitonin concentration below its critical micellular concentration. The explanation for the progressive increase in binding with decreasing concentrations of digitonin has not been explored but it should be kept in mind when designing experiments.

Polyethylene glycol is used during the solubilization procedure to precipitate proteins that tend to slowly precipitate on their own and thereby cause problems during the subsequent affinity chromatography step. It was found that the presence of polyethylene glycol during the affinity chromatography increased the binding of the receptor to the SKF 101253 – Sepharose affinity gel (4). This is also the case with respect to yohimbinic acid – Sepharose. Although polyethylene glycol decreases the stability of the α_2-adrenergic receptor, it is removed when the affinity column is washed after the application of the soluble receptor preparation.

Table 1 shows the yields and fold purifications for each of the chromatographic steps. Affinity chromatography gives the highest fold purification, and although it is not as great as one might hope for, it is three- to fourfold better than the results obtained with our previous affinity gel, SKF 101253 – Sepharose. More extensive washing decreases the fold purification that is obtained. This is because the receptor

Table 1. Yields and fold purifications for the chromatographic steps.

Step	Specific activity (nmol/mg protein)	Fold purification	Step yield (%)	Yield (%)
soluble prep.	0.0006 – 0.0008	1	1	1
1st affinity	0.08 – 0.1	130	39 ± 5	39
heparin	0.5 – 0.7	7	66 ± 10	24
2nd affinity	1.5 – 2.5	3	63 ± 14	14
WGA	5 – 8	3	65 ± 18	10

slowly elutes during the wash whereas the non-specific protein, after falling sharply during the initial wash, elutes at a steady rate for a very long time. Thus, following the initial wash, the specific activity of the eluate rises and then falls as the total amount of receptor remaining on the gel decreases. Washing with higher concentrations of salt results in a more rapid elution of receptor activity and the fold purification is decreased. These effects of salt on the binding of the α_2-adrenergic receptor were also observed with the SKF 101253 affinity gel and suggest that they may result from a property of the receptor. It is noted, however, that we have observed changes in the sensitivity of the receptor – gel interaction to salt depending upon how the affinity gels are prepared. It is worthwhile to do some trial runs on an analytical scale to determine the best washing conditions before working on a preparative scale.

Heparin – agarose gives a good purification of the receptor and has been one of the most consistent steps in the purification scheme. A recent publication (7) reports a 50-fold purification of the neonatal rat lung α_2-adrenergic receptor from a crude soluble prep. Whether a similar fold purification could be obtained after an affinity chromatographic step was not reported. The mechanism of the interaction of the receptor with heparin – agarose is probably ionic. The gene for the human platelet α_2-adrenergic receptor has been cloned and its deduced amino-acid sequence shows that it contains a large hydrophilic region with many basic residues (1). This region, which has been proposed to form a cytoplasmic loop, would potentially interact with the sulphate and carboxylate groups of heparin.

The second affinity chromatographic step is used in the inverse. By equilibrating the receptor and the column with phentolamine, contaminating proteins adsorb to the column while the receptor flows through. The timing of this step is important. If the initial application of the concentrated heparin eluate is left on the column too long, the yield of receptor in the flow-through is decreased. Doing traditional affinity chromatography at this step gives a similar purification but a lower yield.

The specific activity after WGA-agarose chromatography is approximately one half of that expected for a pure preparation of the α_2-adrenergic receptor (15.6 nmol/mg protein, assuming a mass of 64 000 Da). Thus, it is assumed that

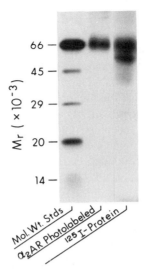

Figure 6. Photoaffinity labelling and radioiodinated protein after WGA-agarose chromatography. [^{125}I]Bolton–Hunter labelling was done as described in the text. After Sephadex G50 chromatography, a 2.5 μl sample was mixed with 50 μl of SDS-PAGE sample buffer and was electrophoresed on a 12% acrylamide gel (4). The gel was dried and an autoradiograph was obtained after a 16 h exposure at −80°C with intensifying screens. For photoaffinity labelling, 1.8 nmol of the α_2-adrenergic photoaffinity ligand, [^{3}H]SKF 102229 (80 Ci/mmol) (2) in 150 μl of carrier solvent (ethanol) was placed in a 12 × 75 mm polypropylene tube and was vacuum evaporated to near dryness with a SpeedVac (Savant Instruments). [Do not dry completely because the ligand is volatile.] The WGA – agarose eluate (2 ml; 140 pmol of receptor-binding activity) was added and the sample was left on ice for 1 h. It was then transferred to a 60 × 15 mm tissue culture dish on ice and was exposed to a hand-held 6 watt UV light (254 nm) for 1 – 2 min from a distance of 3 – 4 cm. Covalent labelling of the receptor appeared to be quantitative. A 20 μl aliquot (1.4 pmol) was vacuum evaporated and was redissolved in 50 μl of SDS-PAGE sample buffer. It was then electrophoresed as above, prepared for fluorography, and dried. An autoradiograph was obtained after a three-day exposure at −80°C.

the preparation is not pure. This seems to be supported by the results obtained following SDS-PAGE of the radioiodinated protein. *Figure 6* shows these results after labelling with [^{125}I]Bolton–Hunter reagent. The upper band represents the receptor since it could be specifically labelled with a photoaffinity probe for the ligand-binding site. The nature of the lower band is unclear; however, preparations such as this were used to generate peptides which were then isolated and sequenced. Of the four peptides sequenced, all were found to be present in the deduced amino-acid sequence of the human platelet α_2-adrenergic receptor. It seems possible, therefore, that the lower molecular weight band represents a degradation product which is incapable of binding ligands.

6. Assay procedures

6.1 Receptor binding

[³H]Yohimbine or [³H]rauwolscine are used for the measurement of α_2-adrenergic receptor binding activity. Both ligands are available with specific activities in the range of 70−80 Ci/mmol (NEN-Dupont, Amersham). These ligands are diastereomers, have very similar binding properties, and have been used to measure α_2-adrenergic receptor activity in a variety of tissues (reviewed in reference 8). For human platelet α_2-adrenergic receptors, [³H]yohimbine has slightly higher affinity and is preferred. During the routine measurement of binding activity, stock solutions of 60 nM [³H]yohimbine and 100 μM phentolamine will be needed. For saturation curve analysis, stock solutions of 120, 80, 60, 40, 20, 10, 5, 1, and 0.5 nM [³H]-yohimbine are required. For competition curve analysis, stock solutions of 10 nM [³H]yohimbine and 10 mM to 1 nM, at half-log intervals, of the competitor will be needed. Dilutions are made with the binding buffer except for the 10 mM solutions of the competitors, which are generally made in 1 mM HCl- or, as required, with ethanol, DMSO, or mixed solvent systems.

6.1.1 Membranes

Protocol 9. Assay of membrane-bound α_2Rs

1. The binding buffer consists of 50 mM Tris-HCl, 10 mM $MgCl_2$, and 1 mM EDTA, pH 7.5.
2. Mix the membrane preparations (up to 400 μl suspended in binding buffer) with 50 μl of [³H]yohimbine and either 50 μl of binding buffer (total binding) or 50 μl of phentolamine (non-specific binding).
3. Incubate for 1 h at room temperature (22−25°C) and then filter under vacuum through Whatman GF/B filters.
4. Wash the filters with four 4 ml rinses of cold (4−6°C) binding buffer.
5. Determine the radioactivity in each filter by liquid scintillation counting. Specific binding is the difference between total and non-specific binding.

6.1.2 Soluble

Protocol 10. Assay of soluble α_2Rs

1. The binding buffer consists of 50 mM Tris-HCl, 1 mM EDTA, 1 mM DTT, 0.05% digitonin, pH 7.2.
2. Mix the soluble receptor preparations (up to 400 μl in binding buffer) with 50 μl of [³H]yohimbine and either 50 μl of binding buffer (total binding) or 50 μl of phentolamine (non-specific binding). (If phentolamine is present in

the soluble receptor preparation it needs to be removed by Sephadex G50 chromatography, see *Figure 4.*)

3. Incubate for 1 h at room temperature, or 2 h in the cold (4−6°C). For the purified receptor, a more accurate determination of the K_d by saturation curve analysis is obtained with an overnight incubation in the cold.

4. Terminate the assays in the cold by Sephadex G50 chromatography as described in *Figure 4.*

5. Measure the radioactivity in the void volume by liquid scintillation counting.

6.2 Protein

Protein determinations for membrane preparations are done using the micro-Bradford assay according to the manufacturer's instructions (Bio-Rad). For soluble preparations the Amido Schwarz assay is employed (9). A description of this assay is given as follows.

Protocol 11. Amido Schwarz protein assay

1. Prepare a stock solution containing 100 μg/ml BSA (Fraction V, fat free) by making a 1 to 100 dilution, with distilled water, of a solution containing 10 mg/ml BSA in 50 mM Tris-HCl, pH 7.5.

2. Prepare the assay standards containing 0, 0.2, 0.4, 0.6, 0.8, 1, 1.5, 2, 4, and 6 μg of BSA in a final volume of 270 μl of water.

3. Maximum sample volumes are 270 μl. Dilute smaller sample volumes with water.

4. Add 30 μl of 1 M Tris-HCl/1% SDS (pH 7.5) to the samples and standards. Mix thoroughly and then add 60 μl of 90% TCA and mix again.

5. Incubate at room temperature for 5 min and filter over nitrocellulose (0.45 μm pore size). For filtration we use a Teflon microfiltration device that was made by our surgical instrument shop (Duke University Medical Center). This device was made according to the same basic dimensions as the commercially available Minifold (Schleicher and Schuell). The latter is available in both acrylic and Delrin versions. We found that the TCA decomposed the Delrin Minifold to the point that it was unusable. TCA etched the acrylic Minifold but it continued to function during the time that we had borrowed it.

6. To filter, place the nitrocellulose over a piece of blotting paper (Whatman 17 or equivalent) and place in the filtration device. Turn on the vacuum, wet the filter with water, and clamp into place.

7. Apply the samples to the sample wells, one row at a time. Rinse the sample tubes with 400 μl of 10% TCA and filter. After all of the rinse has been filtered apply another 400 μl of 10% TCA directly to the sample wells.

8. Remove the nitrocellulose and stain for 2 min in 0.1% Naphthol Blue Black (also known as Amido Schwarz reagent, Naphthalene Black 12B, Acid Black 1, Amido Black, Amido Black 10B, and Aniline Blue Black. Its C.I. number is 20470) dissolved in 45% methanol/10% acetic acid/45% water.

9. Rinse the filter in water for 30 seconds and for 1 min each in three 200 ml volumes of destaining solution (90% methanol/2% acetic acid/8% water), and finally for 2 min in water.

10. For very small amounts of protein (<1 μg) the nitrocellulose can be examined visually and an accurate estimate of the protein content can be obtained. Thus, the spot corresponding to the unknown is compared with the standards and is judged to be equal to or between a given pair of the standards. It is important, when such small amounts of protein are involved, to have an appropriate sample of the original buffer solution (for example, if 2 μl of the heparin−agarose column eluate are sampled for protein, another sample containing 2 μl of the elution buffer should be prepared).

11. For larger amounts of protein, the dye can be eluted from the nitrocellulose and measured spectrophotometrically. Cut out the spots (an 8 mm corkborer works fine) and place in a test-tube with 600 μl of 25 mM NaOH/50 μM EDTA/50% ethanol. Vortex 2−3 times over 10 min and read the absorbance at 630 nm.

The sensitivity of this assay to visual quantitation in the range of 0.2−2 μg of protein is remarkable. It is not affected adversely by digitonin concentrations in the range used in these studies. For Lubrol-containing solutions up to 0.7%, the SDS concentration should be increased to 10% and the samples incubated for 15 min at 37°C prior to the addition of TCA. Solutions containing Tween 60 form an emulsion upon the addition of TCA, but this does not seem to affect either the filtration or staining. Suggested assay volumes for the $α_2$-adrenergic receptor purification procedure are as follows: soluble prep., 2.5 μl; affinity eluate, 25 μl; heparin eluate, 2 μl; WGA eluate, 40 μl.

6.3 Radioiodination

[125I]Bolton−Hunter reagent (10) and chloramine-T with Na125I (11) have been used to radiolabel $α_2$-adrenergic receptors (8). In general, better results have been obtained with the [121I]Bolton−Hunter reagent. A protocol for the use of this reagent is as follows.

Protocol 12. Radioiodination using 125I − Bolton − Hunter reagent

1. Place approximately 100 μCi of [125I]Bolton−Hunter reagent (4000 Ci/mmol, NEN-Dupont) in a clean, dry, glass test-tube (12 × 75 mm) and evaporate the carrier solvent (benzene) with a stream of nitrogen gas.

2. Add 100 μl of the WGA−agarose eluate (5−10 pmol of receptor) directly to the test-tube and place on ice for 3 h.

3. Dilute the sample to 500 μl with 50 mM Tris-HCl/1 mM EDTA/1 mM DTT/0.05% digitonin (pH 7.2) and chromatograph over Sephadex G50 (see *Figure 4*). The [125I]-labelled receptor will be present in the void volume.

Tris buffers interfere with [125I]Bolton−Hunter labelling because they contain primary amines. We have also observed that phentolamine interferes with labelling of the α_2-adrenergic receptor, perhaps by inducing a conformational change in the receptor (8). Both Tris and phentolamine can be removed by Sephadex G50 chromatography prior to the radioiodination reaction (see *Figure 4*).

A note of caution concerns the use of BSA to stabilize the receptor after radio-iodination. We have observed passive labelling of BSA by 125I after the removal of the unreacted [125I]Bolton−Hunter reagent. *Figure 7* 'control' shows a receptor preparation that was approximately 60% pure on the basis of its specific binding

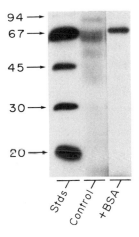

Figure 7. Passive radioiodination of bovine serum albumin (BSA) following storage of [125I]-Bolton−Hunter-labelled receptor preparations in the presence of 0.01% BSA. Partially pure α_2-adrenergic receptors (approximately 5 pmol receptor-binding activity, 0.5 μg protein) were radiolabelled with the [125I]Bolton−Hunter reagent as described in the text. After Sephadex G50 chromatography a sample of the void volume was taken for SDS-PAGE (control) and then BSA was added to the remaining radiolabelled protein to give a final concentration of 100 μg/ml. After 2 weeks at 4−6°C, a sample of this radiolabelled receptor preparation was removed for SDS-PAGE (+ BSA). Both samples were electrophoresed on 10% acrylamide gels which were then dried and autoradiographs obtained. Under similar conditions, storage of receptor preparations in the absence of BSA resulted in degradation of the M_r 64 000 band to one of 41 000. The bands at 80 000 and 55 000 were essentially unaffected unless the preparations were stored in polystyrene test-tubes, in which case the M_r 55 000 band disappeared. In this experiment the samples were stored in polypropylene tubes.

activity and that was radiolabelled with [^{121}I]Bolton−Hunter reagent and run on SDS-PAGE. It can be appreciated that in addition to the receptor band at M_r 64 000, there are additional bands with M_r's of 80 000, 55 000, and 41 000. The latter band is a degradation product of the receptor which accumulates during storage. In an effort to stabilize the receptor, BSA was added after radioiodination to a final concentration of 100 μg/ml (0.01%). After 2 weeks of storage at 4−6°C, the radio-iodinated receptor preparation was again run on SDS-PAGE. '+ BSA' in *Figure 7* shows that now, instead of four bands, there is only one band whose M_r is not the same as any of the original bands. This band of M_r 67 000 corresponds to BSA. The iodination was so extensive that the other bands were not visualized due to the need to shorten the exposure time for autoradiography. It is important to note that the BSA was added after the removal of the unreacted ^{125}I as evidenced by the fact that a band of M_r 67 000 was not present in the original control preparation.

A documented explanation for this passive labelling does not exist. It is assumed that the BSA, present in far greater concentrations than any of the endogenous proteins, acts as sink for iodine radicals and other radiolabelled and chemically reactive species that might be generated during storage of the receptor preparation. It may be significant that the [^{125}I]Bolton−Hunter reagent used for these experiments was diiodinated. Thus, radioactive decay of the first iodine atom might result in the disruption of covalent bonds which could lead to the release of the second iodine atom. A similar, although much less extensive, labelling of BSA took place after using the chloramine-T method of radioiodination (Regan and Nakata, unpublished results). It was noted by Greenwood *et al.* (11) that the presence of BSA during the gel-filtration step of the chloramine-T procedure resulted in the radioiodination of BSA even though the reaction had already been quenched by the addition of sodium metabisulphite.

Acknowledgements

We sincerely thank Bob Lefkowitz and Marc Caron for their generous and enthusiastic support and Hiroyasu Nakata for his help in providing the results shown in *Figures 5* and *7*. We also thank Jeff Benovic, Mark Hnatowich, and Kiefer Daniel for reviewing this manuscript.

References

1. Kobilka, B. K., Matsui, H., Kobilka, T. S., Yang-Feng, T. L., Francke, U., Caron, M. G., Lefkowitz, R. J., and Regan, J. W. (1987). *Science, 238*, 650.
2. Regan, J. W., Kobilka, T. S., Yang-Feng, T. L., Caron, M. G., Lefkowitz, R. J., and Kobilka, B. K. (1988). *Proc. Natl Acad. Sci. USA, 85*, 6301.
3. DeMarinis, R. M., Krog, A. J., Shah, D. H., Lafferty, J., Holden, K. G., Hieble, J. P., Matthews, W. D., Regan, J. W., Lefkowitz, R. J., and Caron, M. G. (1984). *J. Med. Chem., 27*, 918.

4. Regan, J. W., Nakata, H., DeMarinis, R. M., Caron, M. G., and Lefkowitz, R. J. (1986). *J. Biol. Chem.*, **261**, 3894.
5. Repaske, M. G., Nunnari, J. M., and Limbird, L. E. (1987). *J. Biol. Chem.*, **262**, 12381.
6. Lanier, S. M., Hess, H. J., Grodski, A., Graham, R. M., and Homcy, C. J. (1986). *Mol. Pharmacol.*, **29**, 219.
7. Lanier, S. M., Homcy, C. J., Patenaude, C., and Graham, R. M., (1988). *J. Biol. Chem.*, **263**, 14491.
8. Regan, J. W. (1988). In *The alpha-2 adrenergic receptors*. Limbird, L.E. (ed.), Humana Press, Clifton, NJ, p. 15.
9. Schaffner, W. and Weissmann, C. (1973). *Anal. Biochem.*, **56**, 502.
10. Bolton, A. E. and Hunter, W. M. (1973). *Biochem. J.*, **133**, 529.
11. Greenwood, F. C., Hunter, W. M., and Glover, J. S. (1963). *Biochem. J.*, **89**, 114.
12. Kremmer, T. and Boross, L. (1979). *Gel chromatography theory, methodology, applications*. Wiley-Interscience, New York.

7

Purification of nicotinic acetylcholine receptors

SIGRID HERTLING-JAWEED,
GIAMPIERO BANDINI, and FERDINAND HUCHO

1. Introduction—the nicotinic acetylcholine receptor as a model receptor

Receptors are membrane proteins receiving extracellular signals—hormones, growth factors, neurotransmitters—and transducing them to the interior of receiving cells (1). Despite their diversity they have many properties in common, some of which have been discovered through pioneering work on the nicotinic acetylcholine receptor (nAChR), which therefore has become a model for much of the present-day receptor research (2). All membrane-bound receptors are glycoproteins embedded in and partially protruding from the plasma membrane. All are composed of one or several polypeptide chains threaded through the lipid bilayer of the membrane by means of one, four, or seven possibly α-helical hydrophobic sequences. The growth-factor receptors have one membrane-spanning helix; receptors rigidly coupled to ion channels have four, and the receptors transiently coupled to effector enzymes via G-proteins have seven putative transmembrane helices. Most, if not all, receptors can be phosphorylated at sites located on sequences oriented towards the cytoplasm. Receptors exist in interconvertible functional states. They can be considered to be allosterically regulated proteins (3).

Recently, with the advent of recombinant DNA techniques, a wealth of structural information has become available facilitating generalizations and detection of homologies. For example, it became evident that the ion-channel-regulating receptors for acetylcholine (nicotinic), glycine, and GABA comprise one superfamily (4) and the G-protein-coupled receptors for adrenaline, acetylcholine (muscarinic), dopamine, serotonin, and probably many others comprise another superfamily (5). Members of each family exhibit homologies in their primary, secondary, quaternary, and probably in their tertiary structures as well.

1.1 The nAChR molecule

The nicotinic acetylcholine receptor (nAChR) from *Torpedo californica* electric tissue is the best-characterized receptor (2,6). It is the prototype for receptors with an intrinsic ionic conductance. In particular, it is thought to be very similar to the nAChR of the neuromuscular synapse of higher vertebrates. Ganglionic- and CNS-acetylcholine receptors have been mainly characterized through cloned cDNAs and are not yet accessible as proteins (5). This chapter therefore deals exclusively with nAChR from electric fish.

The nAChR from *Torpedo* species is a heteropentameric glycoprotein with an $\alpha_2\beta\gamma\delta$ quaternary structure. All five subunits are integral membrane proteins spanning the plasma membrane, all are glycosylated, only β, γ, and δ are phosphorylated. The total relative molecular mass, including the carbohydrate moieties, is 290 kDa. The apparent molecular weights of the subunits as determined by SDS-PAGE are: α, 40 kDa; β, 48 kDa; γ, 60 kDa; δ, 68 kDa. The binding sites for acetylcholine and other agonists and competitive antagonists are located on the two α-subunits. Additional binding sites for certain local anaesthetics, neurotoxins, and channel-blocking ligands are located within the receptor's ion channel, which is formed by all subunits together.

A further polypeptide chain with M_r 43 kDa is associated in stoichiometric amounts with the AChR. This peripheral protein can be easily removed. The functions of the 43 kDa protein and the precise role of the β, γ, and δ subunits are unclear. In addition to the above-mentioned proteins, which form a complex with stoichiometric composition, several other proteins are associated with the receptor in substoichiometric amounts. Among these are enzymes of unknown physiological functions, several protein kinases (7), a calcium-dependent protease (8) and a Ca^{2+}- and thiol-dependent transglutaminase (9). The latter two enzymes, in particular, can influence the quality of a receptor preparation significantly. The receptor from *Torpedo* electric tissue is isolated as a dimer connected through a disulphide bridge. This dimerization has not been observed with nAChR from other tissues. Its functional significance is unclear. For a comprehensive review of nAChR biochemistry see reference 6.

1.2 Open questions

Although more than 1000 publications on the AChR have appeared, many basic questions are still unanswered, justifying interest in the preparation of this model receptor. Omitting questions pertaining to development and gene regulation, and restricting ourselves to just a few problems to be solved with the receptor protein at hand, we will name only a few topics of future research.

(a) It is not yet clear what the basis of the different effects of agonists and antagonists is. As an initial step this question requires mapping of the agonist and antagonist binding sites on the receptor protein.

(b) The mechanism of activation and desensitization of AChR is unknown.

(c) The regulatory role of the various post-translational modifications is unknown.

(d) The mechanism of selection and transport of ions through the receptor's ion channel is unknown.

(e) The exact secondary, tertiary, and quaternary structures are unknown, i.e. crystals for X-ray diffraction analysis are needed.

1.3 Methods available and methods described here

Different methods of AChR purification are employed by different laboratories. This chapter describes in detail a rapid method for obtaining large amounts of membrane-bound and detergent-solubilized nAChR, including methods for assaying and for an initial attempt at crystallizing nAChR (10). Besides the method below, several affinity chromatography columns using acetylcholine (11), flaxedil (12), and snake venom neurotoxin (13) derivatives as immobilized ligands have been described. Furthermore, a counter-current phase separation protocol and lectin columns have been described. Each of these methods has its merits and special applications. For example, the advantage of the protocols described in Sections 3.2, 3.3.2, 3.4, and 3.5 is their efficiency for large-scale purifications, avoiding tortuous procedures likely to denature parts of the AChR molecules or to produce a heterogeneous population of functional and structural states.

2. Assays and quantification of nAChR

The nAChR is a ligand-gated ion channel. Its function, the controlled passive transport of cations through membranes, cannot easily be assayed after removal of the receptor protein from its lipid environment. A pore is measurable only after reconstitution into an artificial membrane system. Such reconstitution assays are available, but a more direct approach commonly applied in receptor research is to measure ligand binding.

Two methods are available:

(a) equilibrium binding of quasi-irreversible high-affinity ligands such as snake venom toxins, which have dissociation constants in the range of nanomolar to 10 picomolar, or of ligands such as the agonist acetylcholine, with dissociation constants of the order of 10 nanomolar;

(b) measurements of binding kinetics.

The latter gives insight into the mechanism of receptor function, and since only the former is of importance for receptor purification it will be described in detail here.

2.1 Iodination of α-bungarotoxin (BgTX)

Both radioiodinated neurotoxins and tritiated acetylcholine are commercially available (NEN, Amersham; see reference 24, Chapter 1). Because of the short half-live of [^{125}I] and because of the instability of the iodinated toxin, it is recommended to prepare [^{125}I]-toxin regularly fresh. The procedure described here for BgTX (*Bungarus multicinctus* toxin, Sigma) is applicable for other peptide neurotoxins similarly. Iodination is accomplished most gently by the lactoperoxidase (LPO) method (15) as follows.

Table 1. Pipetting scheme for the titration of receptor sites.

[^{125}I]BgTX solution (μl)	Wash buffer (μl)	Protein solution (μl)
1	149	50
5	145	50
10	140	50
50	100	50
150	—	50

Protocol 1. Iodination of α-bungarotoxin

1. Mix the following reagents: 200 μl BgTX solution, 1 mg/ml in H_2O; 300 μl H_2O; 25 μl 1 M K_2HPO_4/KH_2PO_4, pH 7.4; 5 μl ^{125}I (KI from NEN, 350 mCi/ml, 17 Ci/mg, in NaOH, pH 8-10, free of reducing agents); 5 μl of a 1:1000 dilution of H_2O_2 in H_2O (H_2O_2 perhydrol 30%, Merck); 5 μl lactoperoxidase 1 mg/ml in H_2O (lactoperoxidase from milk, 83 U/mg, Sigma).

2. The reaction is started by the addition of lactoperoxidase and should proceed for 10 min at room temperature.

3. Separation of the iodinated toxin from unbound KI is carried out by chromatography on a Sephadex G25 column (20 cm $\times$ 0.6 cm). Elution buffer is 50 mM K_2HPO_4/KH_2PO_4 pH 7.4; with 0.5 ml fractions the toxin peak typically appears in fractions 7−9 thus yielding about 1.5 ml toxin solution at approximately 0.05 mg/ml. For stability add BSA (bovine serum albumin) to 1 mg/ml. Store at 4°C.

4. Determination of specific radioactivity:
   (a) Measure absorbance per cm at 280 nm.
   (b) Divide $A_{280\,nm}$ by 1.15 for toxin concentration in grams protein per litre.
   (c) Divide the latter by 7.98×10^3 to obtain the concentration of toxin in moles per litre.
   (d) Count 10 μl of a 1:10 dilution of toxin in a gamma counter.
   (e) Calculate the specific activity of the labelled toxin in cpm per mole (typically about 10^{15} cpm per mole).

2.2 BgTX-binding assay

A filter binding assay using DE81 filters has proved to be reproducible (16). Receptor-bound [^{125}I]toxin is retained by the ion-exchange filter due to the low pK value of the nAChR. The following protocol is applicable to membrane-bound as well as to detergent-solubilized receptor protein.

Table 2. Pipetting scheme for the titration of the ligand.

Protein solution (μl)	Wash buffer (μl)	^{125}I-BgTX solution (μl)
10	140	50
20	130	50
50	100	50
100	50	50
150	—	50
—	150	50

Protocol 2. Assay of nAChRs by [^{125}I]BgTX binding

1. Take 50 μl of [^{125}I]BgTX solution [diluted 1:200 in wash buffer (composition 10 mM Na_2HPO_4/NaH_2PO_4 pH 7.4, 50 mM NaCl, 0.1% Triton X-100)].

2. Mix with 150 μl buffer (10 mM Na_2HPO_4/NaH_2PO_4 pH 7.4, 50 mM NaCl, 0.4% Triton X-100) with varying concentrations of receptor protein.

3. The receptor concentration in the assay is most critical and has to be explored depending on the preparation to be investigated. If the receptor concentration is too high, ligand-binding sites are irreproducibly covered up by aggregation. For example, the crude extract described below (see Section 3.1) should be diluted to a final concentration in the incubation mixture of 0.01–0.1 μg/200 μl.

4. Transfer 50 μl of the incubation mix to a DE81 (Whatman) filter disc (diameter 2 cm). Set discs upon three needle tips mounted on a plywood board to prevent contact with the underlying surface. After a few seconds wash the filters in a large Petri dish three times with 30 ml wash buffer per filter and count directly in a gamma counter.

5. For titration to determine the concentration of binding sites, follow the scheme given in *Table 1*. Use an incubation time of 30 min at room temperature (at room temperature equilibrium should be reached in less than 5 min), then apply to filters and wash as described in *step 4*.

6. Conversely the toxin can be titrated with receptor protein to determine the percentage of active ligand (as quality control of the [^{3}H]BgTX), following the scheme given in *Table 2*. Use an incubation time of 30 min at room temperature, then apply to filters and wash as described above.

7. For each BgTX-concentration a 'blank' filter (without protein, but washed like the regular samples), to give non-specific binding to the filter, and a 'total AChR' filter (unwashed), to give the total amount of radioactivity in the assay, have to be included in the counting.

8. When the effect of cholinergic ligands on toxin binding is to be determined the receptor protein is pre-incubated for 10 min with effectors before the addition of [^{125}I]BgTX.

167

9. To determine non-specific binding to the membranes add 8 μl of a 25 μM solution of cold BgTX to each sample of a parallel series (final concentration of 1 μM in the assay).

10. Calculate the number of moles of α-BgTX binding sites per mg of protein using the specific radioactivity of α-BgTX as determined in Section 2.1, *step 4* (after subtracting non-specific binding).

2.3 Acetylcholine-binding assay

For the investigation of [^{3}H]acetylcholine binding by equilibrium centrifugation the membrane suspension (0.1 mg/ml, see Section 3.2) is pre-incubated with eserine (Sigma). Eserine is an inhibitor of acetylcholinesterase and thus blocks hydrolysis of the radioactive ligand. ACh binds to its receptor with a K_d of 2×10^{-8} M (desensitized state).

Protocol 3. Binding of [^{3}H]ACh to nAChRs

1. Dissolve the [^{3}H]acetylcholine in Ringer's solution (0.160 M NaCl, 5 mM KCl, 2 mM MgCl$_2$, 2 mM CaCl$_2$, 3 mM Na$_2$HPO$_4$/NaH$_2$PO$_4$, pH 7.4) to make up a 10^{-4} M solution.

2. Further dilute the [^{3}H]ACh to give samples with final concentrations in the range from 10^{-7} to 10^{-5} M, e.g. 4×10^{-7} M, 8×10^{-7} M, 2×10^{-6} M, 4×10^{-6} M, 8×10^{-6} M, 2×10^{-5} M.

3. Preincubate the membrane preparation with 10^{-4} M eserine for 30 min at room temperature, then dilute it 1:100.

4. The assay volume for each ACh concentration is 200 μl:
   160 μl protein solution
    20 μl effector solution
    20 μl [^{3}H]ACh dilution.

5. In an airfuge centrifuge tube incubate the assay solution at room temperature for 5 min.

6. The 'total ACh' concentration is determined by counting duplicate 50 μl aliquots of the above mix directly before centrifugation.

7. Centrifuge at 135 000 g in the airfuge (Beckman) for 10 min.

8. For Scatchard analysis the concentration of the free ligand is determined by counting 50 μl of the supernatant. The concentration of bound ligand is obtained from the difference between total and free ACh.

Non-specific binding is negligible and the [^{3}H]ACh remains stable for over a year. [^{3}H]ACh-chloride is supplied by Amersham as a freeze-dried solid; 88.8 GBq/mmol; 2.4 Ci/mmol.

---

3. Receptor purification

3.1 Rapid preparation of crude receptor extract

Protocol 4. Crude receptor extract

1. In a 100 ml beaker homogenize 5 g *Torpedo californica* electric tissue (frozen in liquid nitrogen: Pacific Biomarine, California) in 10 ml H_2O (in an ice bath) with an Ultraturrax for 1 min at half maximal speed.
2. Centrifuge the homogenate at 1000 g, 4°C, 10 min (e.g. Sorvall SS34 rotor at 3000 r.p.m.).
3. Discard the pellet.
4. Centrifuge the supernatant at 27 000 g, 4°C, 20 min (e.g. Sorvall SS34 rotor at 15 000 r.p.m.).
5. Discard the supernatant.
6. Resuspend the pellet in 5 ml Ringer's solution, 2% Triton X-100 (Serva) by forcing it through a syringe (23 gauge needle).
7. Incubate for 30 min at room temperature, stirring slowly.
8. Centrifuge at 27 000 g, 4°C for 20 min.
9. The supernatant contains the solubilized receptor protein.

3.2 Pure AChR-rich membranes

Protocol 5. (17) nAChR-rich membranes

1. Homogenize 1 vol. frozen *T. californica* tissue in 2 vols of preparation buffer I (0.4 M NaCl, 20 mM Na_2HPO_4/NaH_2PO_4 pH 7.4, 2 mM Na_4EDTA, 0.1 mM PMSF) in a Waring blender (2 litre capacity) for 3 min at speed setting 'high'.
2. Centrifuge the homogenate at 27 000 g, 4°C, 90 min (e.g. Sorvall SS34 rotor at 15 000 r.p.m.).
3. Rehomogenize the pellet in 1 vol. of preparation buffer II (20 mM Na_2HPO_4/NaH_2PO_4 pH 7.4, 2 mM Na_4EDTA, 0.1 mM PMSF) with a Waring blender, 3 min at speed setting 'low'.
4. Centrifuge at 37 000 g, 4°C, 90 min (e.g. Sorvall SS34 rotor at 18 000 r.p.m.).
5. Rehomogenize the pellet in 1 vol. of preparation buffer II with a Waring blender, 3 min at speed setting 'low'.
6. Centrifuge at 37 000 g, 4°C, 90 min (eg. Sorvall SS34 rotor at 18 000 r.p.m.).
7. With a spatula carefully scrape the soft pellet off the denser bottom part.

8. Resuspend the soft pellet in buffer II (1 ml/3 g tissue) using a syringe or short homogenization in the blender.

9. Centrifuge at 1000 g, 4°C for 10 min (e.g. Sorvall SS34 rotor at 3000 r.p.m.).

10. Load the supernatant onto a gradient of 25−50% sucrose made up in H_2O containing 0.02% NaN_3. The homogenate from 100 g tissue typically would fit into 3 tubes of a Beckman SW25 containing 50 ml, or 6 tubes of a Beckman TST28 rotor containing 25 ml, of sucrose solution.

11. Centrifuge the gradient at 59 000 g, 4°C, 8 h (SW25 or TST28 22 000 r.p.m.).

12. Fractionate and pool the peak fractions (for SW25 rotor tubes: 2 ml fractions at 2 ml/min, for a TST28 1 ml fractions. Normally fractions 5−15 will contain the AChR peak—this can be checked by SDS-PAGE).

13. Dilute the pooled fractions 1:5 with H_2O.

14. Centrifuge at 39 000 g, 4°C, 30 min (e.g. Sorvall SS34 at 18 000 r.p.m.).

15. Resuspend the pellet in 50 mM Tris pH 7.4, 10 mM Na_4EDTA, to about 1 mg/ml.

A typical electron micrograph of purified nAChR-rich membranes is shown in *Figure 1*.

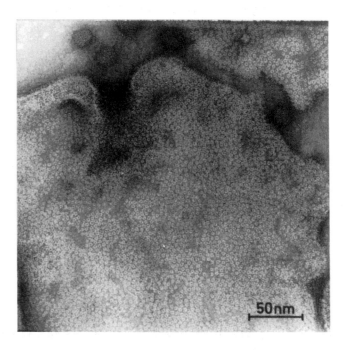

Figure 1. Electron micrograph of membranes from *Torpedo* electric tissue rich in nAChR, negatively stained with phosphotungstic acid (Giersig and Hucho, unpublished).

3.3 Extraction of peripheral proteins

3.3.1 pH 11-extraction

Protocol 6. (18) High pH extraction

1. Carefully adjust the pH of the membrane preparation (at about 2.5 mg protein per ml) to 11.0 with 0.1 M NaOH.
2. Incubate the preparation at room temperature for 1 h, stirring occasionally.
3. Centrifuge at 100 000 g, 4°C, 30 min (e.g. Kontron TFT65 rotor at 40 000 r.p.m.).
4. The supernatant contains the 43 kDa and 100 kDa polypeptides.
5. The receptor is retained in the membrane pellet.
6. This method is less reliable and causes more denaturation of nAChR than the lithium diiodosalicylate (LIDS) method described below.

3.3.2 Extraction with lithium diiodosalicylate

Protocol 7. (19) Extraction with LIDS

1. Resuspend membranes to about 1 mg/ml in 10 mM lithium diiodosalicylate (LIDS) in 20 mM Tris pH 8.1, 10 mM EGTA.
2. Incubate with slow shaking at room temperature for 1 h.
3. Centrifuge the suspension at 100 000 g, 4°C, for 15 min.
4. The supernatant contains the 43 kDa polypeptide.

3.4 Detergent extraction

Protocol 8. Detergent solubilization

1. Resuspend LIDS-extracted membranes (see Section 3.3.2) in 50 mM Tris-HCl pH 7.4, 10 mM EGTA, 2% n-octyl-β-D-glucopyranoside (β-OG, Bachem). The detergent solubilizes the AChR and the opaque solution becomes clear (20).
2. Centrifuge the suspension at 100 000 g, 4°C, 15 min.
3. The receptor protein is contained in the supernatant.

3.5 Chromatography on a Cibacron Blue Sepharose column

The rationale behind this step is to remove ATPase which is a major contaminant. This occurs without removing lipids essential for receptor stability (14).

Protocol 9. Blue Sepharose chromatography

1. Apply the supernatant (see Section 3.4) to a Cibacron Blue Sepharose CL-6B (Pharmacia) column (1 × 4 cm) (see *step 4* for preparation of the column).

2. The elution buffer is 50 mM Tris pH 7.4, 1 mM EGTA, 2% β-OG (about 10 ml for a 2 ml column at 8 ml/h).

3. The receptor protein appears with the void volume (with 1 ml fractions usually in fractions 2 − 10) whilst a proportion of the 100 kDa polypeptide is retained.

4. The Cibacron Blue column is prepared as follows. Swell the gel in H_2O for 15 min, then wash with about 200 ml H_2O per gram of dry powder (1 gram freeze-dried material gives a final volume of approximately 3.5 ml). Equilibrate with 50 mM Tris pH 7.4, 1 mM EGTA (10 column volumes) and wash with 2% β-OG in the same buffer.

5. To regenerate the column, remove the retained polypeptides as follows. Wash the column with Ringer's solution containing 2% Triton X-100, and then re-equilibrate with Tris buffer without detergent.

3.6 Affinity chromatography

3.6.1 Acetylcholine – agarose column

Protocol 10. ACh−agarose affinity chromatography

1. Solubilize the receptor by resuspending the pelleted AChR-rich membranes in 2% Triton X-100 in Ringer's solution to approximately 1 mg/ml.

2. Shake gently for 1 h at 4°C.

3. Centrifuge at 37 000 g, 30 min, 4°C.

4. Discard the pellet.

5. Add the supernatant to 10 ml ACh−agarose (see Section 3.6.3).

6. Shake overnight at 4°C.

7. Pour the gel into a column (1.6 cm diameter).

8. Wash the gel with 10 ml of wash buffer (0.1% Triton X-100, 50 mM NaCl, 10 mM NaH_2PO_4/Na_2HPO_4 pH 7.4).

9. Elute the gel with 10 ml of 50 mM carbamoylcholine in wash buffer.

10. Dialyse the pooled fractions against four changes of wash buffer (2000 ml each).

11. Centrifuge the dialysed preparation through a Centricon 30 microconcentrator (Amicon).

3.6.2 Preparation of bromacetylcholine bromide

Protocol 11. (21) Synthesis of bromacetylcholine bromide

1. Slowly add 12.1 g bromacetyl bromide (Aldrich) to 9.2 g choline bromide (Aldrich) over a period of 40 min in an ice bath, with stirring.
2. Stir for another 90 min in the ice bath.
3. Add 37.5 ml of absolute ethanol.
4. Let the product crystallize at 4°C overnight.
5. Recrystallize twice from 100 ml isopropanol.
6. The theoretical yield is 11.3 g, and the melting point 135−136°C.

3.6.3 Synthesis of acetylcholine – agarose

Protocol 12. (22) Preparation of acetylcholine – agarose

1. Add 10 ml of 10 mM dithioerythritol in 200 mM Tris pH 8 to 10 ml Bio-Rad Affi-gel-401 (preswollen).
2. Shake the gel for 1 h at room temperature.
3. Wash slowly with 1000 ml cold H_2O.
4. Wash with 50 ml of 50 mM NaH_2PO_4/Na_2HPO_4, pH 7.
5. Determine free SH groups in the wash buffer by adding 100 μl wash buffer (see Protocol 10, *step 8* for details) to 1 ml 0.33 mM of dithionitrobenzoic acid in 0.2 M Tris-HCl pH 8. Measure the absorbance at 412 nm. The extinction coefficient is 13 600 cm^2 per mol for the resulting thionitrobenzoate anion.
6. After complete removal of free SH groups from the wash buffer, determine the content of SH groups in the gel by analogous method. This should lie in the range of 2−5 $\mu mol/g$ wet gel.
7. Add 10 ml of 20 mM bromacetylcholine in 0.1 M NaCl, 50 mM NaH_2PO_4/Na_2HPO_4 pH 7 to the sulphydryl-derivatized gel.
8. Shake for 1 h at room temperature.
9. Wash the gel with 300 ml of buffer and determine the level of residual SH groups in the gel. Usually there will be complete substitution. If blocking of the remaining SH groups should be necessary, add 20 mM iodoacetamide and incubate for 1 h at room temperature.
10. Equilibrate the gel with 50 mM NaCl, 100 mM NaH_2PO_4/Na_2HPO_4 pH 7.4 containing 0.1% Triton X-100.

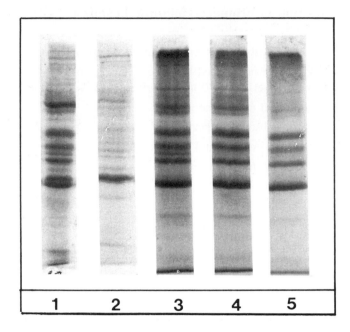

Figure 2. Protein composition at different stages of a receptor preparation. SDS-PAGE (10% gels) stained with Coomassie Brilliant Blue. Lanes: 1, receptor-rich membranes after density gradient centrifugation; 2, peripheral membrane proteins extracted with lithium diiodosalicylate (the predominant band represents the 43 kDa protein); 3, the remaining integral membrane proteins; 4, proteins extracted with 2% β-octylglucoside; 5, protein after Cibacron Blue Sepharose chromatography. The band above the four receptor subunits probably represents a subunit of an ATPase (96 kDa).

3.6.4 Toxin-affinity chromatography

Protocol 13. (13) Toxin affinity chromatography

1. CNBr-activated Sepharose 4B can be obtained from Pharmacia. Swell and wash the gel according to supplier's directions.

2. Couple *Naja naja siamensis* toxin (Sigma) at 10 mg per 40 ml wet gel by dropwise addition of a 10 mg/ml solution in 0.1 M NaH_2PO_4/Na_2HPO_4 pH 7.4, 4°C, agitating gently. Coupling should be complete in one or two hours, this can be monitored by measuring the absorbance at 280 nm.

3. Block excess active groups with 1 M glycine overnight at room temperature with gentle agitation.

4. Wash according to supplier's directions.

For nAChR-adsorption:

1. Wash 10 ml toxin gel with H_2O (100 ml).

2. Mix with Triton X-100-solubilized receptor protein (from 100 g frozen tissue) in 0.1 M NaH_2PO_4/Na_2HPO_4, pH 7.4.

3. Stir gently overnight at 4°C.

4. Pour the gel into a column.

5. Wash the gel with 50 ml of 0.1 M NaH_2PO_4/Na_2HPO_4 pH 7.4, 0.1 M NaCl, 1% Triton X-100.

6. Elute the nAChR with 0.5 M carbamoylcholine in the above buffer.

7. Dialyse the eluted receptor overnight against 500 ml of 0.02 M NaH_2PO_4/Na_2HPO_4 pH 7.4, 0.01 M NaCl, 1% Triton X-100.

8. Over the following 24 h replace the dialysis buffer another 3 times.

This method reportedly has been used for purification of nAChR from bovine, rat, and human muscle, though with modifications to account for the low concentration of receptor in muscle and greater proteolytic susceptibility (23).

4. Conclusion

As mentioned above, a combination of the protocols given in Sections 3.2, 3.3.2, 3.4, and 3.5 is especially useful for rapid large-scale purification (*Figure 2*) of stable nAChR. From such a preparation, crystals (*Figure 3*) can be obtained by slow precipitation with methylpentanediol in the presence of heptanetriol (10).

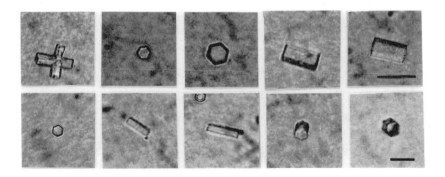

Figure 3. Crystals obtained from a nAChR preparation by precipitation with methylpentane diol in presence of heptane triol. The calibration bar represents 20 μm.

References

1. Hucho, F. (1986). *Neurochemistry, Fundamentals and Concepts*, VCH, Weinheim.
2. Changeux, J. -P. (1981). *Harvey Lect.*, **75**, 85.
3. Changeux, J. -P., Devillers-Thiery, A., and Chemouille, P. (1984). *Science*, **225**, 1335.

4. Barnard, E., Darlison, M., and Seeburg, P. (1987). *TINS*, **10**, 502.

5. Maelicke, A. (ed.), (1989). *Molecular Biology of Neuroreceptors and Ion Channels*, NATO ASI Series H. Springer Verlag.

6. Hucho, F. (1986). *Eur. J. Biochem.*, **158**, 211.

7. Steinbach, J. and Zempel, J. (1987). *TINS*, **10**, 61.

8. Verdenhalven, J., Bandini, G., and Hucho, F. (1982). *FEBS Lett.*, **147**, 168.

9. Hucho, F. and Bandini, G. (1986). *FEBS Lett.*, **200**, 279.

10. Hertling-Jaweed, S., Bandini, G., Müller-Fahrnow, A., Dommes, V., and Hucho, F. (1988). *FEBS Lett.*, **241**, 29.

11. Karlin, A., McNamee, M. G., and Weill, C. L. (1976). In *Methods in Molecular Biology*, Vol. 9 *Methods in Receptor Research*, Blecher, M. (ed.) Marcel Dekker, New York, Part I, p. 1.

12. Olsen, R., Meunier, J. -C., and Changeux, J. -P. (1972). *FEBS Lett.*, **28**, 96.

13. Karlsson, E., Heilbronn, E., and Widlung, L. (1972). *FEBS Lett.*, **28**, 107.

14. Jones, O., Euchanks, J., Earnest, J., and McNamee, M. (1988). *J. Biol. Chem.*, **27**, 3733.

15. Hubbard, A. L. and Cohn, Z. (1972). *J. Biol. Chem.*, **55**, 390.

16. Schmidt, J. and Raftery, M. A. (1973). *Anal. Biochem.*, **52**, 349.

17. Schiebler, W., Lauffer, L., and Hucho, F. (1977). *FEBS Lett.*, **81**, 39.

18. Neubig, R., Krodel, E., Boyd, N., and Cohen, J. (1979). *Proc. Natl. Acad. Sci. USA*, **76**, 690.

19. Porter, S. and Froehner, S. C. (1983). *J. Biol. Chem.*, **258**, 10034.

20. Paraschos, A., Gonzalez-Ros, J. M., and Martinez-Carrion, M. (1982). *Biochim. Biophys. Acta*, **691**, 249.

21. Damle, V. N., McLaughlin, M., and Karlin, A. (1978). *Biochem. Biophys. Res. Comm.*, **84**, 845.

22. Schmidt, J. and Raftery, M. A. (1972). *Biochem. Biophys. Res. Comm.*, **49**, 572.

23. Einarson, B., Gullick, W., Conti-Tronconi, B., Ellisman, M. and Lindstrom, J. (1982). *Biochem.*, **21**, 5292.

24. Hulme, E. C. (ed.) (1990). *Receptor−Ligand Interactions, A Practical Approach*, IRL Press, Oxford, in press.

8

Purification and molecular characterization of the γ-aminobutyric acid$_A$ receptor

F. ANNE STEPHENSON

1. Introduction

γ-Aminobutyric acid (GABA) is the major inhibitory neurotransmitter in the mammalian central nervous system. GABA mediates its effects via the specific interaction with GABA receptors of which two major subclasses have been defined on both a pharmacological and a functional basis. GABA$_A$ receptors are ligand-gated chloride-ion channels and are sensitive to the competitive GABA$_A$ receptor antagonist, bicuculline, whereas GABA$_B$ receptors are bicuculline-insensitive but baclofen, a GABA$_B$ receptor agonist, sensitive. GABA$_B$-receptor activation results in the interaction with a guanine nucleotide binding protein and subsequent modulation of adenylate cyclase activity, of Ca^{2+} and/or K^+ channel activity. GABA$_A$ receptors have a complex pharmacology, which has greatly facilitated the understanding of their structure, whereas to date there is little information available on GABA$_B$ receptors other than, by analogy with other receptors that act via second messenger systems, that they are proteins consisting of a single polypeptide chain.

It is well established, however, that the GABA$_A$ receptor is an integral multisubunit membrane glycoprotein. As indicated above, it is the site of action of several classes of therapeutically important drugs. These include the anxiolytic drugs, the benzodiazepines (e.g. valium and librium); the central nervous system depressant barbiturates, some steroids, e.g. derivatives of progesterone, and the anthelminthic avermectins (reviewed in 1, 2). All of these compounds are allosteric potentiators of GABAergic neurotransmission. Additionally, there exist non-competitive antagonists of GABA$_A$ responses, e.g. picrotoxin and the cage convulsant compounds such as t-butylbicyclophosphorothionate (TBPS) and t-butylbicyclo-phosphate (TBOP) which are thought to act directly at the chloride channel-gating site. *Figure 1* summarizes in diagrammatic form the distribution of these sites within the receptor complex.

This chapter will describe in detail the biochemical characterization of this receptor protein using the methods that have been developed and used routinely in this

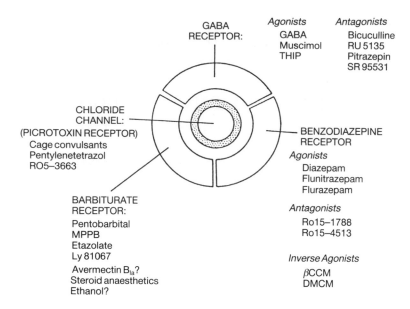

Figure 1. A diagrammatic representation of the GABA_A receptor and the site of action of the drugs that are known to mediate their effects through this neurotransmitter receptor protein. The figure is reprinted from reference 18 with the kind permission of Plenum Press, New York.

laboratory. These studies contributed towards the recent elucidation of the primary structures of the GABA_A receptor α and β polypeptide chains as deduced from molecular cloning (3,4) and, where appropriate, the results obtained from each discipline will be compared.

2. GABA_A receptor solubilization and purification

2.1 Solubilization

The first attempts to solubilize the GABA_A receptor from brain membrane preparations were carried out in the late 1970s and the early 1980s by several groups (reviewed in references 2,5). To summarize briefly, it was found that both GABA- and benzodiazepine-binding sites were readily extracted by a variety of non-denaturing detergents. The efficiency of solubilization and the stability of the resultant extract varied with the detergent employed and, perversely, it was shown that in the presence of the ionic detergent sodium deoxycholate, which solubilized the maximum number of GABA- and benzodiazepine-binding sites, these respective binding sites were unstable. Furthermore, the chloride channel-gating sites and the allosteric interactions involving these sites are inactivated in the presence of sodium deoxycholate. When the zwitterionic detergent, 3-(3-cholamidopropyl)dimethylammonio-1-propane-sulphonate (CHAPS, see Chapter 1, Section 2), was used for the solubilization of

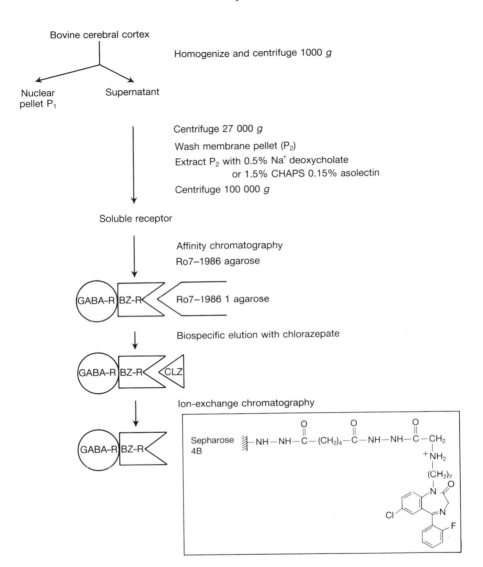

Figure 2. A schematic representation of the method employed for the isolation of the GABA$_A$ receptor. The inset shows the structure of the Ro7-1986/1 benzodiazepine affinity column. The figure is reproduced from reference 5 with the kind permission of Alan R. Liss, Inc., New York.

the GABA$_A$ receptor, it was shown that although the efficiency of solubilization was one-third of that of sodium deoxycholate, all the binding sites and the allosteric interactions between them that had been shown to exist in membrane preparations were now found in the soluble state. Thus we have used two different soluble extracts

Table 1. Purification of the GABA/benzodiazepine receptor complex.

Sample	Total protein (mg)	Specific binding activity (pmol/mg protein)		Yield of [^{3}H]flu-nitrazepam binding
		[^{3}H]muscimol	[^{3}H]flunitrazepam	
Homogenate	14 000	0.04	0.44	100
Crude synaptic membranes	2170	0.28	1.26	44
Soluble receptor	1010	4.4	1.46	24
Affinity column eluate	0.45	1500	—	—
Ion-exchange eluate	0.15	2560	800	2

as the starting material in our purification scheme, the choice of the detergent being dictated by the experiment.

Protocol 1 describes the synthesis of the benzodiazepine affinity column and Section 2.2. gives the detailed procedure that is routinely employed for the purification of the GABA$_A$ receptor by benzodiazepine affinity chromatography, taken from the work of Sigel *et al.* (6,7) and Sigel and Barnard (8). *Figure 2* is a diagrammatic representation of the purification scheme and *Table 1* shows the results of a typical isolation experiment where the starting material was 100 g bovine cerebral cortex.

Protocol 1. The synthesis of Ro7-1986/1 agarose

1. Wash adipic dihydrazide agarose (Pharmacia) (30 ml) with 10 vols ice-cold H$_2$O to remove residual salts.

2. Dissolve sodium iodoacetate (6.3 g) in 15 ml H$_2$O at 4°C at pH 5.0. Add this to the washed adipic dihydrazide agarose on ice. To the gel slurry, add 1.5 g 1-ethyl 3-(3-dimethylaminopropyl)carbodiimide on ice and adjust to pH 5.0. Let the reaction proceed with gentle shaking for 3 h at 23°C.

3. Wash the activated gel on a Buchner filter funnel under vacuum with (i) 3 vols H$_2$O and (ii) 3 vols 100 mM sodium carbonate, pH 9.0 (at this stage a sample of gel can be retained as a control and is subsequently treated identically to the test gel except that all incubations are in the absence of Ro7-1986/1).

4. Dissolve the benzodiazepine Ro7-1986/1, (*Figure 2*) (216 mg) in 3 ml ethanol. Dilute to 30-ml with 10 mM sodium carbonate, pH 9.0, to give a clear yellow solution. Add the ligand to the washed gel slurry and incubate overnight at 23°C with gentle shaking and maintaining a pH of 9.0.

5. Add 1.2 ml of a 10% (v/v) β-mercaptoethanol solution to the gel to block any remaining activated groups, and incubate for 15 min at 23°C.

6. Collect the gel by vacuum filtration as above and wash the gel with 10 vols

each of the buffers used for the regeneration of the affinity column. In this case it is:

(a) high salt and high detergent, 5 mM Hepes pH 7.5 containing 0.5 M NaCl, 0.5 mM EDTA and 2% (v/v) Triton X-100;

(b) a wash containing a protein denaturing agent, 50 mM acetic acid pH 5.5 containing 6 M urea and 1% (v/v) Triton X-100; and

(c) the column equilibration buffer, which is 5 mM Hepes pH 7.4 containing 0.15 M KCl, 0.1 mM EDTA, 0.1% (v/v) Triton X-100, and 0.02% (w/v) NaN$_3$.

7. The concentration of immoblized Ro7-1986/1 can be determined by the difference spectrum between the control and the benzodiazepine gel, provided that each gel is packed in 2 mm path length quartz cells. Knowledge of the extinction coefficient for Ro7-1986/1 enables the determination of the concentration of the immobilized Ro7-1986/1 using the standard Beer−Lambert Law, i.e.

$$A = \Sigma cl$$

where A = absorbance at λ = nm as selected;
Σ = extinction coefficient;
c = concentration;
l = path length.

2.2 Isolation of the GABA$_A$ receptor

2.2.1 Membrane preparation

Protocol 2. Preparation of membranes from bovine cerebral cortex

1. All procedures are carried out at 4°C. Homogenize 100 g of bovine cerebral cortex in 1.4 l 0.32 M sucrose in 10 mM Hepes, pH 7.4 containing 1 mM EDTA, 1 mM benzamidine hydrochloride, 0.02% sodium azide, 0.5 mM phenylmethylsulphonyl fluoride (PMSF), 1 mg/100 ml soya bean trypsin inhibitor, 1 mg/100 ml chicken ovomucoid trypsin inhibitor (Sigma). The PMSF is prepared as 0.17 M stock solution in propan-2-ol and is added to the above buffer, termed buffer 1, immediately prior to use.

2. Centrifuge at 900 g for 10 min, remove the supernatant and centrifuge it at 20 000 g for 40 min.

3. Resuspend the pellet from *step 2* in 10 mM Hepes, pH 7.4, containing 1 mM EDTA, 0.02% NaN$_3$, 1 mg/100 ml soya bean trypsin inhibitor, 1 mg/100 ml chicken ovomucoid trypsin inhibitor. This is referred to as buffer 2.

4. Centrifuge at 20 000 g for 40 min. Discard the supernatant and resuspend the pellet in 170 ml of buffer 2.

2.2.2 Solubilization

Protocol 3. Solubilization of GABA$_A$ receptors

1. Na$^+$-deoxycholate extraction (Method A):
   To the pellet suspension from Protocol 2, *step4* add 18 mg of bacitracin already dissolved in 0.5 ml buffer 2, 0.36 ml of 0.17 M PMSF, 7.9 ml of 3.5 M KCl, and 4.5 ml of 20% (w/v) sodium deoxycholate. This will result in final concentrations of 150 mM KCl, 0.5% sodium deoxycholate, 1 mg/10 ml bacitracin and 0.34 mM PMSF. Gently stir the suspension for 10 min at 4°C.

2. CHAPS extraction (Method B):
   Proceed exactly as for the Na$^+$ deoxycholate extraction above but replacing the detergent with 6.8 ml of 40% CHAPS (Serva, Heidelberg or Sigma) containing 4% soya bean asolectin (Associated Concentrates, New York), to give a final concentration of 1.5% CHAPS, 0.15% asolectin during solubilization. [The stock CHAPS/asolectin mixture is prepared by dissolving 0.8 g asolectin in 20 ml diethyl ether in a glass tube. A thin film is prepared by the evaporation to dryness of the ether under nitrogen gas with rotation of the tube. Add CHAPS (8 g; see Chapter 1, Section 2) and H$_2$O to a volume of 20 ml, followed by 10 μl β-mercaptoethanol. Dissolve the CHAPS and the thin film by vortexing.] Gently stir the suspension for 30 min at 4°C.

3. Centrifuge the suspensions at 100 000 *g* for 75 min. Collect the supernatant by aspiration followed by filtration through washed glass wool. The volume of the respective soluble extracts is 150 ml.

2.2.3 Affinity chromatography

Protocol 4. Affinity chromatography of GABA$_A$ receptors

1. Apply the supernatant from *Protocol 3*, *step 3* at 60 ml/h to an affinity column of the benzodiazepine, Ro7-1986/1, coupled to adipic dihydrazide agarose (*Figure 2*). For the Na$^+$-deoxycholate extraction (Method A), wash overnight with 600 ml of 10 mM potassium phosphate pH 7.4, containing 200 mM KCl, 0.1 mM EGTA, 2 mM magnesium acetate, 10% (w/v) sucrose, 0.02% sodium azide, and 0.2% (v/v) Triton X-100. For CHAPS solubilization (Method B), replace the Triton X-100 with 0.6% (w/v) CHAPS/0.06% asolectin, which is diluted from the stock solution as described in *Protocol 3, step 2*.

2. For Method A, wash with 20 ml of 20 mM potassium phosphate pH 7.4, containing 10% (w/v) sucrose, 2 mM magnesium acetate, 0.02% sodium azide, 0.2% Triton X-100. For Method B, replace the Triton X-100 with 0.6% CHAPS/0.06% asolectin diluted from the stock solution as before.

3. Specific elution of the GABA$_A$ receptor from the affinity column is achieved by passing 20 ml of 10 mM clorazepate (Boehringer Ingelheim, or Sigma),

a water-soluble benzodiazepine in the respective buffers used in *step 2* above, but with no added KCl, at a flow rate of 20 ml/h. Follow this with 30 ml of the buffer, again no KCl, and corresponding detergent alone, again at 20 ml/h. During the elution, fractions of 5 ml are collected.

4. The receptor-containing fractions which have been pre-determined are pooled; the volume is 35 ml. The pH of this pool is adjusted to pH 6.5 with 100 mM H_3PO_4. Apply to a DEAE-Sephacel ion-exchange column (1 × 0.5 cm) equilibrated with 20 mM potassium phosphate pH 6.5, containing 10% (w/v) sucrose, 2 mM magnesium acetate, 0.02% sodium azide, and for Method A, 0.2% Triton X-100 or, for Method B, 0.6% CHAPS/0.06% asolectin. The sample is applied at a flow rate of 40 ml/h then the column washed for 2 h at 40 ml/h with the equilibration buffer above.

5. Elute the column with 0.8 M KCl in the respective ion-exchange equilibration buffers (*step 4*) at a flow rate of 10 ml/h. 1 ml fractions are collected for Method A and 0.5 ml fractions for Method B. The fractions are assayed for $GABA_A$ receptor activity and the peak fractions are pooled and stored at −20°C. Receptor activity is measured by the polyethylene glycol precipitation assay for solubilized receptors (see Appendix; further details are given in reference 25), with the separation of bound and free ligand carried out by either rapid filtration or centrifugation. A summary of the standard incubation conditions is given in *Protocol 5*.

Protocol 5. Radioligand-binding assays for soluble and purified $GABA_A$ receptors

1. This is the standard assay and listed below are the assay conditions for the individual ligands that have been used in the study of the $GABA_A$ receptor. Incubate on ice:
   - 160 µl soluble or purified receptor diluted in 20 mM potassium phosphate pH 7.4, 0.1% (v/v) Triton X-100, 0.1 mM EDTA as required;
   - 20 µl non-radioactive drug for non-specific binding determination, or 20 µl buffer as above;
   - 20 µl radioactive ligand.
   Total volume = 200 µl.

2. Add: 15 µl 3.3% (w/v) bovine γ-globulin;
        85 µl 36% (w/v) polyethylene glycol.
   Total volume = 300 µl.
   Vortex vigorously (the solution should go cloudy) and stand on ice for 12 min.

3. Filter under vacuum onto GF/C glass-fibre filters, 2.5 cm diameter and wash with 3 × 3 ml 7.5% (w/v) polyethylene glycol in the above buffer. Dry, add scintillation fluid and count for radioactivity.

4. For the determination of [^{3}H]muscimol (Amersham) binding, use final assay concentrations of 15 nM [^{3}H]muscimol and 0.1 mM GABA with an assay incubation time of at least 15 min at 4°C.

5. For the determination of [^{3}H]flunitrazepam (Amersham) binding activity, use final assay concentrations of 10 nM [^{3}H]flunitrazepam, 10 μM diazepam with an assay incubation time of at least 30 min at 4°C.

6. For the determination of [^{35}S]TBPS (NEN) binding activity, use final assay concentrations of 40 nM [^{35}S]TBPS, 0.1 mM picrotoxin prepared in a 10% (v/v) ethanol solution, with an assay incubation time of 90 min at 4°C. The final assay concentration of KCl should be kept at 200 mM and it is essential that in this case, the detergent employed is CHAPS at a final assay concentration of 0.3% (w/v).

The recovery of purified GABA$_A$ receptor is low, ~2% of the total benzodiazepine-binding sites present in the brain homogenates. The major loss of activity occurs at the stage of the specific elution of the benzodiazepine affinity column. It is worth noting that in the procedure of Stauber *et al.* (9) the affinity column described therein is eluted by another water-soluble benzodiazepine, flurazepam. The inclusion of 6 M urea in the flurazepam elution increased the yield of purified GABA$_A$ receptor eightfold with respect to elution with flurazepam alone to 48% recovery. The urea was subsequently removed by dialysis with the recovery of benzodiazepine-binding activity (9).

Other points to note in the isolation procedure that may be of general interest are as follows. First, the Na$^+$-deoxycholate is exchanged for Triton X-100 conveniently at the stage of the affinity column wash because the GABA- and benzodiazepine-binding sites are more stable in the presence of Triton X-100 and because, in the buffer system used, Na$^+$ deoxycholate precipitates out of solution at 4°C (see Chapter 1, Section 2). Secondly, although the channel-gating site and the associated allosteric interactions are detectable in CHAPS/asolectin extracts, they are unstable. Bristow and Martin (10) showed that the replacement of the asolectin (a soya bean lipid extract) with a natural brain lipid extract supplemented with cholesterol hemisuccinate stabilized these sites. Thirdly, the wash buffer for the affinity column was chosen specifically to minimize both hydrophobic and electrostatic non-specific interactions; notably, a reduction in the KCl concentration in this medium resulted in an impure preparation. Sucrose was added to all buffers following the application to the affinity column to stabilize the ligand-binding acivities with storage at 4°C and at −20°C. The ion-exchange chromatography provides both a concentration step (tenfold) and a method of removing the ligand used for affinity column elution. It is also a second convenient point for detergent-exchange. No purification factor is achieved by this additional step and losses of activity up to 50% can be incurred. It is noteworthy therefore, that in the determination of the partial amino-acid sequence of the GABA$_A$ receptor polypeptides, where both the amount and the purity of the receptor are of paramount importance, the receptor was concentrated directly from the clorazepate eluate by the chloroform/methanol precipitation method (see *Protocol 8*) and this material was employed for cyanogen bromide cleavage, peptide separation, and subsequent amino-acid sequence determination (3,11; see also Chapter 10, Section 3).

3. Molecular characterization of the purified GABA$_A$ receptor

3.1 Assessment of purity by isoelectric focusing

It is generally accepted that at least two criteria of purity should be satisfied in order to prove the isolation to homogeneity of a receptor protein of interest. These are the specific activity of the pure receptor and an assessment of protein homogeneity in a further separation system which employs non-denaturing conditions. The former of these, the specific activity determination, although required, often falls short of theoretical calculations. This is because invariably there is detergent inactivation of specific ligand-binding sites during the isolation procedure, and also the accurate determination of the protein content of the pure protein is subject to large errors due to the anomalous properties of integral membrane proteins. In the case of the GABA$_A$ receptor, even though the complete primary structures of the α and β polypeptide chains are now known, the absolute ligand-binding site stoichiometry per receptor oligomer has not been determined definitively. Thus, the calculated specific activity of the GABA$_A$ receptor is uncertain. The homogeneity of the protein isolated by the method outlined in Section 2.2 was assessed therefore by narrow-range isoelectric focusing. The purified material was radioiodinated prior to focusing by a mild chloramine-T procedure (12). The methods are given in *Protocols 6* and *7* for both the iodination procedure used and the isoelectric focusing in polyacrylamide disc gels.

Protocol 6. Radioiodination of purified GABA$_A$ receptor

1. Dialyse the purified GABA$_A$ receptor (100 μl from the pool of the ion-exchange fractions) against 1000 volumes of 10 mM potassium phosphate pH 7.4, 0.1% Triton X-100 for 2 h at 4°C, to remove salts. See Chapter 10, Section 6 for pretreatment of dialysis tubing, but use Triton X-100 instead of SDS.

2. Add the dialysed receptor (100 μl) to a microfuge tube on ice and add sequentially, 10 μl of 1 M sodium phosphate pH 7.4, 2.5 μl carrier-free Na[^{125}I] (0.25 mCi; Amersham), 4 μl of 1 mg/ml chloramine-T (Pierce). Incubate on ice for 20 min.

3. Stop the reaction by the addition of 20 μl of 10 mM dithiothreitol. Incubate for 5 min and add 20 μl of 10 mg/ml sodium iodide followed by 10 μl of 2.5 mM tyrosine. Mix and leave on ice for 20 min.

4. Apply the reaction mixture to a Sephadex G75 column (18 × 0.5 cm) pre-

equilibrated with 10 mM potassium phosphate, pH 7.4, containing 0.1% Triton X-100, 0.02% NaN$_3$, and 5 mg/100 ml bovine serum albumin to separate [^{125}I]GABA$_A$ receptor and free ^{125}I. Collect fractions of 10 drops. Count 5 μl samples from each fraction, pool those containing [^{125}I]GABA$_A$ receptor and store at 4°C.

Protocol 7. Narrow range isoelectric focusing in polyacrylamide gels of [^{125}I]GABA$_A$ receptor

1. Mix (scale-up ad lib):
   - 219 μl 28% acrylamide/1.6% NN^1-methylenebisacrylamide (BDH Electran grade, or similar);
   - 82.5 μl Ampholines range pH 4−6.5 (Pharmacia) and pH 6.5−9 mixed at 1:1.25 ratio;
   - 192 μl Glycerol;
   - 82.5 μl 20% Triton X-100;
   - 5.5 μl 1:10 dilution of $NNN'N'$-Tetramethylethylenediamine;
   - 1000 μl H$_2$O.

2. Degas the solution and add 41.25 μl 1.5% (w/v) ammonium persulphate. Mix and pour immediately into silanized glass tubes (11 $\times$ 0.5 cm internal diameter) covered at the bottom with dialysis tubing held in place by an O-ring. Overlay with H$_2$O and leave to polymerize for 1 h at room temperature.

3. Set up the disc electrophoresis apparatus with 10 mM H$_3$PO$_4$, pH 2.3, containing 0.2% Triton X-100 at the anode; and 20 mM NaOH, pH 11, containing 0.2% Triton X-100 at the cathode. Wash the top of each gel and overlay with cathode buffer.

4. Pre-run the gels at 1 mAmp/gel for 20 min at 4°C.

5. Prepare the samples by mixing:
   (a) 20 μl pool [^{125}I]GABA$_A$ receptor; 65 μl H$_2$O; 25 μl glycerol;
   (b) 85 μl H$_2$O; 3 mg myoglobin as standard marker;
   (c) 100 μl H$_2$O
   Layer each sample into a rod gel using a fine pipette tip.

6. Focus at 400 V constant voltage for 5 h at 4°C.

7. Remove the gels from the gel tubes by carefully injecting water between the gel and the tube using a syringe and fine-gauge needle until the gel slides free, or can be expelled by gentle pressure from a Pasteur pipette bulb.

8. Slice each gel at 2 mm intervals and either count in a γ-counter for sample (a) or measure the OD at $\lambda = 406$ nm for sample (b). For sample (c), incubate each slice with 200 μl degassed 10 mM KCl, overnight at 4°C. Measure the pH of each corresponding solution. Gel slicers are available commercially.

9. Note that isoelectric focusing can also be performed in free solution. A suitable apparatus, the Rotofor is marketed by Bio-Rad. An information leaflet is available.

10. Tubes are silanized by immersion in 5% dichlorodimethyl-silane (Pierce) in chloroform, followed by draining, washing in deionized water and baking at 100°C for 30 min.

11. This protocol can be applied to the isoelectric focusing of other receptors. The Triton X-100 can be replaced by other suitable non-ionic detergents; such as digitonin. A potential problem that of ammonium persulphate-catalysed disulphide bond formation can be avoided by removing the polymerized gels from the tubes as in *step 7* above, soaking them in 5 mM dithiothreitol in solution 1, made up with water replacing the acrylamide solution, for 5 h, sucking the gels back into the tubes, and then proceeding as in *step 3*.

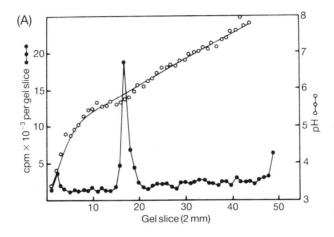

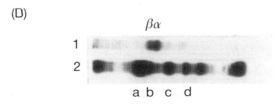

Figure 3. Narrow-range isoelectric focusing and SDS-polyacrylamide-gel electrophoresis of purified bovine [125I] GABA$_A$ receptor. (A) Isoelectric focusing of the [125I]GABA$_A$ receptor isolated from bovine cerebral cortex (an estimated 86 fmol [3H]muscimol-binding sites), carried out exactly as described in Protocol 7. The [125I]GABA$_A$ receptor focused as a single sharp band with an isoelectric point, pI 5.6. (B) SDS-PAGE under reducing conditions in a 10% polyacrylamide gel of the [125I]GABA$_A$ receptor, carried out as described in Protocol 8. Lane 1 corresponds to 25 fmol [3H]muscimol-binding sites of [125I]GABA$_A$ receptor. Lane 2 contains [125I]standard proteins radioiodinated as a pool by the method in Protocol 6: a, bovine serum albumin, M$_r$ 67 000; b, catalase, M$_r$ 36 000; c, ovalbumin, M$_r$ 43 000; d, lactate dehydrogenase, M$_r$ 36 000. Reprinted from reference 24 with the kind permission of Raven Press, New York.

Figure 3 shows the results of the isoelectric focusing and SDS-PAGE under denaturing conditions (see *Protocols 7, 8*) of the [^{125}I]GABA$_A$ receptor. The [^{125}I]GABA$_A$ receptor focused as a single sharp band with pI 5.6. It is interesting to note that the α and the β polypeptide chains are basic. The charges as calculated from the deduced respective primary structures of the mature subunits at pH 6.8 are $+15$ for α1 and $+11$ for β (3). Thus the carbohydrate moieties of the α and the β subunits must dictate the overall acidic properties of the oligomer.

3.2 Molecular weight determination of the GABA$_A$ receptor

The accurate determination of the molecular weight of the pure GABA$_A$ receptor oligomer has been difficult and the reasons for this are twofold. First, in common with other receptors described in this volume, it is an integral membrane protein and in detergent solution, it is in the form of a protein−detergent or protein−detergent−phospholipid complex. The bound molecules of detergent and phospholipid are less dense than the protein and thus produce a buoyant effect on the hydrodynamic behaviour of the protein such that the receptor behaves anomalously in solution. The molecular weight therefore cannot be extrapolated directly from the sedimentation coefficient ($s_{20,w}$) or the Stokes' radius (a) as for soluble globular proteins. Thus the method of choice is the determination of the sedimentation coefficient and the partial specific volume of the protein−detergent complex using sucrose density gradient centrifugation in media of two different densities, i.e. H$_2$O and ^{2}H$_2$O, as first described for the nicotinic acetylcholine receptor by Meunier *et al.* (13). This technique, with the knowledge of the partial specific volume of the non-denaturing detergent used, permits the percentage contribution by weight of the detergent to the protein−detergent complex (see below), hence the molecular weight of the receptor if the Stokes' radius is known.

Secondly, the pure GABA$_A$ receptor has a strong tendency to form very high molecular weight aggregates. In initial experiments with the isolated receptor, the ligand-binding activities were found in the void volume in gel-filtration studies

Figure 4. Sucrose density gradient centrifugation of the GABA$_A$ receptor purified from bovine cerebral cortex by the Na$^+$ deoxycholate/Triton X-100 method. (A) The heterogeneous profile of [^{3}H]flunitrazepam-binding activity of the bovine GABA$_A$ receptor purified by the Na$^+$ deoxycholate/Triton X-100 method following sucrose density gradient sedimentation in the presence of 0.5 M NaCl in the gradient media. Marker enzymes (↓) are from left to right: cytochrome *c*, $s_{20,w}$ = 1.9S; yeast alcohol dehydrogenase, $s_{20,w}$ = 7.4S; catalase, $s_{20,w}$ = 11.4S; and *E. coli* β-galactosidase, $s_{20,w}$ = 16.0S. (B) The profile of [^{3}H]flunitrazepam (■ − ■) and [^{3}H]muscimol (● − ●) specific ligand-binding activity of the bovine GABA$_A$ receptor following sucrose density gradient sedimentation in the presence of 1.0 M NaCl in a D$_2$O gradient. In this case, the receptor was purified by the Na$^+$ deoxycholate/Triton X-100 method from membranes which had been pre-extracted with 0.05% Triton X-100. The markers (↓), from left to right, are haemoglobin, $s_{20,w}$ = 4.5S; yeast alcohol dehydrogenase, catalase, and β-galactosidase. The sedimentation coefficient, $s_{20,w}$, for the pure GABA$_A$ receptor determined for the conditions described in (B) was used to calculate the molecular weight of the receptor as shown in Section 3.2. The figure is reprinted from reference 14 with the kind permission of Raven Press, New York.

(i.e. $a >> 6.8$ nm) and the profile of binding following sedimentation in sucrose density gradients was heterogeneous (*Figure 4*), thus making the accurate determination of the $s_{20,w}$ of the receptor monomer difficult. Two factors were found to reduce the aggregation state of the receptor. These were the inclusion of

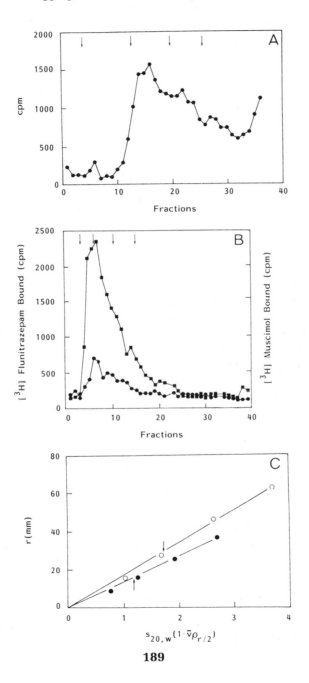

Table 2. Hydrodynamic parameters of the Na$^+$-deoxycholate-solubilized GABA$_A$ receptor [from Mamalaki et al. (14)].

Triton X-100 pretreatment of membranes[a]	NaCl in gradient medium[b]	$s_{20,w}$ (S)	$\bar{v}$ (R-Det) (ml/g)	Stokes' radius (nm)	$M_{\text{R-Det}}$	Mol. wt of receptor
−	50 mM	10.9	−	−	−	−
+	50 mM	10.5	−	−	−	−
−	1 M	7.8	0.73	7.25	236 000	236 000
+	1 M	7.7	0.77	7.3	307 000	240 000

[a] Indicates that brain membranes were pretreated with 0.05% Triton X-100 before membrane protein solubilization.
[b] The concentration of NaCl present in the sucrose density gradient.

high concentrations of NaCl, i.e. ≥0.5 M, in the separation media and, secondly, the pretreatment of the brain membranes with 0.05% Triton X-100 at 37°C for 1 h prior to membrane protein solubilization (*Figure 4*). Under these conditions (reported fully in reference 14), the GABA$_A$ receptor was non-aggregated and a molecular weight was determined in the range of 230 000−240 000 Da. A point to note is that in H$_2$O sucrose density gradients of the soluble receptor in the low salt the $s_{20,w}$ = 10.9 ± 0.3S whereas in the sucrose density gradients of the soluble receptor in high salt, the $s_{20,w}$ = 7.8 ± 0.5S. When the soluble receptor was sedimented in both media in H$_2$O and ^{2}H$_2$O, the calculated molecular weight of the receptor was identical. This demonstrates that the gradient media can influence the $s_{20,w}$ and thus the deduced molecular weight of the protein, and highlights the dangers of the deduction of a molecular weight from one parameter. Thus, the molecular weight of the GABA$_A$ receptor oligomer in Triton X-100 detergent was determined from the following procedure and the experimental results are summarized in *Table 2*.

The GABA$_A$ receptor was isolated according to the Method A outlined earlier (Protocol 3, *step 1*) and immediately prior to use was dialysed against 20 mM K$^+$-phosphate pH 7.4, 0.02% (w/v) NaN$_3$, 0.5% (w/v) Triton X-100 (10 volumes) to remove sucrose. Sucrose density gradient centrifugation in 5−20% sucrose gradients of H$_2$O and ^{2}H$_2$O and gel-filtration chromatography were carried out according to standard procedures (e.g. reference 14; see also Chapter 2). The buffer used for these separation procedures was 20 mM K$^+$-phosphate pH 7.4, 0.5% Triton X-100, 1.0 M NaCl. Martin and Ames (15) showed that for density gradient centrifugation the distance r_i travelled from the meniscus by any macromolecule in medium (i) is

$$r_i = k_i \, s_{20,w} \, (1 - \bar{v}\varrho_i)$$

where k_i is a constant in a medium of a given average density ϱ_i, for all macromolecules with the same partial specific volume ($\bar{v}$). Sucrose density gradient centrifugation was carried out in H$_2$O (i = 1) and ^{2}H$_2$O (i = 2) to determine, by the solution of the pair of simultaneous equations, $s_{20,w}$ and $\bar{v}$ for the protein–detergent complex.

These equations were obeyed for the GABA$_A$ receptor protein, both in crude extracts and in the pure state. The molecular weight of the receptor–detergent complex ($M_{R\text{-Det}}$) was calculated from the following equation with the substitution of the experimentally determined parameters:

$$M_{R\text{-Det}} = \frac{6\pi\eta_{20,w}Ns_{20,w}a}{(1 - \bar{v}_{R\text{-Det}}\varrho_{20,w})}$$

where N is Avogadro's number, $\eta_{20,w}$ is the viscosity of water at 20°C, and $\varrho_{20,w}$ is the density of water at 20°C. The Stokes' radius a was that determined by gel-filtration under identical conditions to those used for the sucrose density gradient centrifugation. The molecular weight of the receptor was calculated after the solution of:

$$\bar{v}_{R\text{-Det}} = x\bar{v}_{Det} + (1 - x)\bar{v}_R$$

where $\bar{v}_{Det}$ and $\bar{v}_R$ are the partial specific volumes of the detergent and the receptor protein, respectively, and x is the weight fraction of detergent bound. See Chapter 2, Section 6 for a detailed protocol for these procedures.

3.3 Subunit composition of the GABA$_A$ receptor

In the early studies of the determination of the subunit composition of the GABA$_A$ receptor, the receptor was subjected to SDS-PAGE under denaturing and reducing conditions in 10% slab gels. The standard procedure, outlined in *Protocol 8*, included GABA$_A$ receptor precipitation and gel-electrophoresis followed by protein detection by the silver stain or Coomassie Blue method.

Protocol 8. SDS-PAGE under denaturing conditions of the GABA$_A$ receptor

1. Precipitate the protein using either (a) or (b).
   (a) *Trichloroacetic acid (TCA) method:*
       Add an equal volume of 24% (w/v) TCA to the receptor, vortex, and incubate on ice for 15 min. Centrifuge for 15 min in an Eppendorf microfuge, discard the supernatant, and add with mixing 0.5 ml 12% (w/v) TCA. Centrifuge as above, discard the supernatant. Add 0.5 ml acetone, vortex, centrifuge, then repeat the acetone wash. Evaporate to dryness under vacuum.

   (b) *Chloroform/methanol precipitation method:*
       This is the method of Wessel and Flügge (22). Add methanol (4 vols) to the receptor (1 vol.), vortex, and centrifuge for 10 sec in an Eppendorf microfuge. Add chloroform (1 vol.), vortex, and centrifuge for 10 sec as above. Add H$_2$O (3 vols) vortex vigorously, centrifuge for 10 sec.

Carefully remove the upper layer (a protein precipitate should be visible at the interface of the two phases). Add methanol (3 vols), vortex gently, centrifuge for 4 min. Discard the supernatant and evaporate the pellet to dryness under vacuum.

2. Each sample is dissolved in standard gel sample buffer in the presence of 10 mM dithiothreitol and boiled for 5 min. The samples are applied to a 10% polyacrylamide slab gel and electrophoresis carried out at 10 mAmp, constant current. This is a standard method and a good reference is found in the previous volume of this series, *Gel Electrophoresis of Proteins: A Practical Approach* (26); see also Chapter 10, Section 3.

3. At the end of the run, the gel is rinsed superficially with H_2O and silver staining of the gel carried out. We have used two silver stain methods; one is a slow method, which includes a periodic acid oxidation step, whereas the second method is rapid and can be completed within 2 h and utilizes a microwave oven. Both methods are given here. Gloves must be worn throughout both procedures to prevent marking of the gel.

4. *Silver stain—slow method:*

   (a) Fix the gel overnight at room temperature with 25% (v/v) propan-2-ol, 10% (v/v) acetic acid in H_2O.

   (b) Incubate with 7.5% (v/v) acetic acid in H_2O for 30 min at room temperature.

   (c) Dissolve 400 mg periodic acid in 200 ml H_2O and incubate on ice for 1 h.

   (d) Wash the gel with H_2O for 3 h at room temperature, changing the H_2O every 30 min.

   (e) Incubate the gel with 0.5 mg dithiothreitol in 100 ml H_2O for 30 min at room temperature.

   (f) Do not rinse the gel, and incubate with 100 mg silver nitrate in 100 ml H_2O for 30 min at room temperature.

   (g) Prepare the developing solution by dissolving 9 g sodium carbonate and 750 μl of 8% formaldehyde solution in 300 ml H_2O. Wash the gel rapidly once with H_2O, 2 × 100 ml with developer then incubate with the developer until the silver stain appears in the gel.

   (h) Stop the reaction by the addition of 48% (w/v) citric acid (10 ml) and dry the gel for storage.

5. *Silver stain—fast method:*

   (a) Fix the gel with 50% (v/v) methanol, 10% (v/v) acetic acid in H_2O, 2 × 100 ml, 15 min each volume at room temperature.

   (b) Incubate with 5% (v/v) methanol, 7% (v/v) acetic acid in H_2O (100 ml) for 1 min in a microwave oven. Use a low setting to avoid overheating and subsequent 'yellowing' of the gel.

(c) Wash 3 × H$_2$O, each 1 min incubation in the microwave.

(d) Incubate the gel with 100 ml of 5 μg/ml dithiothreitol in H$_2$O for 1 min in the microwave oven.

(e) Agitate the gel in 200 ml of 0.1% (w/v) silver nitrate solution for 30 min at room temperature.

(f) Wash rapidly, twice with H$_2$O and once with developer, which is 3% (w/v) sodium carbonate and 75 μl of 37% formaldehyde solution (150 ml). Develop in the remainder until required.

(g) As above, stop the reaction by the addition of 48% (w/v) citric acid (10 ml), rinse with H$_2$O, and dry the gel.

It was found that the subunit composition of the GABA$_A$ receptor purified by either Method A or Method B (*Protocol 3*) was identical (7,8). Thus it was shown to be heterologous and consisted of two polypeptide chains, the α-subunit with M_r 53 000 and the β-subunit with M_r 57 000 (*Figure 5*). The same pattern was found

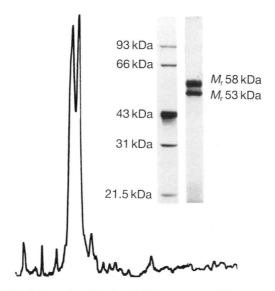

Figure 5. SDS-PAGE of the purified bovine GABA$_A$ receptor. The receptor was purified by the Na$^+$-deoxycholate/Triton X-100 method, and SDS-PAGE under reducing conditions was carried out as described in *Protocol 8*. The inset, of a silver-stained gel of molecular weight standards on the left and the purified receptor on the right, shows clearly the α-subunit with M_r 53 kDa and the β-subunit with M_r 58 kDa. The remainder of the figure is a densitometric scan of photographic negatives in the linear response conditions of the silver-stained gel of the purified GABA$_A$ receptor. Computer integration of the overlapping peaks seen of the α- and β-subunits yielded in replicate a mean ratio of 0.93:1 (α:β). Results are taken from reference 16.

Table 3. A summary of the properties of the $\alpha1$, $\alpha2$, and $\alpha3$ subunits of the bovine GABA$_A$ receptor.

α-subunit	$\alpha1$	$\alpha2$	$\alpha3$
Number of amino-acid residues	429	423	464
Molecular weight	48 000	48 000	52 000
Net charge at pH 6.8	+15	+11	+6
Number of potential glycosylation sites (Asn-X-Thr/Ser)	2	3	4

for the pure receptor, using the periodic acid Schiff's base reagent, showing that both the polypeptides contain carbohydrate. Also, the same pattern was found if the gels were run in the absence of reducing agents, indicating no disulphide bonding between adjacent subunits.

Attempts to determine the ratio of the $\alpha:\beta$ subunits by quantitative densitometric scanning of Coomassie-stained gels gave either $3\alpha:1\beta$ (7) or, later, $1\alpha:1\beta$ (16). This led to the proposal, together with the determination of the molecular weight of the receptor (see above), that it is a heterologous tetramer with an $\alpha_2\beta_2$ composition. This method of the quantification of the respective subunits is subject to error because it depends on the ability of each polypeptide to bind the protein stain to the same extent and this has been shown not to be always the case, one example is the δ-subunit of the nicotinic acetylcholine receptor. A more rigorous quantification of the subunit ratio could be determined by the N-terminal sequence analysis of the GABA$_A$ receptor α- and β-subunits. This has not been possible to date, however, since numerous attempts by several groups to determine the N-terminal amino-acid sequence by Edman degradation have yielded no information, due to chemically blocked N-termini. From the sequences of the α- and β-subunits, it is now known that the blocked N-termini are not intrinsic properties of the polypeptides but may be caused by post-translational modification or by reaction during the purification procedure.

A further complication in the determination of the subunit stoichiometry of the GABA$_A$ receptor by classical protein chemistry methods was the discovery, by molecular cloning, of isoforms of the (to date) α-subunit (4). Three forms of the α-subunit, $\alpha1$, $\alpha2$, and $\alpha3$, have been identified, and their properties are summarized in *Table 3*. It could not be excluded from the determination of the partial amino-acid sequence of the purified native receptor from cerebral cortex that all the three forms of the α-subunit are present in the isolated material and are not resolved by one-dimensional electrophoresis in 10% polyacrylamide gels. Because the respective α-subunit mRNAs have been shown to have different brain regional expression (17), they may even be present in varying amounts between GABA$_A$ receptor preparations from the same 'gross' brain regions. With the knowledge of the existence of the isoforms of the α-subunits from molecular cloning approaches, it has become possible to identify the presence of multiple bands in the purified receptor using higher resolution methods, such as polyacrylamide gradient gels. (18; see also Chapter 10, Section 3). The use of subunit-specific antibodies, together with recombinant DNA technology, will be important in the further elucidation of the receptor structure at the subunit level. Protocol 9 gives one method that we have used successfully for the production of anti-GABA$_A$ receptor sequence-specific polyclonal antibodies.

Protocol 9. Production of sequence-specific GABA$_A$ receptor polyclonal antibodies

This is the glutaraldehyde-coupling method which cross-links the peptide to the carrier protein by reaction with primary amines such as the N-terminal amino acid of the peptide or any lysines which may be present in the peptide sequence.

1. Dissolve the peptide, e.g. the bovine GABA$_A$ receptor N-terminal sequence of the α1 subunit, α1 1−15, sequence QPSLQDELKDNTTVF (5 mg), and the carrier protein keyhole limpet haemocyanin (5 mg) in 0.1 M NaHCO$_3$ to give a carrier concentration of 2 mg protein/ml.

2. Thaw a fresh vial of glutaraldehyde (Sigma, ultra-pure grade 1, 25% (v/v) stock) and add 5 μl to the peptide-carrier solution to give a final concentration of 0.05% (v/v) which is pale yellow in colour. Mix overnight at room temperature.

3. Add 0.277 ml of a 1 M glycine ethyl ester pH 8.0 stock solution to give a final concentration of 0.1 M glycine ethyl ester and leave the mixture to stand for 30 min at room temperature.

4. Add ice-cold acetone (12.4 ml, i.e. 4.5 vols) to the peptide-carrier conjugate and leave to precipitate for 30 min at −70°C. Pellet the conjugate by centrifugation at 10 000 g for 20 min. Decant the supernatant and air-dry the pellet. Resuspend the peptide−carrier conjugate in 0.9% (w/v) NaCl pH ∼7.0 to a carrier concentration of 1 mg/ml. Store frozen at −20°C until use.

5. For the production of polyclonal antibodies, for the primary immunization, emulsify the peptide−carrier conjugate (0.75 ml equivalent to 0.2 μmol peptide) with 0.35 ml Freund's Complete adjuvant. Inject a Dutch-belted rabbit at two sites intramuscularly with the emulsion.

6. For the second and subsequent immunizations, at 1 month intervals, emulsify 0.35 ml peptide−carrier with 0.35 ml Freund's Incomplete adjuvant. Inject the rabbit at two sites intramuscularly and bleed from the ear 7 days following the booster injections. Collect the serum by standard methods and assay for the production of anti-peptide antibodies and for their reactivity with denatured receptor in immunoblots, or for reactivity against native receptor by either enzyme-linked immunoadsorbent assay (ELISA) or soluble immunoprecipitation assay. Further details can be found in references 20 and 21.

3.4 The carbohydrate properties of the GABA$_A$ receptor

The carbohydrate moieties of the GABA$_A$ receptor have been studied by three different methods. These are:

(a) The ability of the receptor protein to bind to various lectin affinity columns;

(b) lectin binding to the α and β polypeptides by a technique based on Western blotting; and

(c) the effect of endoglycosidase digestion on the GABA$_A$ receptor.

Methods (a) and (c) can be applied to the receptor in its crude soluble form and give information on both the effect of carbohydrate removal on ligand-binding activity, the carbohydrate specificity of the receptor, and, if photoaffinity-labelled soluble receptor is used, the contribution by weight of the sugar portion to the receptor polypeptides. The advantge of method (b) is that, using the pure receptor, it gives information on the carbohydrate specificity of the individual polypeptides, which may be useful for further structural analyses of the receptor as were employed by Kubalek *et al.* (19) in the study of the nicotinic acetylcholine receptor.

Protocol 10 outlines the method that has been used in this laboratory for the determination of the molecular weight of the core α- and β-proteins following carbohydrate removal by *N*-glycanase treatment. The enzyme used was from Genzyme Corporation (agents: KochLight) and it hydrolyses asparagine-linked oligosaccharides from glycoproteins to give the free sugar and the peptide containing aspartic acid at the glycosylation site. A protocol used for N-glycanase digestion of mAChR peptides is given in Chapter 10, Section 5.

Protocol 10. *N*-Glycanase treatment of the purified GABA$_A$ receptor

1. Precipitate 3×200 μl samples of the purified GABA$_A$ receptor by the trichloroacetic acid preciptation method described in *Protocol 8*. Air-dry the protein pellet.

2. Add to the pellet in the following order with gentle mixing:
   - 100 μl sodium phosphate pH 8.6;
   - 11 μl 10% sodium dodecyl sulphate;
   - 50 μl 100 mM phenanthroline in methanol;
   - 50 μl 10.25% NP40;
   - 289 μl sodium phosphate, pH 8.6.

   Total volume = 500 μl.

3. Add 5 μl H$_2$O to 4°C and 37°C control samples. Add 2.5 units *N*-glycanase to test sample.

4. Incubate overnight at 37°C.

5. Precipitate the sample by the addition of an equal volume (505 μl) 24% (w/v) trichloroacetic acid and wash as described in *Protocol 8*.

6. Dissolve the protein pellet in gel-electrophoresis buffer and carry out SDS-PAGE under denaturing conditions as described in *Protocol 8*.

7. In cases where the effect of *N*-glycanase treatment on the ligand-binding properties of purified receptor were studied, the GABA$_A$ receptor was not precipitated and the *N*-glycanase reaction was carried out as above except in the absence of SDS.

The results obtained are depicted in *Figure 6* where the products of the reaction are visualized by a monoclonal antibody, 1A6, that was raised by Cleanthi Mamalaki

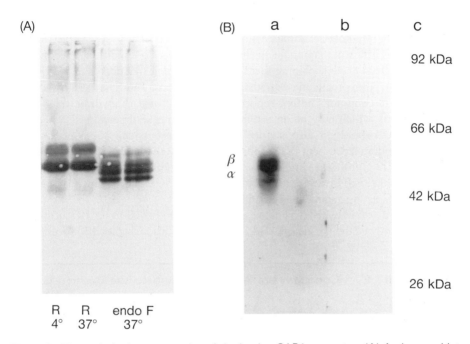

Figure 6. The carbohydrate properties of the bovine GABA$_A$ receptor. (A) An immunoblot before and after deglycosylation of the isolated GABA$_A$ receptor. The receptor was treated with *N*-glycanase as described in *Protocol 10*, subjected to SDS-PAGE under denaturing conditions, as described in *Protocol 8*, then probed with the monoclonal antibody 1A6, which recognizes both the α and β polypeptides (16), by immunoblotting. R4° and R37° are control samples incubated in the absence of the enzyme overnight at 4°C and 37°C, respectively. Endo F 37° are duplicate samples treated with *N*-glycanase. Note the change in the mobility of the α and β polypeptides. The three lower molecular weight bands correspond to non-deglycosylated α-subunit (M_r 53 kDa) and two deglycosylated products of the α-subunits with M_r 48 – 49 kDa and M_r 44 – 46 kDa, respectively, demonstrated by using a polyclonal α-subunit-specific GABA$_A$-receptor antibody (16). The top band corresponds to deglycosylated β-subunit with M_r 55 kDa. (B) [^{125}I]Concanavalin A lectin overlay of the purified bovine GABA$_A$ receptor as described in *Protocol 11*. The first lane (a) shows labelling with [^{125}I]Concanavalin A with radioactivity incorporation seen in the α, the β, and the α' polypeptides, where α' is a known proteolytic product of the α-subunit. Lane b shows labelling with [^{125}I]Concanavalin A in the presence of 0.2 M α-methyl mannoside and demonstrates the specificity of the lectin reaction. Lane c shows the position of the molecular weight standards.

and which recognizes both the α and the β polypeptide chains. The molecular sizes of the deglycosylated α and β polypeptides are M_r 44 000−46 000 (final product) and M_r 48 000−49 000, respectively. This compares to values for the mature α polypeptide of M_r 48 800 (429 residues, α1) and for β, M_r 51 400 (449 residues) (3). It should be noted that we find two products of α identified with an α-specific polyclonal antibody as a result of N-glycanase treatment. The amount of the intermediate product is dependent on the efficiency of the deglycosylation reaction. Also note from *Protocol 10* that the reaction is carried out under denaturing conditions. For the study of the effect of the removal of carbohydrate on the ligand-binding activities of the GABA$_A$ receptor, the reaction must be carried out in the absence of sodium dodecyl sulphate. In our experience firstly, the deglycosylation was even less efficient than under the standard denaturing conditions. Secondly, it was evident from protein staining that for control samples incubated at 37°C in the absence of enzyme, proteolytic degradation had occurred. Thus it is essential to carry out all control conditions before reporting any effect of carbohydrate removal on ligand binding which may be an artefact due to protease digestion.

Genzyme Corporation also supply an O-glycanase which removes O-linked oligosaccharides. Treatment of the purified GABA$_A$ receptor with neuraminidase followed by O-glycanase, which is the protocol recommended by Genzyme, gave no change in the mobility of α or β-subunits in SDS-PAGE under reducing conditions (F. A. Stephenson, unpublished results).

Protocol 11, outlines the procedure that we have followed in [^{125}I]Concanavalin A lectin overlay of the purified GABA$_A$ receptor (see also Chapter 10, Section 3.7).

Protocol 11. [^{125}I]Concanavalin A lectin binding to the purified GABA$_A$ receptor

The method of Bartles and Hubbard (23) is followed.

1. Iodinate 100 μl of 5 mg/ml Concanavalin in 10 mM Hepes, pH 7.4, by the same method described in *Protocol 5* but separate [^{125}I]Concanavalin A and [^{125}I] by gel filtration on a Sephadex G25 column (18 × 0.5 cm) equilibrated with 10 mM Hepes pH 7.4 containing 1 mM calcium chloride, 0.02% NaN$_3$, and 5 mg/100 ml bovine serum albumin. Store [^{125}I]Concanavalin A in aliquots at −20°C.

2. Precipitate purified GABA$_A$ receptor and carry out SDS-PAGE under denaturing conditions in 10% polyacrylamide slab gels as described in *Protocol 8*.

3. Transfer the protein to nitrocellulose filters by Western blotting. The transfer buffer is 25 mM Tris pH 8.3, containing 192 mM glycine, 0.1% (w/v) SDS, and 20% (w/v) methanol. The blotting is carried out at 30 V overnight at 4°C followed by 1 h at 50 V at 4°C (see Chapter 10, Section 3 for a detailed protocol).

4. Cut in half the nitrocellulose and rinse each for 10 min at 37°C with

physiological buffered saline (PBS). Pyrex oven dishes are useful for this and subsequent steps.

5. Incubate each filter for 30 min at room temperature with 2% (w/v) polyvinyl-pyrrolidine in PBS.

6. Aspirate the blocking buffer and pre-incubate one half with 2% (w/v) polyvinylpyrrolidine in PBS and the second with 0.2 M α-methylmannoside in 2% (w/v) polyvinylpyrrolidine/PBS for the control binding.

7. Add [^{125}I]Concanavalin A to both solutions such that the final concentration is ~6 × 10^5 cpm/ml. Incubate each for 4 h at room temperature with gentle shaking.

8. Wash each filter for 5 × 5 min with 2% (w/v) polyvinylpyrrolidine in PBS. Air-dry the filters and expose to X-ray film.

9. After the development of the X-ray film, the nitrocellulose filter can be stained for total protein. A sensitive stain method is the colloidal gold total protein stain from Bio-rad Laboratories. Thus rinse the filters with PBS followed by incubation with the colloidal gold protein stain until the protein bands appear. Rinse and air-dry.

Figure 6 shows the results of such an experiment where both the α and the β polypeptide chains were shown to bind specifically to the lectin and thus both possess the mannose/glucose residues. This method can also be adapted to other lectins with different carbohydrate-binding specificities (e.g. WGA). Available now are biotinylated lectins which, when used in combination with streptavidin-peroxidase, can be an alternative method to detect the specific lectin binding to the receptor of interest. An advantage of the lectin overlay procedure is that, in peptide-mapping studies of the receptor, it can be used to identify specifically which of the resultant cleavage products are glycopeptides. Since the sequence of the α and β polypeptide chains of the receptor and the glycosylation consensus sequence sites within these are now known, they are a useful tool in, for example, the location of specific ligand-binding domains.

4. Conclusions

The work described in this chapter has focused on the characterization of the molecular properties of the purified GABA$_A$ receptor of bovine cerebral cortex. It has considered primarily the determination of those properties of the receptor that are requisite both for the identification of the receptor complementary DNAs and biosynthetic reconstitution by the now conventional methods of partial protein-sequence determination followed by oligonucleotide probe screening of complementary DNA libraries and *in vivo* translation of the respective RNAs in the *Xenopus* oocyte (3). The wheel goes full circle though because the results of the molecular cloning of the GABA$_A$ receptor have shown that multiple isoforms of the

GABA$_A$ receptor polypeptides exist in the brain, and it would appear that the protein chemistry is far more complex than was predicted from the early purification studies. The availability of the respective subunit sequences, however, provides an important additional tool with which to study further the microstructure of the GABA receptors.

Acknowledgements

The work described in this chapter has been a team effort and I would like to acknowledge particularly Erwin Sigel, Cleanthi Mamalaki, Stefano Casalotti, Michael Duggan, and Eric Barnard who have all contributed to the work described herein. FAS holds a Royal Society University Research Fellowship in the Department of Pharmaceutical Chemistry, The School of Pharmacy, University of London.

References

1. Olsen, R. W. and Venter, J. C. (eds) (1986). *Benzodiazepine/GABA Receptors and Chloride Channels: Structural and Functional Properties.* Alan R. Liss, New York.
2. Stephenson, F. A. (1988). *Biochem. J.*, **249**, 21.
3. Schofield, P. R., Darlison, M. G., Fujita, N., Burt, D. R., Stephenson, F. A., Rodriguez, H., Rhee, L. M., Ramachandran, J., Reale, V., Glencorse, T. A., Seeburg, P. H., and Barnard, E. A. (1987). *Nature*, **328**, 221.
4. Levitan, E. S., Schofield, P. R., Burt, D. R., Rhee, L. M., Wisden, W., Kohler, M., Rodriguez, H., Stephenson, F. A., Darlison, M. G., Barnard, E. A., and Seeburg, P. H. (1988). *Nature*, **335**, 76.
5. Stephenson, F. A. and Barnard, E. A. (1986). In *Benzodiazepine/GABA Receptors and Chloride Channels: Structural and Functional Properties*, (ed. R. W. Olsen and J. C. Venter), p. 261. Alan R. Liss, New York.
6. Sigel, E., Mamalaki, C., and Barnard, E. A. (1982). *FEBS Lett.*, **147**, 45.
7. Sigel, E., Stephenson, F. A., Mamalaki, C., and Barnard, E. A. (1983). *J. Biol. Chem.*, **258**, 6965.
8. Sigel, E. and Barnard, E. A. (1984). *J. Biol. Chem.*, **259**, 7219.
9. Stauber, G. B., Ransom, R. W., Dilber, A., and Olsen, R. W. (1987). *Eur. J. Biochem.*, **167**, 125.
10. Bristow, D. R. and Martin, I. L. (1987). *J. Neurochem.*, **49**, 1386.
11. Rodriguez, H. (1985). *J. Chromatog.*, **350**, 217.
12. Froehner, S. C., Reiness, C. T., and Hall, Z. W. (1977). *J. Biol. Chem.*, **252**, 8589.
13. Meunier, J.-C., Olsen, R. W., and Changeux, J.-P. (1972). *FEBS Lett.*, **24**, 63.
14. Mamalaki, C., Barnard, E. A., and Stephenson, F.A. (1989). *J. Neurochem.*, **52**, 124.
15. Martin, R. G. and Ames, B. N. (1961). *J. Biol. Chem.*, **236**, 1372.
16. Mamalaki, C., Stephenson, F. A., and Barnard, E. A. (1987). *EMBO J.*, **6**, 561.
17. Wisden, W., Morris, B. J., Darlison, M. G., Hunt, S. P., and Barnard, E. A. (1988). *Neuron,* **1**, 937.
18. Olsen, R. W., Bureau, M., Ransom, R. W., Deng, L., Dilber, A., Smith, G., Khrestchatisky, T., and Tobin, A. J. (1988). In *Neuroreceptors and Signal Transduction.* (ed. S. Kito, T. Segawa, K. Kuriyama, M. Tohmaya, and R. W. Olsen), p. 1. Plenum Press, New York.

19. Kubalek, E., Ralston, S., Lindstrom, J., and Unwin, N. (1987). *J. Cell Biol.*, **105**, 9.
20. Duggan, M. J. and Stephenson, F.A. (1989). *J. Neurochem.*, **53**, 132.
21. Stephenson, F. A., Duggan, M. J., and Casalotti, S. O. (1989). *FEBS Lett.*, **243**, 358.
22. Wessel, D. and Flügge, U. I. (1984). *Anal. Biochem.*, **138**, 141.
23. Bartles, J. R. and Hubbard, A. L. (1984). *Anal. Biochem.*, **140**, 284.
24. Stephenson, F. A., Casalotti, S. O., Mamalaki, C., and Barnard, E. A. (1986). *J. Neurochem.*, **46**, 854.
25. Hulme, E. C. (ed.) (1990). *Receptor−Ligand Interactions, A Practical Approach*, IRL Press, Oxford, in press.
26. Hames, B. D. and Rickwood, D. (1986). *Gel Electrophoresis of Proteins, A Practical Approach*, IRL Press, Oxford.

9

Purification of the epidermal growth factor receptor from A431 cells

GEORGE N. PANAYOTOU and MARY GREGORIOU

1. Introduction

The EGF receptor is a 170 kDa membrane-spanning glycoprotein. An extracellular, EGF-binding domain is connected to the cytoplasmic catalytic domain by a single transmembrane link of 23 amino acids (1). Binding of EGF results in a variety of events, including clustering and internalization of EGF−receptor complexes, phosphorylation of cellular proteins and of the receptor itself at tyrosine residues, changes in ion fluxes, increase in protein synthesis, and finally, upon prolonged stimulation, DNA synthesis and cell division (2,3).

Advances in our understanding of the structure and function of the EGF receptor have been greatly facilitated by the availability of purified protein from the human epidermoid carcinoma cell line A431, which over-expresses the EGF receptor to approximately 2×10^6 per cell (4). A431 cells are grown *in vitro* and detergent extracts of the cells or crude membrane preparations are used for the purification of the receptor in a single affinity step. A method for the purification of the EGF receptor from A431 cells was first described by Cohen *et al.* (5,6) using EGF coupled to Affi-Gel-10 as an affinity matrix. More recently, with the availability of anti-EGF-receptor monoclonals, mostly raised against the human receptor, immunoaffinity methods have been developed and can be used instead of EGF affinity chromatography for the isolation of the human EGF receptor (7−9). Obviously, both epitope and species specificity of the monoclonal determine its applicability as a general affinity ligand. Methods have also been published for the purification of mouse EGF receptor from liver (10) and of human receptor from placenta (11).

Here we describe the purification of the A431 EGF receptor by affinity chromatography using EGF as an affinity ligand.

2. Preparation of the affinity matrix

2.1 Purification of EGF

EGF can be purified from male mouse submaxillary glands following the procedure described by Savage and Cohen (12) with minor modifications.

Protocol 1. Purification of EGF

1. Remove the submaxillary glands from 150−200 sacrificed adult male mice (>30 g body weight), freeze immediately in dry ice, and store at −70°C until required.

2. Thaw at 4°C and homogenize the tissue (30 g) in 120 ml 0.05 M acetic acid, pH 3.0, for 3 min at 4°C using an Ato-mix blender (or similar equipment, e.g. a polytron) at high speed.

3. Transfer the extract to a round-bottom flask and shell-freeze in an isopropanol/dry-ice bath. Thaw in a 37°C water bath (taking care that the temperature does not rise above 4°C) and centrifuge at 100 000 g for 30 min.

4. Filter the supernatant through glass wool to remove fat, and lyophilize overnight.

5. Dissolve in 10 ml 1 M HCl, dilute with 32 ml 0.05 M HCl and adjust the pH to 1.5 with NaOH. If necessary, clarify the extract by centrifugation at 100 000 g for 30 min at 4°C.

6. Equilibrate 100 g of Biogel P-10 (200−400 mesh, Bio-Rad) in 0.05 M HCl, 0.15 M NaCl, and pack into a 5 × 90 cm column at 4°C.

7. Apply the extract and develop in equilibration solution at a flow rate of 1 ml/min, collecting 16 ml fractions. EGF elutes after approximately two column volumes, in a small broad peak of the absorbance profile at 280 nm. The fractions containing EGF activity can be identified by displacement of [^{125}I]EGF binding on A431 cells (13).

8. Pool fractions, adjust the pH to 5−7 with ammonia and dialyse in 38 mm SpectraPor 6 dialysis tubing (2000 MW cut-off) against 1 mM ammonium acetate, pH 5.7, with two buffer changes overnight.

9. Lyophilize EGF, dissolve in a small volume of distilled water, and dialyse against 20 mM ammonium acetate.

10. Equilibrate a 2 × 25 cm DE52 column in 20 mM ammonium acetate at 4°C, apply the sample, and wash with the same buffer until no more protein is detected (approximately two column volumes).

11. Elute EGF with a linear gradient of 20−200 mM ammonium acetate (400 ml total volume), collecting 10 ml fractions.

12. Lyophilize the major EGF-containing peak and store at 4°C. The amount of protein recovered can be estimated by amino-acid analysis. Typically, 10−15 mg EGF are obtained (from 30 g of tissue) of purity higher than 98%.

2.2 Preparation of EGF – Affi-Gel

Affi-Gel-10 (Bio-Rad), a *N*-hydroxysuccinimide ester of a cross-linked agarose gel bead support, is the matrix of choice for preparation of an EGF-affinity column, as it shows little non-specific protein binding and is convenient to use.

Protocol 2. Immobilization of EGF on Affi-Gel-10

1. Dissolve EGF (10 mg) in 4 ml 10 mM HCl and adjust the pH to 8 with 1 M NaHCO$_3$. Dialyse against 3 × 1 l changes of 0.1 M NaHCO$_3$, pH 8.2, to remove low molecular weight compounds with free amino groups. Estimate the amount present either by amino-acid analysis or from the absorption of the solution at 280 nm.

2. Wash Affi-Gel-10 quickly with three bed volumes of isopropyl alcohol and three volumes of cold distilled water and equilibrate by washing with 10 volumes 0.1 M NaHCO$_3$, pH 8.2, at 4°C. This procedure should be completed in less than 20 min and the gels should not be allowed to dry during washing.

3. Weigh 5 g of wet gel and transfer to a screw-cap tube (e.g. ELKAY tube) containing the EGF solution. Mix thoroughly and rotate in an end-over-end mixer overnight at 4°C.

4. Collect unreacted EGF by filtering the suspension through a small Buchner funnel and washing the gel twice with 10 ml of 0.2 M glycine-HCl, pH 2.5. Estimate the amount of uncoupled EGF from the absorbance of the solution (pH 2.5) at 280 nm.

5. Wash the gel with 100 ml each of 0.5 M NaCl, 5 mM ethanolamine, pH 9.7, and 0.1 M NaHCO$_3$, pH 8.2.

6. Transfer the gel to a clean tube, add 0.1 ml of 1 M ethanolamine, pH 8.2, per ml of suspension and shake gently at room temperature for 2 h to block any unreacted sites on the gel.

7. Wash free from ethanolamine with at least 1 litre of Tris-buffered saline (TBS: 0.05 M Tris, 0.15 M NaCl, pH 7.5) and store the gel at 4°C in the presence of 0.05% sodium azide.

3. Solubilization and purification of the EGF receptor

3.1 Culture of A431 cells

Stock cultures of A431 cells are routinely maintained in DMEM (Dulbecco's Modified Eagle's Medium) supplemented with 5% newborn calf serum, penicillin (30 mg/l), and streptomycin (75 mg/l). The cells are seeded at a cell density of 2 × 10^6 cells per 75 cm^2 flask in 15−20 ml medium and kept under 5% CO$_2$−95% air in a humidified 37°C incubator for 4−5 days. The cells are subcultured at confluence. Stock cultures can be expanded to six 140 cm^2 dishes at 5 × 10^6 cells/dish, which in turn can be subcultured into roller bottles at 5 × 10^7 cells/bottle in 100 ml medium. The medium is changed after 3−4 days and the cells harvested at confluence (6−7 days). For consistent yields, glass roller bottles should be cleaned with trypsin, washed with detergent, thoroughly rinsed in distilled water, sterilized by autoclaving, and coated with 0.1% gelatin at 4°C before use. Typically, 3−5 × 10^8 cells can be obtained per roller bottle. Details of cell-culture techniques are given in reference 20.

Cells are harvested in the following way:

Protocol 3. Harvesting cultured A431 cells

1. Decant the medium into disinfectant, rinse the cells twice with 50 ml of 5 mM EDTA/PBS, followed by a 10−15 min exposure to fresh EDTA/PBS solution. This procedure results in cell detachment and also protects the EGF receptor from Ca^{2+}-activated proteolysis.

2. Pool cell suspensions into 250 ml centrifuge tubes, collect cells by pelleting at 1000 g for 10 min at 4°C, and wash once with PBS.

A crude membrane fraction can be prepared at this point using the method of Thom *et al.* (14), except that calcium is omitted from all buffers:

Protocol 4. Membrane preparation from A431 cells

1. Wash the cells once with harvesting solution (50 mM boric acid, 150 mM NaCl, pH 7.2) and scrape into the same solution using a rubber policeman.

2. Collect cells by centrifugation at 300 g for 10 min.

3. Resuspend in two pellet volumes of harvesting solution, add slowly with stirring to 100 volumes of extraction buffer (20 mM boric acid, 0.2 mM EDTA, pH 10.2), leave stirring for 10 min, add 8 volumes of 0.5 M boric acid, pH 10.2, and stir for a further 5 min.

4. Filter through nylon gauze and centrifuge the suspension at 450 g for 10 min at 4°C. Discard the pellet and centrifuge the supernatant at 12 000 g for 30 min at 2°C.

5. Resuspend the pellet in a small volume of phosphate-buffered saline, layer on top of a sucrose solution cushion [35% (w/w) sucrose in PBS, 2 ml], and centrifuge at 24 000 g for 1 h at 2°C.

6. Remove the membrane fraction from the sucrose−PBS interface, resuspend in PBS, and collect membranes by centrifugation at 100 000 g for 10 min at 2°C. Store the membranes at −70°C until used.

Either whole cells or the membrane fractions can be used for the subsequent solubilization and purification of the EGF receptor.

3.2 Extraction of the EGF receptor

The EGF receptor can be solubilized from A431 cells or membranes using a variety of detergents, e.g. Triton X-100, NP40, sodium deoxycholate, and *n*-octylglucoside (see Chapter 1, Section 2). Membrane preparations require smaller quantities of solubilizing agent (10 mg detergent per mg protein) and are therefore a preferable starting material when using expensive detergents. Triton X-100 has been used widely for the purification and characterization of the properties of the receptor, and is required at a minimum concentration of 0.2% for EGF-binding activity (15) and at 0.2−0.5% for maximal kinase activity (16).

Protocol 5. Solubilization of EGF receptor

1. Resuspend the cell pellet in one volume of TBS and add gradually with stirring to solubilization buffer [10 mM Tris pH 7.4, 1% (v/v) Triton X-100, 1 mM EDTA, 115 mM NaCl, 25 mM benzamidine, and 10% (v/v) glycerol] at a ratio of 2×10^6 cells/ml (*c.* 100 pellet volumes).

2. Stir the suspension for 20 min at room temperature and remove debris by centrifugation at 3000 *g* for 5 min at 4°C.

3. Clarify the supernatant at 100 000 *g* for 1 h at 2°C. The supernatant can be used immediately or stored at −70°C for several months without loss of activity.

3.3 Affinity chromatography using EGF – Affi-Gel-10

Protocol 6. Affinity chromatography using EGF – Affi-Gel-10

1. Mix the A431 cell extract (200 ml prepared from 2×10^8 cells) with the EGF-Affi-Gel beads (5 g) and agitate for 1 h at room temperature in a polypropylene centrifuge tube.

2. Wash the beads in a sintered-glass Buchner funnel with 100 bed volumes of wash buffer (10 mM Tris pH 7.5, 1% Triton X-100, 2 mM EDTA, 25 mM benzamidine, 10% glycerol) at 4°C and then with 20 bed volumes of a 0.2% Triton, 10% glycerol solution adjusted to pH 7.5 with NaOH.

3. Transfer the gel to a 10 ml Bio-Rad column and elute the receptor with six column volumes of 5 mM ethanolamine, 0.2% Triton, 10% glycerol, pH 9.7 at 4°C.

4. Immediately bring the eluates to pH 8.0 using a 2 M Tris-HCl solution, pH 6.8 (final concentration 20 mM Tris), and add benzamidine to 25 mM.

5. Pool the eluted material and concentrate to the required volume by pressure ultrafiltration (XM-50 membrane filters, Amicon).

6. Store the receptor at −70°C.

7. Wash the EGF−Affi-Gel column with at least 50 volumes of PBS and store in PBS/0.05% sodium azide.

Elution of the EGF receptor can also be accomplished in the presence of detergents having a higher critical micellar concentration, such as *n*-octylglucoside, to enable reconstitution of the receptor in artificial lipid bilayers by detergent dialysis (17). In this case, *step 3* above should be modified as follows:

3a Wash the gel with at least five bed volumes of 40 mM *n*-octylglucoside, 10% glycerol, pH 7.5. Elute with a solution of 5 mM ethanolamine, 40 mM *n*-octylglucoside, 10% glycerol, pH 9.7.

Detergent exchange and concentration of the receptor to a small volume can also be accomplished in one step using anion-exchange chromatography:

Protocol 7. Detergent exchange by DEAE chromatography

1. Mix the eluted receptor with 0.5 ml DEAE anion-exchange gel (e.g. DE52) equilibrated with 20 mM Tris-HCl pH 7.5, 0.2% Triton X-100, 10% glycerol and agitate for 20 min at 4°C.

2. Collect the exchanger by centrifugation (600 *g*, 3 min) and pack into a 1 ml disposable syringe plugged with siliconized glass wool.

3. Exchange detergent, if required, as described in *step 3a* above.

4. Recover the receptor by eluting with 0.4 M NaCl in 20 mM Tris pH 7.5, 10% glycerol, 25 mM benzamidine and the appropriate concentration of detergent, as above.

It should be noted that the combination of benzamidine and sodium deoxycholate causes precipitation of material and one or the other should be avoided. Deoxycholate should be used at pH 8.0.

The amount of receptor obtained can be most conveniently estimated by radioimmunoassay, exactly as described by Gullick *et al.* (18). The same method can be used to estimate the affinity of the purified material for EGF:

Protocol 8. Estimation of EGF receptor concentration

1. Prepare dilutions of the EGF receptor to a final volume of 0.2 ml in 20 mM Tris pH 7.5, 0.2% Triton, 150 mM NaCl, 10% glycerol buffer containing final concentrations of 200 nM [^{125}I]EGF and 10 μg/ml monoclonal antibody R1 (both [^{125}I]EGF and R1 are available from Amersham). For dissociation constant estimations use a range of [^{125}I]EGF concentrations (0.2−100 mM).

2. Incubate for 1 h at room temperature and then add 20−30 μl of a 1:1 suspension of protein A−Sepharose and mix for a further 30 min at room temperature.

3. Wash the beads three times with ice-cold assay buffer and count the associated radioactivity. Determinations should be made in triplicate and non-specific binding values, obtained in the presence of a 200-fold excess of unlabelled EGF, should be subtracted.

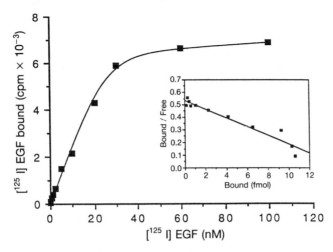

Figure 1. The binding of [^{125}I]EGF to EGF-affinity purified receptor was measured by radioimmunoassay. The specific activity of ligand was 1.1 × 10^5 c.p.m./ng. Background values were obtained for each point by incubating with a 100-fold excess of unlabelled EGF. Insert: Scatchard analysis of the binding data.

Table 1. Purification of the EGF receptor

Number of cells:	2 × 10^8
Theoretical amount of EGF receptor:	120 μg
EGF receptor in 100 000 g lysate:	95 μg
EGF receptor bound on EGF − Affi-Gel:	(17 μg)^a
EGF receptor eluted:	10 μg
% yield:	8.3%

^{a} The amount of EGF receptor bound on the column is an extrapolated figure, based on calculations from a different experiment: EGF receptor was prepared from cells biosynthetically labelled with [^{35}S]methionine, using a small amount of EGF − Affi-Gel. After elution, the radioactivity of the purified receptor was measured by scintillation counting and compared with the radioactivity remaining on the gel.

Determination of protein content by adaptations of the Lowry assay or dye binding assays are not precise because of detergent interference and the low levels of purified protein usually available.

Figure 1 shows a representative binding curve obtained using radioimmunoassay. Scatchard analysis gives a single apparent dissociation constant (K$_d$) of 3.2 × 10^{-8} M. The yield of EGF receptor varies in different preparations, but typically is 10 μg from 2 × 10^8 A431 cells. *Table 1* summarizes the results of a typical purification. The relatively low yield of receptor appears to be due mainly to incomplete binding on the EGF−Affi-Gel, rather than to inefficient elution. Non-adsorbed material can be used for a second round of purification; however, the EGF receptor obtained in this way is usually less pure, with the appearance of proteolytic

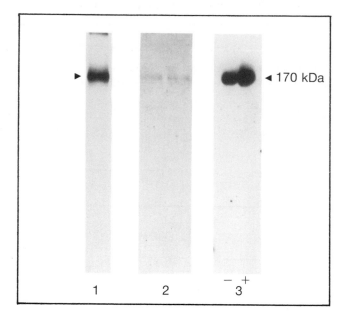

Figure 2. Purification of the EGF receptor by EGF-affinity chromatography. Lane 1, Fluorograph of purified, biosynthetically labelled EGF receptor after SDS-PAGE (7.5% gel); lane 2, Coomassie Blue staining of purified EGF receptor; lane 3, autophosphorylation of the purified EGF receptor in the absence or presence of EGF [see *Receptor-Effector Coupling, A Practical Approach* (21) for a description of the methodology of tyrosine phosphorylation assays].

degradation products (150 kDa form) and therefore this procedure is not recommended.

The purity of the EGF-receptor preparation can be estimated by densitometric scanning of either Coomassie Blue stained gels or of fluorographs of biosynthetically labelled receptor ([^{35}S]methionine) after SDS-PAGE (see Chapter 10, Section 3 for protocols). Using the above purification method, the receptor is found routinely to be homogeneous by both techniques (*Figure 2*).

A major disadvantage of the EGF-affinity purification protocol is the significant loss of tyrosine protein kinase activity of the receptor usually observed. This is presumably due to the high pH conditions during elution and therefore minimization of the exposure time to these conditions is required for recovery of kinase activity (*Figure 2*, lane 3). An alternative elution procedure can be followed to obtain more active EGF receptor kinase:

Protocol 9. Elution of active EGF receptor tyrosine kinase

1. Complete all washing procedures as described above.
2. Incubate the gel with one bed volume of 0.2% Triton, 10% glycerol, pH 7.5, containing 1 mg/ml EGF.

3. Shake for 2 h at 4°C and collect the eluted material.

This procedure results in a purified EGF-receptor preparation containing large amounts of EGF which may hinder following experiments.

Provided that relatively large amounts of pure ligand are available, the above affinity purification procedure can be modified for the purification of other receptor tyrosine kinases. One relevant example is the purification of the insulin receptor kinase by insulin-affinity chromatography following a wheat-germ agglutinin chromatography step (19).

References

1. Ullrich, A., Coussens, L., Hayflick, J. S., Dull, T. J., Gray, A., Tam, A. W., Lee, J., Yarden, Y., Libermann, T. A., Schlessinger, J., Downward, J., Mayes, E. L. V., Whittle, N., Waterfield, M. D., and Seeburg, P. H. (1984). *Nature*, **309**, 418.
2. Carpenter, G. and Cohen, S. (1979). *Ann. Rev. Biochem.*, **48**, 193.
3. Carpenter, G. (1987). *Ann. Rev. Biochem.*, **56**, 881.
4. Fabricant, R. N., DeLarco, J. E., and Todaro, G. J. (1977). *Proc. Natl Acad. Sci. USA*, **74**, 565.
5. Cohen, S., Carpenter, G., and King, L. (1980). *J. Biol. Chem.*, **255**, 4834.
6. Cohen, S. (1983). *Meth. Enzymol.*, **99**, 379.
7. Parker, P. J., Young, S., Gullick, W. J., Mayes, E. L. V., Bennett, P., and Waterfield, M. D. (1984). *J. Biol. Chem.*, **259**, 9906.
8. Weber, W., Bertics, P. J., and Gill, G. N. (1984). *J. Biol. Chem.*, **259**, 14631.
9. Yarden, Y., Harari, I., and Schlessinger, J. (1985). *J. Biol. Chem.*, **260**, 315.
10. Cohen, S., Fava, R. A., and Sawyer, S. T. (1982). *Proc. Natl Acad. Sci. USA*, **79**, 6237.
11. Hock, R. A., Nexo, E., and Hollenberg, M. D. (1980). *J. Biol. Chem.*, **255**, 10737.
12. Savage, C. R. and Cohen, S. (1972). *J. Biol. Chem.*, **247**, 7609.
13. Haigler, H., Ash, J. F., Singer, S. J., and Cohen, S. (1978). *Proc. Natl Acad. Sci. USA*, **75**, 3317.
14. Thom, D., Powell, A. J., Lloyd, C. W., and Rees, D. A. (1977). *Biochem. J.*, **168**, 187.
15. Carpenter, G. (1979). *Life Sci.*, **24**, 1691.
16. Hanai, N., Nores, G., Torres-Mendez, C. -R., and Hakomori, S. -I. (1987). *Biochem. Biophys. Res. Comm.*, **147**, 127.
17. Panayotou, G. N., Magee, A. I., and Geisow, M. J. (1985). *FEBS Lett.*, **183**, 321.
18. Gullick, W. J., Downward, D. J. H., Marsden, J. J., and Waterfield, M. D. (1984). *Anal. Biochem.* **141**, 253.
19. Petruzzelli, L. M., Herrera, R., and Rosen, O. M. (1984). *Proc. Natl Acad. Sci. USA*, **81**, 3327.
20. Freshney, R. I. (ed.) (1986). *Animal Cell Culture, A Practical Approach*, IRL Press, Oxford.
21. Hulme, E. C. (ed.) (1990). *Receptor-Effector Coupling, A Practical Approach*, IRL Press, Oxford.

10

Peptide mapping and the generation and isolation of sequenceable peptides from receptors

M. WHEATLEY

1. Introduction

As increasing numbers of receptors are purified to homogeneity, research interests are aimed at elucidating structural information about the receptor proteins themselves. This encompasses obtaining general sequence information often used for constructing the probes required for cloning the receptors, and in addition investigating the location within the molecule of functionally important amino-acid residues or moieties. A popular approach to this is site-directed mutagenesis of a residue or region of interest, followed by expression of the protein and its characterization. By virtue of its indirect approach, it is sometimes difficult to relate any observed change in function with a given mutation. A complementary and synergistic approach is to chemically modify the pure receptor directly, and then analyse the protein by peptide-mapping studies and protein-sequencing techniques. The aim of this chapter is to provide the information necessary for peptide mapping receptors utilizing the low-technology, and therefore relatively cheap, high-resolution system of sodium dodecylsulphate polyacrylamide gel electrophoresis (SDS-PAGE). Furthermore, information will be presented pertinent to the preparation and purification of receptor-derived peptides suitable for sequencing. It is not within the scope of this chapter to describe the actual sequencing procedure itself. When appropriate, examples from our research on muscarinic acetylcholine receptors (mAChRs) will be presented.

2. Sample preparation for peptide mapping

2.1 Radioiodination

Throughout the course of analytical, and sometimes preparative, procedures, it can be useful to have a means of labelling the receptor in a manner that facilitates following the protein through various protocols, and allows cleavage of the original protein

213

to generate labelled products. Radioiodination with ^{125}I forms a very sensitive and generally applicable means of doing this. By virtue of this sensitivity, the inadvertent iodination of extraneous protein must be avoided by using reagents and water of the highest grade of purity available (e.g. purchase HPLC-grade water), and by the investigator wearing plastic gloves. Several methods have been developed for radioiodinating proteins. The most popular protocols are described below. It should be noted that although radioiodination is employed as a general label of protein, the incorporation of label is not uniform throughout the protein but is entirely dependent on the distribution of the reacting iodinatable residues (1). For example, the 111-amino-acid protein Thy-1 from rat has only two tyrosine residues to iodinate and thus has proved difficult to radiolabel. In addition, the microenvironments of amino acids influence their reactivity with respect to iodination, and consequently the same residue in different locations within the receptor protein will be iodinated to varying degrees. An extreme example of this is the increased incorporation of ^{125}I into a protein after it has been fully denatured when compared to the incorporation observed with the native conformation of the protein. Obviously in the native conformation potentially reactive amino-acid residues are buried within the hydrophobic core of the molecule, and in consequence are not readily iodinated. Therefore, radioiodination undertaken in this laboratory utilizes receptor protein that has been fully unfolded by the following protocol:

Protocol 1. Denaturing receptor protein

1. Add a 10% (w/v) solution of SDS (use specially purified SDS, AnalaR Biochemical grade, BDH) to the sample to give a final concentration of 1% (w/v). SDS can be recrystallized by the method of Hunkapiller *et al.* (25).

2. Warm the sample at 60°C for 15 min. The receptor samples are not boiled as reported by some laboratories, so as to avoid the possible formation of protein aggregates (2).

Incorporation of ^{125}I into a protein is commonly achieved by two basic methodological approaches viz. oxidizing techniques and conjugation of a prelabelled structure to the protein.

2.2 Oxidizing methods of radioiodination

The main variable parameter in these methods is the oxidizing agent utilized (*Figure 1*); however, the underlying principle remains the same. Essentially, the radioiodide is oxidized to generate cationic iodine which, in the presence of protein, can be incorporated predominantly as iodotyrosine but also as iodohistidine. *Important:* it should be noted that during the radioiodination protocols described below, radioactive iodine vapour can be evolved. Consequently, the procedures should be executed in a fume cupboard and appropriate precautions taken.

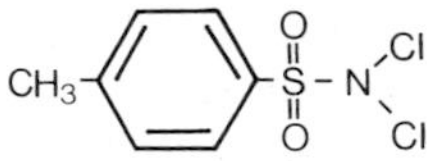

Figure 1. The structures of oxidants routinely employed for radioiodinating proteins.

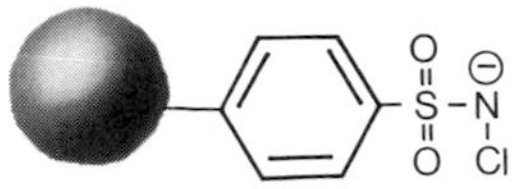

2.2.1 Chloramine-T

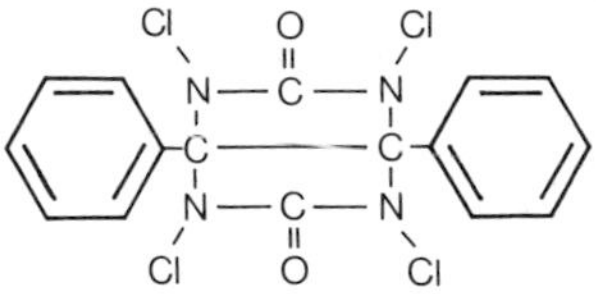

This method is still widely used, and has changed little since its introduction (3). The solution of the commercially available oxidant, chloramine-T (Pierce), should be freshly prepared for each iodination. A basic method is presented, but optimization to the receptor protein under study may be required. Furthermore, it is usually beneficial to keep the overall reaction volume low (100−200 µl).

Protocol 2. Radioiodinating protein using chloramine-T

1. Dispense 250 µCi of Na^{125}I into a 1.5 ml plastic microcentrifuge tube.
2. Add 25 µl of 0.5 M phosphate buffer, pH 7.5.
3. Add the fully denatured receptor sample in a small volume (25−150 µl).
4. Dissolve chloramine-T in 50 mM phosphate buffer, pH 7.5, to give a fresh stock solution of 5 mg/ml. Add 25 µl of this stock solution.
5. Incubate for 2 min at ambient temperature.
6. Stop the reaction by adding 100 µl of a 1.2 mg/ml sodium metabisulphite solution.

The addition of reducing agent, at *step 6*, inactivates the oxidizing agent, and causes the conversion of [^{125}I]iodine to [^{125}I]iodide, thereby stopping the reaction.

2.2.2 Iodo-beads

This protocol utilizes the oxidant *N*-chloro-benzene sulphonamide immobilized onto 2.8 mm diameter polystyrene beads (4). These are available commercially from

Pierce Chemical Company under the trade name 'Iodo-beads'. It can be seen from the structures presented in *Figure 1* that Iodo-beads essentially provide an immobilized form of chloramine-T. This endows this procedure with increased ease of execution, as to initiate iodination the reactants are exposed to an Iodo-bead, and to terminate the reaction the reactants can be rapidly withdrawn from the bead using a pipette. In addition, the need for fresh solutions and reducing agent is averted. We have found Iodo-beads to be the method of choice for routine radioiodination of receptor.

Protocol 3. Radioiodinating protein using Iodo-beads

1. Dispense 250 μCi of Na^{125}I into a 1.5 ml plastic microcentrifuge tube.
2. Add 50 μl of 100 mM phosphate buffer, pH 7.5.
3. Add one Iodo-bead. Two Iodo-beads can be added to each tube if it is found that this increases incorporation of label. Incubate for 5 min, with occasional stirring, at room temperature.
4. Add the denatured receptor sample (50 – 150 μl).
5. Incubate at ambient temperature for 15 min with occasional stirring.
6. Draw the sample off the Iodo-bead using a pipette.

2.2.3 Iodo-gen

By virtue of its hydrophobic properties 1,3,4,6-tetrachloro-3α,6α-diphenylglycoluril (*Figure 1*) is virtually insoluble in water, and can be used to coat a reaction vessel, thereby functioning as a solid-phase oxidant. This compound is marketed by Pierce Chemical Company under the trade name 'Iodo-gen'. Iodination is terminated by simply withdrawing the reactants from the coated vessel. As contact between the protein being labelled and the oxidant is kept to a minimum, it has been proposed that oxidative damage to the protein is minimized. However, it has also been reported (5) that Iodo-gen is solubilized by detergents. This is significant when studying receptor proteins which are transmembrane, integral proteins, and therefore require detergent to remain in solution.

Protocol 4. Radioiodinating protein using Iodo-gen

Reaction vessels can be prepared in batch form in advance as follows:

1. Dissolve 1 mg/ml of Iodo-gen in dichloromethane or chloroform.
2. Aliquot 50 μl into plastic microcentrifuge tubes and allow the tubes to dry.

A basic iodination procedure is:

3. Add 50 μl of 100 mM phosphate buffer, pH 7.5.
4. Add 250 μCi of Na^{125}I.

5. Add the denatured receptor protein (50–100 µl) and incubate at room temperature with occasional mixing for 15 min.

6. Withdraw the reaction mixture to terminate the iodination.

A general complication of using oxidative procedures for radioiodination is that the oxidant can damage the receptor protein, and amino acids can be oxidized. This can have severe ramifications for subsequent peptide mapping and sequencing strategies. One such modification is the oxidation of methionine residues. As shown in *Figure 2*, the sulphur atom in methionine can be oxidized to form methionine sulphoxide. This reaction can occur in the presence of the oxidizing agents routinely used for the radioiodination of proteins. The sulphoxide can be further oxidized to methionine sulphone; however, this peroxidation reaction usually requires a more potent oxidant than is provided in the protocols described. The effect of this oxidation of methionine on peptide-mapping protocols utilizing cyanogen bromide is detailed in Section 3.2.

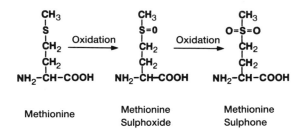

Figure 2. Oxidation of methionine.

2.3 Conjugation of a prelabelled structure

The most commonly used agent in this type of radioiodination is the Bolton–Hunter reagent (*Figure 3*), which can be obtained preradiolabelled from Amersham International plc. This provides an alternative strategy for radioiodinating receptors and does not require oxidizing agent (6). It is often utilized for labelling proteins which do not possess the reactive residues required for the procedures presented in the preceding section (Section 2.2), or when these residues are inaccessible or not very reactive. Alternatively, it is a useful protocol if the receptor protein being studied is particularly sensitive to oxidative damage. Essentially, the label is incorporated by coupling to amino groups in the protein, which cleave the ester linkage of the reagent (*Figure 3*).

Protocol 5. Radioiodinating protein using Bolton – Hunter reagent

1. Dispense 40 µl of the [^{125}I]Bolton–Hunter reagent (in benzene) into a microcentrifuge tube and dry under a stream of nitrogen. *Care:* benzene is carcinogenic so perform the drying in a fume cupboard.

2. Gel-filter the receptor protein into 0.1% (w/v) SDS, 25 mM sodium tetraborate, pH 8.5 with HCl. The detergent is required to keep the hydrophobic receptor soluble.

3. Add 10% (w/v) SDS solution to give a final concentration of 1% (w/v), and denature the receptor by incubating at 60°C for 15 min.

4. Add the sample in 250 μl to the reagent tube, and incubate at room temperature for 3 h.

5. Gel-filter the sample into buffer containing 0.1% (w/v) SDS to remove excess reagent (see following section).

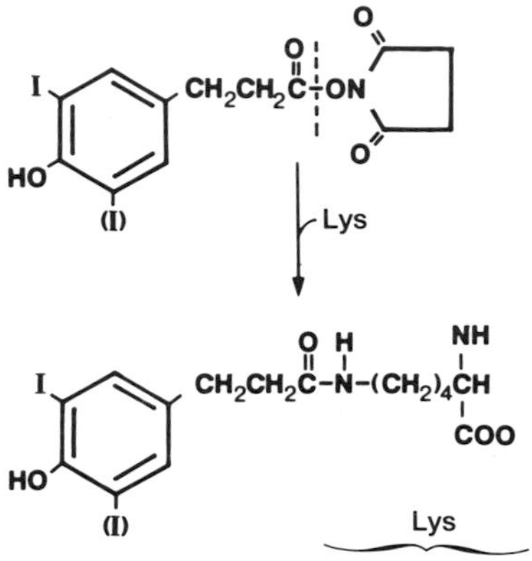

Figure 3. [^{125}I]Bolton – Hunter reagent labelling a Lys residue. The dotted line indicates the bond cleaved during the reaction.

2.4 Removal of free radioiodine

After the procedures given in Section 2.2, it is necessary to remove the non-incorporated iodide. This is achieved quickly and conveniently using a 2 ml Sephadex G50 fine column (diameter approximately 8 mm) pre-equilibrated with phosphate buffer, pH 7.4, containing 0.1% (w/v) SDS. The gel-filtration steps described in this and subsequent procedures are conveniently carried out using disposable polypropylene or polycarbonate columns. We have found the disposable 10 ml plastic columns marketed by Kontes very useful after inserting a sinter cut from porous polythene (Amicon Ltd). We cut the sinter using a tool resembling a cork borer mounted in a power drill. These columns can be effectively cleaned prior to use by soaking in 35% (v/v) nitric acid, followed by thorough washing with deionized water. In general, it seems that problems of adsorption of receptor proteins to surfaces

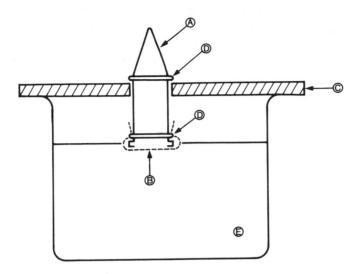

Figure 4. Arrangement for removing free iodine from a small sample by dialysis. (A) microcentrifuge tube; (B) dialysis membrane; (C) test-tube rack; (D) ring of flexible plastic tubing; (E) dialysis buffer.

is avoided if plastic rather than glassware is utilized. The radioiodinated protein elutes from these columns in the void volume. Because some free iodide co-elutes with the protein, possibly due to its partitioning into detergent micelles, dialysis can be utilized instead of, or supplementary to, gel-filtration. The dialysis should be against a large excess (1000-fold if possible) of buffer containing 0.1% (w/v) SDS, and this should be changed 3 or 4 times. Obviously, waste buffer from the dialysis will be radioactive so appropriate disposal procedures should be implemented. The sample can be dialysed in a bag of dialysis tubing; however, handling losses can be reduced if the sample, or protein fraction from the gel-filtration, is collected in a microcentrifuge tube, and this is arranged as in *Figure 4*.

Protocol 6. Dialysis of small samples

1. Remove the cap from the microcentrifuge tube containing the sample, ensuring that no sharp edges result.
2. Cover the top of the tube with a small piece of dialysis membrane (see Section 6 for pretreatment of dialysis tubing). Fold the membrane back over the tube and secure with a ring made by cutting the end off flexible plastic tubing of an appropriate diameter to ensure a tight fit around the microcentrifuge tube. This ring is most conveniently located by first stretching it over a test-tube of a slightly larger diameter than the microcentrifuge tube, placing the microcentrifuge tube into the mouth of the test-tube, then easing the ring off the test-tube and around the sample tube.

3. Insert this sample tube into a suitable rack, then place another ring onto the tube on the underside of the rack to locate the tube securely.

4. Invert the rack so that the sample is resting on the dialysis membrane. Then place the rack over the dialysis vessel such that the membrane is just under the dialysis buffer. Dialyse for 24 h with stirring. Change the buffer at regular intervals.

5. When dialysis is complete, spin the microcentrifuge tube to collect the sample in the bottom, prior to removing the dialysis membrane.

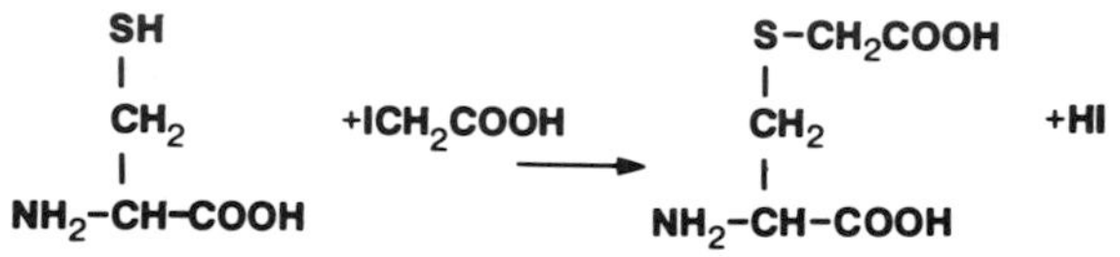

$$\underset{\underset{\displaystyle NH_2-CH-COOH}{\overset{\displaystyle |}{}}}{\overset{\underset{\displaystyle |}{\overset{\displaystyle SH}{}}}{CH_2}} \quad +ICH_2COOH \quad \xrightarrow{\hspace{2cm}} \quad \underset{\underset{\displaystyle NH_2-CH-COOH}{\overset{\displaystyle |}{}}}{\overset{\underset{\displaystyle |}{\overset{\displaystyle S-CH_2COOH}{}}}{CH_2}} \quad +HI$$

Figure 5. S-carboxymethylation of fully reduced cysteine by iodoacetate.

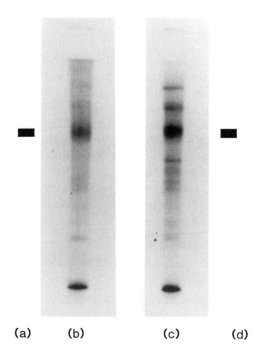

(a)　　(b)　　　　(c)　　(d)

Figure 6. The effect of reduction/S-carboxymethylation on the migration properties of polypeptides on SDS-PAGE. Radiolabelled polypeptides present in a semi-purified preparation of mAChR are analysed on a 10% polyacrylamide gel. Lanes b and c are the same sample without and with reduction/S-carboxymethylation, respectively. Lanes a and d mark the position of fluoresceinated BSA standard.

It should be noted that the employment of the inverted-tube dialysis method is restricted to use with small ($100-200$ μl) samples, where the sample forms a thin layer over the membrane. Larger samples should be dialysed in a bag which allows a greater area of contact with the dialysis buffer.

2.5 Reduction and S-carboxymethylation

It has been our experience during the course of purifying the mAChR, that the handling characteristics of proteins are improved after full reduction and S-carboxymethylation. The rationale for this series of manipulations is to block cysteine residues (see *Figure 5*) which can cause non-specific aggregation of proteins via intermolecular disulphide-bond formation. In the absence of S-carboxymethylation this phenomenon is facilitated by the concentrating or stacking effect of the discontinuous gel system of Laemmli (7), which is routinely used for peptide mapping (see Section 3) and preparing peptides for sequencing (see Section 6). *Figure 6* shows the effects of reduction and S-carboxymethylation on partially purified mAChR. It can be seen that the protein bands are sharper, and the formation of high molecular weight aggregates evident at the top of the gel in the non-S-carboxymethylated sample has been prevented.

Protocol 7. Reduction and S-carboxymethylation of proteins

1. Aliquot the receptor sample ($50-500$ μl) into plastic microcentrifuge tubes.

2. Add 10% (w/v) SDS solution to give a final SDS concentration of 1% (w/v). Warm the sample at 60°C for 15 min to fully denature the protein. Use AnalaR Biochemical or other high grade SDS (BDH Chemicals Ltd, Calbiochem, Pierce).

3. Add 2 M Tris-HCl buffer, pH 7.9 (at room temperature) to give a final concentration of 0.5 M. This ensures a slightly alkaline pH throughout the procedure.

4. Add 100 mM dithiothreitol (DTT) to give a final concentration of 10 mM. This solution should be freshly made each time using high-purity DTT supplied by Calbiochem.

5. Place the sample tube into a 30 ml Universal container with a screw-cap supplied by Sterilin Ltd. After flushing the pot with nitrogen via two hypodermic needles through the cap (see *Figure 7*), seal the vessel with tape. Incubate at 37°C for 2 h.

6. Remove the sample tube and cool it on ice to approximately 4°C. Add 100 mM iodoacetic acid (IAA) to give a final concentration of 22 mM IAA. Incubate in the dark at 4°C for 2 h. The IAA solution should be freshly made using IAA specially purified for biochemical work (e.g. BDH Chemicals Ltd). This provides a small excess of IAA which works well with the mAChR; however, some groups use a larger excess of IAA. The excess employed may have to be empirically determined.

7. The S-carboxymethylated protein can then be separated from the reagents by gel-filtration on Sephadex G50 (fine) equilibrated with buffer containing 0.1% (w/v) SDS.

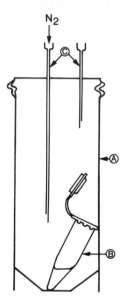

Figure 7. Arrangement for transferring the sample to a nitrogen atmosphere. See the text for further details. (A) Universal container with screw cap; (B) sample in microcentrifuge tube; (C) hypodermic needles.

If you wish to carboxymethylate free sulphydryl groups, but preserve native disulphide bonds, add the iodoacetic acid and Tris-Cl to the receptor sample before the SDS, and omit the reduction step. This procedure can be used to isolate disulphide-linked peptides which can then be separated from one another later on by further reduction and re-carboxymethylation. An alternative to the use of IAA is to use *N*-ethyl maleimide or iodoacetamide.

3. One-dimensional peptide mapping

The underlying principle of both one- and two-dimensional peptide mapping is that the protein under investigation is cleaved in a highly specific fashion to yield reproducibly a series of discrete products, the size of which is directly related to the original protein sequence. In this way proteins can be probed to establish the existence of any sequence homology with suspected related proteins, or to analyse precursor/product relationships. Furthermore, if functionally important residues or moieties can be irreversibly labelled, then peptide mapping can supply information on the location of the residues modified by these introduced labels. The precise position of modified residues within the protein sequence can then be established by sequencing the proteolytic peptides (see Section 6). The mapping procedure as originally developed by Cleveland and colleagues (8) incorporates three processes:

(a) generation of discrete products by limited cleavage of the fully denatured receptor; followed by

(b) separation of these cleavage products to generate a peptide map; then

(c) location or visualization of the peptides in this map.

3.1 Enzymatic cleavage methods

Some proteolytic enzymes have very restricted cleavage-site requirements, making them ideal for peptide mapping. Protocols for the most useful enzymes are presented. However, it must be stressed that conditions such as quantity of enzyme, duration of digest, and number of enzyme additions should be optimized to the requirements of the investigator and determined empirically in each case. If an enzyme digest is being utilized to obtain sequence information about a receptor (Section 6), or if a general protein-stain visualization method is being employed for peptide mapping (Section 3.7), then an endoproteinase autodigest control should be run in parallel to the experimental digestion. This is to ensure that endoproteinase autodigest products are not inadvertently sequenced, or mistaken for products derived from the sample and subsequently interpreted as proof that two proteins are related by virtue of their similar peptide maps. These errors are obviously reduced by keeping the endoproteinase to substrate ratio as low as practically possible. In addition, the absolute amount of enzyme added can be minimized by performing the cleavage in a small volume. It is also advisable to run a 'blank' of undigested receptor.

3.1.1 Trypsin (EC 3.4.21.4)

Trypsin cleaves protein at the carboxy terminus of lysine and arginine residues. However, even 'pure' commercial preparations of trypsin display α-chymotrypsin-like activity. α-Chymotrypsin (EC 3.4.21.1) has a wider cleavage spectrum than trypsin, cleaving at the carboxy terminus of leucine, methionine, phenylalanine, tryptophan and tyrosine residues (9). Fortunately, this α-chymotrypsin activity can be inhibited by treatment with *N*-tosyl-L-phenylalanyl chloromethylketone (TPCK), and TPCK-trypsin is commercially available from several suppliers, including Worthington Biochemical Corporation Ltd. As with all enzymatic cleavage methods, the specificity of cleavage is not absolute. Not all theoretically cleavable sites are cleaved and the rate of bond cleavage is not uniform. Lys−Pro and Arg−Pro bonds are not usually cleaved, and the presence of acidic residues on either side of the Lys or Arg residues reduces the rate of bond cleavage. If the protein is soluble, then trypsinization can be performed after the protein has been denatured.

Protocol 8. Cleavage of protein with trypsin

1. Add guanidinium HCl (ARISTAR grade, BDH Chemicals Ltd) to a final concentration of 6 M in 0.5 M Tris-HCl, pH 7.9, to denature the protein.

2. Dilute the guanidinium HCl to 2 M with 100 mM Tris-HCl, pH 7.9, supplemented with 20 mM calcium chloride. The calcium chloride is included as it helps to stabilize the trypsin.

3. Add TPCK-trypsin (Worthington Biochemical Corporation Ltd) at approximately 1 part to 100 parts protein and incubate for 16 h at 37°C. Make the stock solution of TPCK-trypsin at a concentration of 1 mg/ml in pure water adjusted to pH 3.0 with AnalaR HCl, and store at 4°C. At pH 3.0 the stock solution is far removed from the pH optimum of trypsin, viz. pH 8.0, so that the enzyme is virtually inactive and autodigestion is kept to a minimum. It is possible to check the concentration of the TPCK-trypsin stock by absorbance, as 1 mg/ml in a cuvette with a path length of 1 cm gives an absorbance of 1.4 at 280 nm.

4. Repeat the digestion with a further addition of TPCK-trypsin.

As receptor proteins are integral membrane proteins, they are strongly hydrophobic in nature, and hence solubility and unfolding is best effected with SDS. As trypsin is somewhat sensitive to denaturation by SDS this presents a drawback to its applicability to peptide mapping receptors. However, digestion of receptors with trypsin in SDS has been reported (10), but multiple additions of enzyme might be required.

The specificity of cleavage of trypsin can be further restricted to either Lys residues only, or Arg residues only, by chemical modification of the basic residue at which cleavage is not required. Appropriate protocols have been described (11); however, this methodology has been superseded by the commercial availability of Lys-specific, and Arg-specific, endoproteinases.

It should also be noted that cleavage sites for lysyl bond-cleaving enzymes can be introduced into proteins. Cysteine residues can be converted to a 'Lys look-alike' residue by reacting with ethyleneimine (12) to generate S-(β-aminoethyl)-cysteine. Peptide bonds involving the carboxyl group of this S-aminoethylation product are susceptible to cleavage by Lys-specific endoproteinases; however, the rate of cleavage is often very much slower than for Lys residues. This reduced rate of cleavage can be compensated for by increasing the amount of enzyme utilized, or by employing longer incubations.

3.1.2 Endoproteinase Arg-C

The enzyme is commercially prepared from the submaxillary gland of mice and specifically cleaves proteins at the carboxy terminus of Arg residues. It is supplied as a lyophilysate obtainable from Sigma Chemical Company Ltd as protease XX, and from Boehringer Mannheim Biochemicals (information sheet available).

Protocol 9. Cleavage of protein with endoproteinase Arg-C

1. Resuspend the lyophilysate in 50 mM Tris-acetate, pH 7.8, to give a stock solution with a protein concentration of 2 $\mu g/\mu l$. Dispense into small aliquots and store at $-20°C$ for up to 3 months.

2. Gel-filter the protein sample into 50 mM Tris-acetate, pH 7.8, using Sephadex G50 F.

3. Add endoproteinase Arg-C stock solution to give a final enzyme to protein ratio of 1:30–100, then incubate at 25°C until the digestion has proceeded as far as required. For a total digestion a further addition of enzyme after 10 h may be necessary.

Unfortunately, endoproteinase Arg-C is sensitive to even low [0.1% (w/v)] concentrations of SDS, so its applicability to analysing hydrophobic membrane proteins like receptors in solution is limited. In our experience, Arg-C is not a very aggressive enzyme.

3.1.3 Endoproteinase Lys-C

This endoproteinase specifically cleaves proteins at the carboxy terminus of Lys residues (13). Like most proteinases, it cleaves most rapidly at clusters of vulnerable residues. It is obtained from *Lysobacter enzymogenes*, and is commercially available from Boehringer Mannheim Biochemicals (information sheet available). With respect to studying receptor proteins, Lys-C together with Glu-C (see below) have the great advantage over trypsin and endoproteinase Arg-C, of being fully active in 0.1% (w/v) SDS. Bacterial proteinases tend to be tougher than their mammalian counterparts. This means that even the hydrophobic domains of receptors can be exposed to cleavage by endoproteinase Lys-C after full denaturation by 1% (w/v) SDS provided that the SDS concentration is then lowered to 0.1% (w/v) after the protein has been denatured. The continued presence of SDS in the sample then facilitates the easy manipulation, and analysis by SDS-PAGE, of both hydrophilic and hydrophobic peptides generated by the enzyme. Note that in principle, the binding of SDS to positively charged residues might be expected to restrict the activity of Lys-C. However, in practice, the enzyme works well under the conditions described.

Protocol 10. Cleavage of proteins with endoproteinase Lys-C

1. For a stock solution of endoproteinase Lys-C, resuspend the lyophilizate (3 U of enzyme; 100 μg of protein) in 50 μl of 50 mM Tris-acetate, pH 7.8, to give a stock solution of 2 $\mu g/\mu l$. Dispense small aliquots (5–10 μl) into small microcentrifuge tubes and store at $-20°C$. It is advisable not to use this stock after 3 months.

2. Add stock endoproteinase Lys-C to a denatured receptor protein sample in 50 mM Tris-acetate, pH 7.8, containing 0.1% (w/v) SDS, to give a final enzyme

to receptor ratio of 1:30—100. For analytical digests, we routinely use Lys-C at a final concentration of 16 μg/ml. Higher concentrations are needed for total digestion.

3. Incubate at 25°C until the digestion has proceeded as far as required. If necessary, add more enzyme after 10 h. After 24 h, digestion has usually proceeded more or less to completion. However, vulnerable bonds are usually cleaved much more rapidly.

When constructing a time-course of digestion with endoproteinase Lys-C, it should be noted that although this enzyme is a serine proteinase, it is not inhibited by phenylmethylsulphonyl fluoride (PMSF). Effective termination of digestion is most conveniently achieved by adding a 10% (w/v) SDS solution to give a final SDS concentration of 1% (w/v), followed by heating at 60°C for 15 min. Lys—Pro bonds are often incompletely hydrolysed by endoproteinase Lys-C, and Lys residues flanked by acidic residues can be resistant to cleavage.

It should be noted that Wako Pure Chemical Industries Ltd market a Lys-specific endoproteinase from *Achromobacter lyticus* as a lyophilizate, under the name Lysyl Endopeptidase (*Achromobacter* Protease I). This enzyme is also stable to 0.1% (w/v) SDS but has the advantage of effectively cleaving Lys—Pro bonds, unlike endoproteinase Lys-C. In addition, Lysyl Endopeptidase hydrolyses the carboxy terminus of S-(β-aminoethyl)-cysteine residues at the same rate as Lys residues (cf. trypsin). We have used this enzyme successfully under the conditions described above for Lys-C.

3.1.4 Endoproteinase Glu-C (V8 proteinase) (EC 3.4.21.19)

This is an extracellular endoproteinase from the V8 strain of *Staphylococcus aureus*, which is marketed as a lyophilyzate by Boehringer Mannheim Biochemicals (information sheet available) and displays two pH optima, pH 4.0 and pH 7.8. Houmard and Drapeau (14) reported that in bicarbonate buffer, pH 7.8, endoproteinase Glu-C cleaves specifically at the carboxy terminus of Glu residues, whereas in phosphate buffer, pH 7.8, both Glu and Asp residues are cleaved at the carboxy terminus. However, with the mAChR we have found that only Glu residues were cleaved, regardless of whether phosphate or bicarbonate buffer was employed. The nature of neighbouring residues can affect the rate of cleavage of a bond, and Glu—Pro bonds are not cleaved. In addition, cleavage occasionally occurs at the carboxy-terminal of non-acidic residues. As with endoproteinase Lys-C, endoproteinase Glu-C is fully active in low concentrations of SDS, in fact up to 0.2% (w/v), thereby endowing it with the benefits ascribed to endoproteinase Lys-C (see above), for peptide-mapping receptors. It should be noted that preparations of endoproteinase Glu-C have been reported to be contaminated by another proteinase secreted by *S. aureus* which is inhibited by EDTA (15). Although this contaminant is not usually present in commercial preparations, it is nevertheless advisable to include 0.2 mM EDTA in incubations with endoproteinase Glu-C as a precaution against non-specific cleavage. This is particularly relevant for digests of long duration, or if a high enzyme to receptor ratio is utilized.

Protocol 11. Cleavage of proteins with endoproteinase Glu-C

1. Dissolve the lyophilyzate in 0.2 mM EDTA, 50 mM NH_4HCO_3, pH 7.8 (or 0.2 mM EDTA, 50 mM phosphate buffer, pH 7.8, if trying to select for cleavage at both Glu and Asp residues) to give a stock solution of 2 $\mu g/\mu l$. After aliquoting this stock solution into small tubes it can be stored at $-20\,°C$ for up to 3 months.

2. Using Sephadex G50 F, gel-filter the denatured receptor protein sample into 0.1% (w/v) SDS, 0.2 mM EDTA, 50 mM NH_4HCO_3, pH 7.8 (or phosphate buffer equivalent).

3. Add stock endoproteinase Glu-C to give a final enzyme to receptor protein ratio of 1:30–100. We have used final concentrations in the range 5–20 $\mu g/ml$ for analytical digests. Then incubate at 25 °C, adding additional endoproteinase if required. Incubation times are in the range 1–48 h, depending on the vulnerability of the bonds concerned. Again, there is evidence to suggest that the enzyme cleaves preferentially at clusters of acidic residues, where the net negative charge density is high.

3.1.5 Endoproteinase Asp-N

The marketing of this enzyme has just been initiated by Boehringer Mannheim Biochemicals (who provide an information leaflet). Although largely untried as yet with respect to studying receptor proteins, this enzyme is potentially very useful, as it specifically cleaves the N-terminal bond of Asp residues and $CysSO_3H$ residues. The commercial availability of endoproteinase Asp-N will obviously make it complementary to endoproteinase Glu-C (see above) which has been reported to cleave Asp and Glu residues at their C termini. Endoproteinase Asp-N can probably be used essentially as per the endoproteinase Glu-C protocol with specific conditions being determined empirically. Unfortunately, Asp-N appears not to be a very aggressive enzyme, and it is difficult to obtain a high enough enzyme:receptor ratio to get large-scale cleavage of hydrophobic peptides. The enzyme also appears to be sensitive to SDS. Furthermore, it is expensive.

3.1.6 Oxidation of cysteine to cysteic acid followed by cleavage with Glu-C or Asp-N

Provided that the cysteine residues of a protein have not been carboxymethylated, it is possible to oxidize them to cysteic acid by exposure of the peptide to performic acid in the vapour phase. This introduces fresh cleavage sites for Asp-N or Glu-C, which may then be exploited by digestion as described above. N.B. Avoid buffers containing Cl^- ions, which will be oxidized to Cl_2 by the performic acid.

A protocol for the oxidation of cysteine to cysteic acid (A. Aitken, personal communication) is given below.

Protocol 12. Oxidation of cysteine to cysteic acid

1. Place the peptide sample in an Eppendorf tube and evaporate to dryness in a Speed-Vac concentrator (Savant).

2. Mix 5 ml of 30% hydrogen peroxide (100 volumes, BDH ARISTAR) with 95 ml of 98% formic acid (BDH ARISTAR). Incubate for 2 h on ice.

3. Place the performic acid in a Petri dish in a desiccator. Place the peptide sample in the open Eppendorf tube in a separate container in the desiccator. Note: remove the zinc divider, which will otherwise be attacked by the vapours.

4. Evacuate the desiccator at a water pump, seal and leave for 2−4 h, so that performic acid in the vapour phase can access the peptide.

5. Remove the Petri dish, and evacuate thoroughly using an oil pump and trap to remove residual performic acid.

6. Take the oxidized peptide up in a suitable buffer for cleavage.

3.1.7 Pepsin (EC 3.4.23.1) and thermolysin (EC 3.4.24.4)

Both these enzymes cleave proteins by hydrolysing a wide range of peptide bonds, with the actual cleavage site being virtually impossible to predict. However, by employing short incubation periods and/or low enzyme concentrations, a degree of specificity can be introduced, as only the most susceptible bonds will be cleaved. Although not recommended for primary cleavage of receptors, these enzymes can prove useful for sub-fractionating peptides generated by a more specific cleavage method. This strategy was successfully employed for obtaining peptides from the human platelet α_2-adrenergic receptor (16). This receptor possessed only four methionines, so after an initial cleavage by CNBr (see Section 3.2.1) sequenceable peptides were generated by successive treatment with pepsin and thermolysin. Thermolysin is a very robust enzyme, easily surviving 0.1% SDS, and active at high temperatures (60°C) at which digestion times can be kept short. Both thermolysin and pepsin have their devotees. The pH optimum for pepsin activity is low (*c.* 4.0). At this pH, the NH_2 termini of the peptides generated are protonated, and thus protected against N-blocking, caused, for instance, by traces of aldehydes in the buffers. The N-blocking of peptides during prolonged digestion is a difficult problem, which is certainly mitigated by working under somewhat acidic conditions.

3.1.8 Immobilized proteinases

A number of proteinases are now commercially available immobilized on an inert matrix. Of the peptidases mentioned above, trypsin, α-chymotrypsin, and Glu-C are available in this form from Mobitec (Scotlab) or Pierce. These may allow better control over the cleavage process, and minimize the contamination of the peptide fraction with proteinase autodigest products.

3.2 Chemical cleavage methods

These have the obvious advantage over enzymatic procedures (Section 3.1) that the cleavage process does not involve a proteinaceous agent which is itself degradable. Consequently one source of misinterpretation is eradicated. However, there are not many chemical methods available which exhibit the required specificity of cleavage, and which do not affect amino-acid side chains. Cleavage of methionine, tryptophan, tyrosine, and cysteine residues is possible. More details are to be found in Biological Membranes, a Practical Approach (40), Chapter 6. The most useful methods are those for cleavage of methionine and tryptophan. These are described below.

3.2.1 Cyanogen bromide

This is becoming an increasingly popular method for producing fragments of receptor for obtaining sequence information and, of course, can be utilized for peptide-mapping studies. It is the most useful of the chemical cleavages. Cleavage occurs at the carboxy terminus of Met residues (17) to generate the interconvertible residues homoserine and homoserine lactone (*Figure 8*). The specificity of cyanogen bromide is dependent on pH. At alkaline pH, basic groups in the protein react, whereas at low pH reaction with cyanogen bromide is largely restricted to Met residues, and with cysteine residues

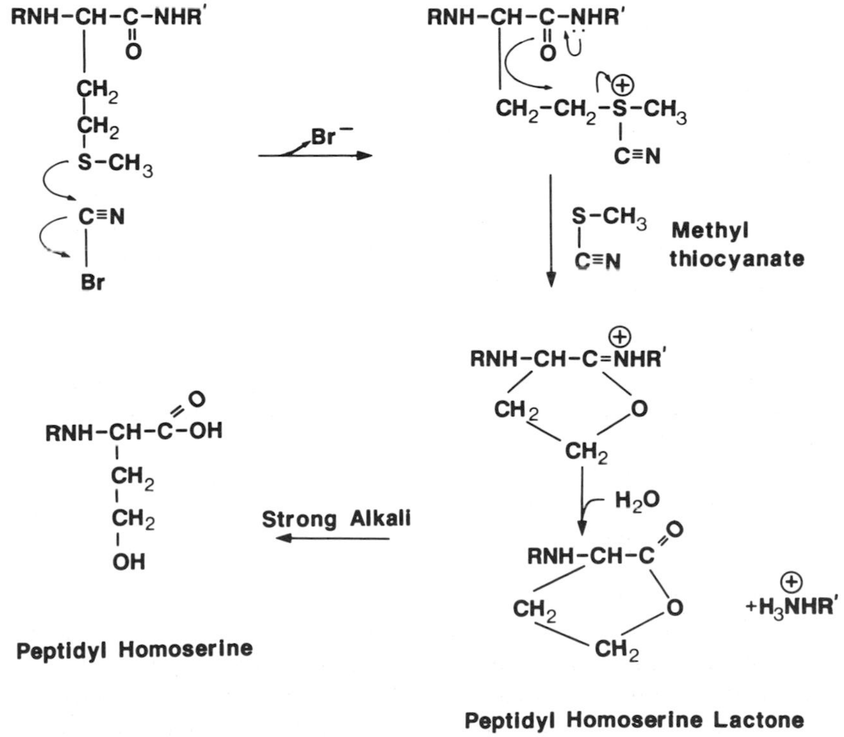

Figure 8. Cleavage of the methionyl peptide bond by cyanogen bromide.

229

which are slowly oxidized to cysteic acid. Cleavage at tryptophan may also occur, but is rare in practice. Reaction with cysteine is, of course, prevented if the protein has been fully reduced and S-carboxymethylated (see Section 2.5). It should be noted that the oxidation products of methionine, viz. methionine sulphoxide and methionine sulphone (see Section 2.2.3), do not react with cyanogen bromide. The sulphoxide can be reduced to methionine, but the sulphone is more stable to reduction. In addition, the acidic conditions used to maintain the specificity of cleavage can cleave Asp−Pro bonds, and cause formylation of Ser and Thr residues. Met−Thr and Met−Ser bonds may be poorly cleaved due to the β-hydroxyl groups altering the course of the normal cyanogen bromide cleavage reaction (18). *Important:* cyanogen bromide and its volatile product hydrogen cyanide are very toxic and appropriate precautions should be taken during its use and disposal. Work on a tray in a fume cupboard, and familiarize yourself with the procedures for treating cyanide poisoning.

Protocol 13. Cleavage by cyanogen bromide

1. If the receptor protein has not recently been reduced and S-carboxymethylated, reduce with 5% (w/v) DTT (high-purity grade, Calbiochem) in 10 mM Tris-HCl, pH 8.0. Then remove the DTT by gel filtration (preferable) or dialysis. Alternatively, the protein can be precipitated by adding AnalaR ethanol to a final concentration of 90−95% (v/v) and incubating at −20°C overnight.

2. Dissolve the protein in 10−50 μl of 70% (v/v) AnalaR formic acid which has been purged by bubbling with nitrogen.

3. Add a 50-fold molar excess of CNBr (nominally equivalent to the weight of protein). It is critical to use fresh CNBr. Absolute concentrations vary from 1−20 mg/mol. We prefer the higher concentrations.

4. Transfer to a nitrogen atmosphere as described for reduction and S-carboxymethylation (see Section 2.5 and *Figure 7*), and incubate at 20°C in the dark for 16 h.

5. Lyophilize the sample. This removes excess reagent and the by-product methylthiocyanate, both of which are volatile. It should be noted that very hydrophobic peptides may adhere very strongly to the tube after lyophilization and may not be resuspended efficiently in *step 7*.

6. Resuspend in 100% ethanolamine for 10 min at room temperature, then lyophilize.

7. Dissolve the cleaved protein in 1% (w/v) SDS for 15 min at 60°C prior to analysis by SDS-PAGE. Alternatively, if reverse phase HPLC (see Section 7.2) is the subsequent procedure, the protein can be redissolved in HPLC solvent, such as 50% (v/v) trifluoroacetic acid (TFA; HPLC grade, e.g. Pierce Chemical Company) followed by addition of water and acetonitrile (HPLC grade, e.g. Romil Chemicals) to give the gradient starting conditions.

We have also used an alternative protocol to cleave peptides prepared by high-resolution gel-filtration on a Superose 12 column. In this procedure, peptides generated by primary cleavage with Lys-C in the presence of 0.1% SDS are fractionated by FPLC gel-filtration on a Superose 12 column in a buffer containing 50 mM Tris-Cl, 0.1% SDS, pH 7.8, at a flow rate of 0.3 ml/min. 250 μl fractions are collected. We have used two procedures for CNBr cleavage of the peptide fractions so generated. These are as follows.

Protocol 14. Cleavage of small samples by cyanogen bromide after gel filtration

1. Analytical scale cleavage. Take $20-50$ μl of the fraction of interest. Add 100 mM dithiothreitol (Calbiochem, high-purity grade) to a final concentration of 5 mM. Incubate the solution at room temperature overnight in an atmosphere of N_2 as described for reduction and S-carboxymethylation (Section 2.5).

2. Make up 100 ml of a solution containing 100 μg/ml tryptophan in 100 mM HCl (BDH, ARISTAR grade). Prepurge the solution by bubbling with nitrogen. The tryptophan is added to prevent tryptophan residues within the peptides from undergoing modification and cleavage.

3. Place a fresh, colourless, dry crystal of CNBr in a pre-weighed Eppendorf tube. Take the crystal from deep down in the bottle: surface crystals are more likely to have deteriorated. Cap and weigh the tube. N.B. perform all manipulations of CNBr in a fume cupboard, and wear gloves.

4. Dissolve the CNBr at a final concentration of 40 mg/ml by adding an appropriate amount of the prepurged HCl to the Eppendorf tube.

5. Add a volume of the CNBr solution equal to that of the fraction to be cleaved. This yields a final free HCl concentration of c. 25 mM, and a pH of c. 1.7. This may be checked by placing 1 μl of the reaction mixture on a piece of low pH-range indicator paper. The large excess of CNBr will react with any residual DTT.

6. Flush out the Eppendorf tube with N_2, and place it in a Sterilin bottle under N_2, as described (Section 2.5). Wrap the bottle in foil, and keep it in the dark for 24 h at room temperature.

7. At the end of the cleavage period, the sample can be neutralized with NaOH (to give a final concentration of 25 mM) and diluted with running buffer containing 20% glycerol to c. 100 μl prior to SDS-PAGE analysis. The presence of the cleavage reagents does not seem to affect the running of the gels. Alternatively, the peptides can be separated by reversed-phase HPLC (see Section 7). Excess CNBr, and contaminated tubes, etc. should be immersed in a solution of bleach in a fume hood prior to disposal.

The procedure can be modified slightly to allow larger amounts of peptide fractions to be processed.

Protocol 15. Cleavage of larger samples by cyanogen bromide after gel filtration

1. Take $100-200$ μl of the fraction in Tris-Cl/SDS buffer and reduce it with DTT as described above.

2. Apply the fraction to a 2 ml column of Sephadex G50 F, equilibrated in 0.01% SDS, with no buffer. Make a first addition to the column of enough 0.01% SDS to bring the total volume applied to 200 μl, then wash in with 2×200 μl of 0.01% SDS. Elute the peptides in 0.5 ml of 0.01% SDS, collecting them in an Eppendorf tube (1.5 ml). The recovery is good, even with quite hydrophobic receptor-derived peptides. The use of Kontes Disposaflex columns with porous polythene sinters is recommended. The gel-filtration removes excess DTT, Tris buffer, and reduces the SDS concentration.

3. Evaporate the peptide sample to c. 20 μl in a Speed-Vac concentrator (Savant).

4. Make up a solution of 60 mg/ml CNBr in 50 mM HCl and 100 μg/ml tryptophan, prepurged with N_2 as described above. Add an equal volume of this solution to the concentrated peptide solution and incubate as described above.

5. The cleaved reaction mixture can be purified further, or, under suitable circumstances, applied to the polybrene-coated disc of a pulsed liquid-phase sequenator, and sequenced directly, after drying to remove the excess CNBr.

These procedures have several virtues.

(a) They avoid lyophilization of the sample. Hydrophobic peptides stay in solution at all times, because of the presence of SDS in the buffers. Therefore, recoveries are excellent.

(b) The procedures avoid exposure of the peptides to high concentrations of formic acid, which may modify or N-block the peptides.

(c) Provided that the CNBr is fresh, the cleavages should proceed to near completion.

(d) Note that the fact that the product of cleavage is a C-terminal homoserine lactone means that the peptides can be directly immobilized on an amino propyl resin, using the method of Horn and Laursen (27) or a variant thereof, for solid-phase Edman degradation and sequencing.

3.2.2 Cleavage at tryptophan

Several protocols have been developed for cleavage at tryptophan residues. While they do not, in general, give yields as good as those obtained from CNBr cleavage at methionine, they may still be found useful. These protocols are as follows.

3.2.3 Dimethylsulphoxide (DMSO)/hydrobromic acid

The method developed by Savige and Fontana (19) which is presented below selectively generates high cleavage yields $(50-70\%)$, and has the advantages of using inexpensive reagents and ease of execution. In addition to cleaving at the carboxy terminus of Trp residues, oxidation of Met and Cys residues to methionine sulphoxide and cystine, respectively, also occurs. Cleavage of Asp−Pro bonds may also be observed.

Protocol 16. Cleavage of proteins with DMSO/HBr

1. Disperse the receptor protein in a mixture of 600 μl AnalaR acetic acid, 300 μl concentrated (~ 12 M) AnalaR HCl, 25 μl DMSO, and 20 mg of phenol.

2. Clear the solution by sonicating for 2 min then stirring for 30 min.

3. Add 100 μl of 50% (w/v) HBr (AnalaR grade, BDH Chemicals Ltd) and incubate for 30 min.

4. Remove most of the acid and water by concentrating on a rotary evaporator at 35°C.

5. Disperse the viscous residue in 450 μl of AnalaR acetic acid, then add 150 μl of water. Separate the protein from the reagents by collecting the void volume of a gel-filtration using Sephadex G10 equilibrated with 70% (v/v) acetic acid.

3.2.4 Iodosobenzoic acid

The reaction is cleavage after oxidative halogenation. The C-terminal spirolactone thus formed is useful for peptide immobilization to an amine-derivatized solid support (36). The protocol is based on that of Mahoney, Smith and Hermodson. (33).

Protocol 17. Cleavage of proteins by iodosobenzoic acid

1. Make up a solution of guanidinium HCl (BDH, ARISTAR; 0.382 g) in 1 ml 80% formic acid (BDH, ARISTAR). The guanidinium HCl dissolves on whirlimixing.

2. Add 16 mg of iodosobenzoic acid (Pierce) and 20 μl p-cresol (to protect tyrosine residues), and bubble with N_2.

3. Add 150 μl of the solution to 10 μl of the fraction to be cleaved. Incubate under N_2 in a foil-wrapped Sterilin bottle, as described above.

4. After 24 h at room temperature, add 16 μl of 10% SDS and sonicate.

5. Gel-filter the peptides on a 4 ml column of Sephadex G25 F in 50 mM Tris-Cl/0.1% SDS, pH 7.8, or use a suitable modification of this procedure (e.g. gel-filtration in 50% acetic acid) to permit further analysis.

In our experience, there is a strong tendency for very hydrophobic peptides to be caused to adhere to the tube by the high guanidinium HCl concentration. However, attempts to modify the protocol have not been successful.

3.2.5 BNPS skatole

The method is based on that of Fontana (34,35).

Protocol 18. Cleavage of proteins by BNPS skatole

1. Make up 20 mg/ml BNPS skatole (Pierce) in glacial acetic acid (BDH ARISTAR) containing 100 μg/ml tyrosine (to protect tyrosine residues).

2. Add 10 μl of sample in 50 mM Tris-Cl/0.1% SDS, pH 7.8. Any yellow precipitate which forms can mainly be redissolved by vortex mixing and sonicating.

3. Incubate under N_2 for 24 h at room temperature in the dark.

4. Add 1 μl of 100 mM DTT, to destroy excess BNPS skatole.

5. To remove excess acetic acid, and get a sample which can be run on an SDS gel, we add 200 μl of 0.05% SDS. The BNS skatole precipitates. The sample is evaporated to *c*. 20 μl. A further 100 μl of deionized water is then added, and the procedure repeated. Any remaining acid is then neutralized by the addition of NaOH (check with indicator paper). The sample can then be supplemented with glycerol and loaded onto an SDS gel. The yellow precipitate does not appear to interfere with the running of the gel, although the presence of some residual salt causes the band to spread. The precipitate can be removed by centrifugation before analysis of the sample, but care must be taken to ensure that the peptide of interest does not adhere to the pellet.

6. Alternative approaches are to remove the excess BNPS skatole by ether extraction or by gel-filtration in aqueous acetic acid. Further details are given in references 34 and 35.

3.3 Separation of cleavage products by slab SDS-PAGE

Once specific cleavage of the receptor has been attained it is then required to separate the products. Many methods are available for protein separation, any one of which may be used. However, it should be noted that the presence of SDS interferes with some, e.g. reverse-phase HPLC (although this may not prove fatal), and can affect or reduce the resolution of peptides by sizing techniques. A means of circumventing this is to utilize SDS-PAGE. This has the advantage that SDS is an integral part of the separation procedure, so its presence in the sample is a prerequisite not a problem. Furthermore, systems are relatively inexpensive to purchase, economical to run, easy to operate, and have excellent resolving power. Further details are given in reference 41, and Bio-Rad Bulletin 1156 is an excellent guide to PAGE technology.

The gel is formed by polymerizing acrylamide ($CH_2 = CH\text{-}CO\text{-}NH_2$) and the cross-linker *N,N'*-methylene bisacrylamide in the presence of a catalyst and an initiator at room temperature. The catalyst is routinely *N,N,N',N'*-tetramethyl ethylene diamine (TEMED) with either ammonium persulphate, or a combination of riboflavin and UV light, acting as the initiator. The porosity of the gel is affected by the concentration of acrylamide. To prevent excessive fragility or brittleness, gels normally contain 2−30% (w/v) acrylamide. High-density gels are used to resolve low molecular weight species, and for high molecular weight proteins low-density gels are employed. It should be noted that, in general, receptors are glycoproteins, and that the carbohydrate moiety causes the receptor protein to migrate as a diffuse band on SDS-PAGE, with an anomalous molecular weight, e.g. the m1 mAChR has a protein molecular weight of 51.5 kDa, but migrates on SDS-PAGE as a 70 kDa protein. The gel system routinely used in this laboratory is a discontinuous system based on that of Laemmli and co-workers (7) and is described below. *Warning:* both acrylamide and *N,N'*-methylene bisacrylamide are neurotoxins and can be absorbed through the skin, so gloves should be worn when these are handled. The polymerized polyacrylamide gel is not neurotoxic.

3.4 Stock solutions

In all cases use the highest grade chemicals and water, and ensure extraneous proteins are prevented from contaminating stocks, e.g. wear gloves when handling solutions and when manipulating the polyacrylamide gel.

Protocol 19. Preparation of SDS-PAGE stock solutions

1. Solution A: 36.3 g Tris, 0.23 ml *N,N,N',N'*-tetramethyl ethylene diamine (TEMED). Dissolve in water and adjust to pH 8.9 with HCl. Final volume 100 ml.

2. Solution B: 5.98 g Tris, 0.46 ml TEMED. Dissolve in water and adjust to pH 6.7 with HCl. Final volume 100 ml.

3. Solution C: 28.0 g acrylamide (Electran, grade 1; BDH Chemicals Ltd), 0.735 g *N,N'*-methylene bisacrylamide (Electran, BDH Chemicals Ltd) in water to a final volume of 100 ml. Deionize by stirring with 5 g/100 ml acrylamide solution Amberlite MB-1 mixed-bed resin (BDH) (optional). It is advisable to filter this solution, e.g. through a hard analytical filter-paper (Whatman no. 1), before storage.

4. Solution D: 10 g acrylamide, 2.5 g *N,N'*-methylene bisacrylamide in water to a final volume of 100 ml.

Solutions A−D can be stored at 4°C for some weeks.

5. Solution E: 4 mg riboflavin in water to a final volume of 100 ml. Prepare fresh each time.

6. Solution F: 40 g AnalaR sucrose in water to a final volume of 100 ml. Prepare fresh each time.

7. Solution G: 0.14 g ammonium persulphate in water to a final volume of 10 ml. Prepare fresh each time.

8. Solution H (the running buffer): 3.0 g Tris, 14.4 g glycine, 1 g SDS in water to a final volume of 1 litre. This solution should possess a pH of 8.2−8.4

9. Solution I: 10% (w/v) SDS in water.

10. Solution J: 8 mg of riboflavin in water to a final volume of 100 ml. Prepare fresh each time.

3.5 Composition of SDS-polyacrylamide gels

Protocol 20. Mixing polyacrylamide gels

1. 7% (w/v) polyacrylamide separating gel:

Solution A:	6 ml;
Solution C:	12 ml;
Water:	27.1 ml.
Degas,then add:	
Solution G:	2.4 ml;
Solution I:	0.5 ml.

Gels with different polyacrylamide concentrations can be made by utilizing a $2\times$ concentration of Solution G, and adjusting the volumes of Solution C, Solution G, and water to produce the desired polyacrylamide concentration. Degassing is necessary, as oxygen inhibits the polymerization. This can be done by swirling the gel solution in a stoppered Buchner flask at the water pump until it bubbles. You may have to work quickly to avoid premature polymerization of the gel. (Some workers do not bother with this step but it is necessary for reproducible polymerization).

2. Stacking gel [2.5% (w/v) polyacrylamide]:

Solution B:	2 ml;
Solution D;	4 ml;
Solution F:	7.9 ml;
Solution E:	2 ml.
Degas, then add:	
Solution I:	0.16 ml.

Table 1. Composition of separating SDS-polyacrylamide gels using riboflavin as the initiator. The volumes stated are to prepare 15 ml total volume in each case.

	Final polyacrylamide concentration			
	10%	**14%**	**18%**	**22%**
Solution A	1.88 ml	1.88 ml	1.88 ml	1.88 ml
Solution C	5.40 ml	7.54 ml	9.64 ml	11.78 ml
Solution J	1.0 ml	1.0 ml	1.0 ml	1.0 ml
Water	6.57 ml	4.43 ml	2.33 ml	0.19 ml
		degas, then add		
Solution I	0.15 ml	0.15 ml	0.15 ml	0.15 ml

In high-percentage polyacrylamide gels, polymerization may be too rapid for convenience. This can be avoided by using riboflavin as initiator (see *Table 1*) and excluding light with aluminium foil until the gel is fully poured and polymerization is required. This can be particularly relevant when pouring gradient gels (Section 3.6) using a gradient maker, as this may be quite a slow process. Note that Bio-Rad market a new cross-linking agent, piperazine di-acrylamide, which can be substituted for methylene bisacrylamide on a weight-for-weight basis, and is supposed to give gels with superior mechanical strength and resolution.

3.6 Pouring and running gels

3.6.1 Routine method

Many laboratories use their own home-made electrophoresis apparatus and obtain good results. We have found the commercially available system Protean II marketed by Bio-Rad very satisfactory, but there are many variants.

Protocol 21. Routine method for SDS-PAGE

1. While wearing gloves, clean the glass plates by immersion in 35% (v/v) nitric acid. Rinse well with water, then acetone, then dry. Clean plates may be stored wrapped in clingfilm.

2. Assemble the plates and spacers to form a mould for the gel. Suitable spacers are 0.7−1.5 mm thick.

3. Mix the separating gel as described in Section 3.5, then pour the gel into the mould. This is conveniently done using a large (60 ml) disposable plastic syringe and needle.

4. Gently overlay the gel with a small amount of water, disturbing the gel solution as little as possible. For example, dispense the water from a 2 ml disposable plastic syringe with a length of firm polypropylene tubing pushed over the needle, or use a peristaltic pump. This excludes air from the polymerizing gel, and produces a straight top to the separating gel. During the polymerization process the water/gel interface will temporarily disappear. Polymerization takes about 60 min; however, it is better to allow the gel to polymerize overnight (see later in this section).

5. After polymerization has occurred decant off the water layer and rinse the top of the gel with water. The remaining water can be soaked up with the corner of a Kleenex while holding the gel at an angle.

6. Mix the stacking gel (see Section 3.5), and pour this on top of the separating gel. Immediately insert the comb to form the sample wells in the stacking gel. Try to insert this at a slight angle to ensure that air bubbles are not trapped on the underside of the comb. There should be 1−1.5 cm of stacking gel between the bottom of the sample wells and the top of the separating gel. Then illuminate the mould, e.g. with a portable fluorescent tube, to polymerize the stacking gel (30−90 min).

7. Remove the comb. Fill the bottom chamber of the electrophoresis tank with running buffer (Solution H). Place the bottom of the gel mould into this at a steep angle, so that air bubbles trapped at the bottom of the gel, between the glass plates, are removed. Jiggle if necessary. If bubbles are not efficiently removed the electric field will be distorted and anomalous migration of proteins will result.

8. Clamp the gel to the tank and fill the top chamber (the cathode) with Solution H supplemented with 0.1 mM thioglycolate as anti-oxidant, as suggested by Koziarz and colleagues (20). This travels at the dye front during electrophoresis and scavenges free-radicals and oxidants that might be present in the gel matrix. If these are not removed they can destructively react with amino-acid residues in the protein samples, especially Met, Trp, and His. As a further precaution, it is advisable to allow the separating gel to polymerize overnight prior to use.

9. Flush any poorly polymerized gel out of the sample wells using the cathode running-buffer in a Pasteur pipette.

10. Samples to be run should be fully denatured (see Section 2.1), reduced, and S-carboxymethylated (see Section 2.5) and contain 0.1% (w/v) SDS. The ionic strength of the sample should be kept as low as possible, as high ionic strength can cause severe distortion of protein migration. In addition, high concentrations of SDS in the sample can cause fuzzy banding. Consequently, if samples have been concentrated prior to electrophoresis, or obtained from ion-exchange columns, they should be dialysed, or gel-filtered, into running buffer (Solution H). A useful trick is to gel-filter the sample into running buffer diluted 1:5 to 1:10 with deionized water, and then concentrate it five- to tenfold as follows:

    (a) Pour a column (Kontes Disposaflex, or similar) containing 2 ml of Sephadex G50 F or G25 F, depending on the molecular weight of the peptide (greater than 10 kDa or less than 10 kDa).

    (b) Equilibrate the column with c. 5 bed volumes of Tris/glycine/SDS running buffer diluted 1:5 − 1:10 with water. The SDS concentration is well below the CMC (0.2%).

    (c) Apply the sample in 200 μl. Allow it to run in.

    (d) Wash the sample into the gel with a further 2 × 200 μl of diluted running buffer.

    (e) Elute the sample with 500 μl of diluted running buffer. Collect the eluate in a weighed microcentrifuge tube.

    (f) Concentrate the eluate five- to tenfold using a centrifugal evaporator (e.g. Speed-Vac), checking the progress of the concentration by weighing the tube.

    (g) The procedure can be scaled up or down as desired.

    Add AnalaR glycerol to each sample to a final concentration of 20% (v/v) to ensure that the sample sinks to the bottom of the well. Add 0.005% (w/v) bromophenol blue to the samples to act as a tracking dye. Load the samples into the wells. A convenient way of doing this is to draw the sample up into a 200 μl disposable pipette tip using an Oxford or other similar hand-held pipette. The sample volume will usually be about 100 μl. Push the pipette tip firmly into the conical inlet of the luer fitting of a 21 gauge, 40 mm, disposable hypodermic needle. Position the end of the needle near to the bottom of the well. Expel the sample from the tip into the bottom of the well via the needle. As the needle has not had contact with the sample before the sample is expelled, potential contamination of the cathode buffer is avoided. Furthermore, this method has the attraction of disposability, with a new tip and needle being employed for each sample loaded. A slight disadvantage is that some sample remains in the needle after loading. A preferable alternative now available is to use flat-ended pipette tips (multiflex tips, P & S Biochemicals).

11. Connect the power pack to the tank's electrodes, with the anode (positive) at the bottom of the gel. Run at 30 mA, 150 V, until the tracking dye is near the bottom of the gel. Do not let this run off the gel as it can be used to locate sample lanes. Furthermore, if samples have been radioiodinated (Section 2.1) any residual free ^{125}I will run at this front. Heat generated during electrophoresis can cause band distortion, hence some systems possess a circulatory water cooling coil, alternatively the apparatus can be cooled by a fan. Do not run the gel in a cold room as the SDS will precipitate out of solution (although the lithium salt of dodecylsulphate can be used at 4°C).

12. When electrophoresis is complete, turn off the power and transfer the gel from the mould into a container (e.g. a plastic sandwich box) of fixative. A convenient fixative is 30% (v/v) methanol, 10% (v/v) glacial acetic acid and water. Alternatively, a solution of glutaraldehyde (10%) in deionized water can be used. If silver staining proteins fix as per Section 3.7.2. Use enough fixative to completely cover the gel, and gently agitate overnight on a shaking water-bath. The fixing process can be beneficial if radioiodinated samples have been run, as free ^{125}I will be removed. The gel can then be stained (see Section 3.7), or dried down with heat under vacuum onto a paper support (Whatman 3 mm; Whatman Ltd) for analysis and storage. Several gel driers are marketed, including the Model 483 Slab Drier (Bio-Rad). To dry a gel, remove the fixative and transfer the gel onto a piece of Whatman 3 mm paper cut slightly larger than the gel. To ensure the gel is flat, place the paper on top of the gel then, whilst holding it in place, invert the vessel containing the gel. Overlay the gel with clingfilm of sufficient size to completely cover the gel and overlap onto the paper support. Dry the gel in the drier following the manufacturer's instructions. It is important to ensure that a good vacuum is attained. This can be aided in the first instance by having a large side-arm flask between the pump and the drier. Close the tubing between the flask and the drier, by folding it back on itself, and allow the flask to evacuate, then on releasing the tubing a vacuum at the drier is effectively established. The duration and temperature required will vary with the density of the gel and the model of drier. We use

80°C for 3 h for uniform gels of 7−14% (w/v) acrylamide, and 60°C for 4.5 h for gradient gels, high-percentage acrylamide gels, or gels containing a fluor (see Section 3.7.4), followed by evacuation at room temperature overnight. Excessive heating or premature release of vacuum will cause gel-cracking.

It is important to note that small peptides (<6 kDa) may not be fixed and may be lost from the gel. Thus, if peptides of this molecular weight range are of interest, dry the gel down directly without fixing. However, if this is done any free ^{125}I is not removed from the gel, so it is important to ensure that free ^{125}I is efficiently removed from the samples (see Section 2.4) prior to electrophoresis.

3.6.2 Gradient gels

In these gels, the concentration of polyacrylamide is not uniform but increases down the gel. This means that as proteins migrate, they experience increasing molecular sieving. This has the effect of expanding the molecular weight range which can be analysed on one gel. In consequence this is the most frequently utilized gel for peptide mapping. The steepness, and range, of the gradient can be varied according to requirements, e.g. when analysing endoproteinase cleavage products of the mAChR we routinely use an 18−22% polyacrylamide gradient gel.

Protocol 22. Pouring gradient gels for SDS-PAGE

1. Calculate the total volume of polyacrylamide required to fill the mould to the required level for a separating gel. This is obviously determined by the dimensions of the mould.

2. Mix the polyacrylamide solutions required for the separating gel as described above, using the compositions in *Table 1*. Then construct the gradient using a gradient-maker connected to a pump to ensure that a uniform gradient is obtained. The figures in *Table 1* are for a gradient gel with a total volume of 30 ml (15 ml of each polyacrylamide solution).

3. To prevent premature polymerization when using riboflavin as initiator, wrap the gradient-maker and pump tubing with aluminium foil. Put the highest percentage acrylamide solution next to the outlet and add a stir-bar to this chamber. Pour the low-percentage acrylamide solution into the other chamber. Whilst mixing on a magnetic stirrer, simultaneously start pumping the acrylamide solution at a rate of about 3 ml/min, and open the tap permitting flow between the two chambers. Keep the outlet tube of the gradient-maker in the middle of the mould and just above the rising level of liquid at all times, ensuring the gradient is disturbed as little as possible. Then gently overlay the gel with water.

4. The stacking gel is prepared as for uniform gels. Gradient gels can be run and processed as already described for uniform gels (Section 3.6.1).

The polyacrylamide gel systems presented above satisfy most laboratories' requirements for peptide separation. In addition, a slab-gel system containing 6 M urea has been developed (21) for separating peptides in the $1-10$ kDa range. The $18-22\%$ gradient gel polymerized with riboflavin, separates in the $2-60$ kDa molecular weight range.

3.7 Detection, visualization, and quantitation of peptides after SDS-PAGE

After the proteolytic cleavage products have been resolved by SDS-PAGE it is necessary to locate them. Several of the most popular methods will be described.

3.7.1 Coomassie Brilliant Blue R-250

This is a frequently used general protein stain suitable for locating proteins after SDS-PAGE. It is a simple procedure but may not provide the required sensitivity if only small amounts of receptor protein are available for peptide mapping.

Protocol 23. Staining protein in gels with Coomassie Brilliant Blue

1. Make a stock solution of a stain by dissolving 2.5 g Coomassie Brilliant Blue R-250 (Sigma) in 900 ml of 50% (v/v) methanol, then add 100 ml of glacial acetic acid. Remove insoluble material by filtration through Whatman no. 1 filter paper.

2. After fixing the gel (see Section 3.6.1), replace the fixative with sufficient stain to cover the gel and agitate gently for 90 min (1.5 mm gels: thin gels stain more rapidly).

3. Discard the stain, then destain with 30% (v/v) methanol, 10% (v/v) glacial acetic acid, in water. Change the destain solution repeatedly and agitate the gel gently. Floating a small piece of sponge in the destain can speed up the process as it absorbs stain. To avoid over-destaining, replace the destain solution with 10% (v/v) acetic acid.

3.7.2 Silver staining

This general protein stain is 100-fold more sensitive than Coomassie Brilliant Blue R-250 and is widely employed. Stain occurs predominantly at the surface of gels, which can be a disadvantage when using thick gels. The method presented is that of Morrisey (22) as modified (23) to give low background staining.

Protocol 24. Staining protein in gels with silver

1. Fix the gel overnight in 25% (v/v) isopropyl alcohol, 10% (v/v) glacial acetic acid, in water.

2. Discard the fixative and soak the gel in 7.5% (v/v) glacial acetic acid with gentle shaking, at room temperature for 30 min.

3. Soak the gel in 0.2% (v/v) aqueous periodic acid with gentle shaking at 0°C for 60 min.

4. Rinse the gel by agitating in 500 ml of distilled water for 30 min. Repeat washing a further 4 times.

5. Soak in 5 μg/ml dithiothreitol for 30 min.

6. Discard the solution, then without rinsing add 0.1% (w/v) silver nitrate. Incubate for 30 min.

7. Rinse once rapidly with a small volume of distilled water, then twice rapidly in a small amount of developer [50 μl of 37% (w/v) formaldehyde in 100 ml of 3% (w/v) sodium carbonate]. Incubate in the developer until the required amount of staining is attained.

8. Stop the staining by adding 5 ml of 2.3 M citric acid directly to the developer, and agitating for 10 min.

9. For storage, soak the gel in 0.03% (w/v) sodium carbonate for 10 min.

Silver-stained gels are best analysed when illuminated from below, consequently it is better to photograph the hydrated gel rather than preserve it dried down onto paper.

3.7.3 Autoradiography

This constitutes a very sensitive technique for locating proteins. In principle, emissions from proteins that have been radiolabelled (see Section 2.1), or probed with radiolabelled antibodies and lectins after electroblotting onto nitrocellulose (see Section 3.7.6), are used to develop photographic film, thereby generating an image of the proteins' location. In the procedure given below *steps 2* and *4* should be performed in a darkroom under safelight.

Protocol 25. Obtaining an autoradiograph of a gel

1. Dry the gel onto paper as described in Section 3.6.1.

2. Place the dried gel (or nitrocellulose paper if appropriate) into a light-proof cassette and place on top of this a sheet of X-ray film (e.g. Fuji RX film). To avoid chemography, the clingfilm can be left in place between gel and film, when using high-energy emitters such as ^{125}I. Close the cassette securely.

3. Store the cassette at −70°C until the film is sufficiently exposed. The duration of exposure is dependent on the specific activity of the radiolabelled proteins.

4. Allow the cassette to warm to room temperature, then develop the film according to the manufacturer's instructions. Standard films are best processed using an automated X-ray film processor. Re-expose the gel for longer or shorter periods to optimize the degree of film development.

If low levels of radioactivity are present then pre-flashing the X-ray film can be used to sensitize it. This can be done conveniently with the film in the cassette. The flash unit should be *c*. 1 metre from the film (39). Sensitivity can also be improved by including an intensifying screen in the cassette (e.g. Cronex lightening plus; Du Pont), such that the film is sandwiched between the gel and the intensifying screen. The pre-flashed side of the film should face the screen. With this arrangement, high-energy emissions from ^{125}I and ^{32}P pass through the film and impinge on the intensifying screen. This generates photons which develop the film, thereby enhancing the final image. Low-energy emissions, e.g. from ^{3}H, do not pass through the film so intensifying screens cannot be employed (however, see Section 3.7.4). It is advisable to mark the paper support of the gel with ink which has been 'spiked' with a small amount of waste radioactivity. This will facilitate correct alignment and orientation of the autoradiograph with the original gel. The yields of the various peptides can be elucidated by determining the extent each peptide band has developed the film using a densitometric scanner. However, it should be noted that at its extremes, development is not linear with respect to exposure.

3.7.4 Fluorography

This is essentially the same technique as autoradiography (see Section 3.7.3) except that it is compatible with low-energy β-emitters such as ^{3}H, ^{14}C, and ^{35}S. This is made possible by impregnating the slab gel with a fluor. Emissions from radiolabelled proteins cause the scintillant to generate photons which develop the film. Intensifying screens should always be employed when doing fluorography. The scintillants can be obtained commercially, e.g. En^{3}Hance (Du Pont). The major drawback with this technique is that films may take weeks to develop.

Protocol 26. Obtaining of fluorograph of a gel

1. Fix the gel (see Section 3.6.1).

2. Impregnate the gel with scintillant following the manufacturer's instructions. Enhancers are also available in spray form, for surface application (e.g. Amplify Spray, Amersham).

3. Dry the gel using medium heat (60°C) in a gel drier. Temperatures above 80°C may cause the fluor to become volatile.

4. Set up a cassette as for autoradiography (Section 3.7.3).

Note that a variety of high sensitivity X-ray films for fluorography and autoradiography are now available from Kodak, or Amersham (Hyperfilm) who also provide a useful technical booklet on high-performance films and autoradiographic standards. These films lack the usual coating on the emulsion, enhancing their sensitivity to low-energy emissions, but making them more sensitive to mechanical damage. For the same reason, they have to be developed by hand. Note that the use of enhancers or enhancing screens will tend to broaden bands. If the highest resolution is being sought, and enough radioactivity is available, it is preferable to use a high-sensitivity film without screens or enhancer. It is best to do some trial runs to determine sensitivity.

3.7.5 Slicing gels

This method can be utilized for one-dimensional peptide maps of proteins radiolabelled with either low- or high-energy emitters. Usually autoradiography is the preferred method for proteins labelled with ^{125}I or ^{32}P; however, gel-slicing offers a reasonably rapid means of determining the position of peptides labelled with low-energy emitters and is routinely used by this laboratory in preference to fluorography. Furthermore, the gel-slicing procedure generates quantitative data (see *Figure 9*).

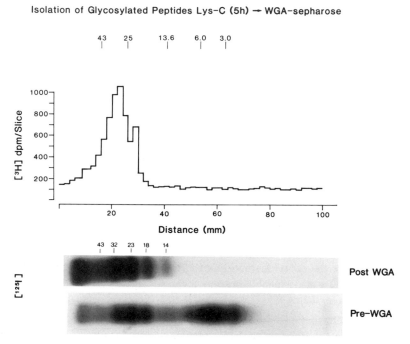

Figure 9. Isolation of radioiodinated and [^{3}H]PrBCM-labelled peptides from a 5 h endoproteinase Lys-C digest using WGA – Sepharose. Either [^{125}I]mAChR, or [^{3}H]PrBCM-labelled mAChR, was digested with endoproteinase Lys-C for 5 h and the proteolytic products isolated by WGA – Sepharose were then analysed on an 18 – 22% gradient gel. The [^{125}I]peptides were visualized by autoradiography, whereas [^{3}H]peptides were detected by gel-slicing.

Protocol 27. Analysing gels by slicing

1. Load the sample for electrophoresis onto the polyacrylamide gel utilizing alternate wells only.

2. When electrophoresis is complete, unclamp the mould from the apparatus and carefully remove one glass plate so that the gel remains supported by the remaining glass plate.

3. Using the position of the sample well and tracking dye as guides, cover the entire sample lane with a strip of clear adhesive tape. This prevents the sample lane distorting due to dehydration, and provides support for the gel which facilitates the cutting process. Mark the position of the sample lane(s) by drawing lines on the back of the glass plate.

4. Place the slab gel, still on its glass plate, on top of a sheet of graph paper such that the top of the separating gel is superimposed on a line on the graph paper, and the sides of the gel are perpendicular to this. Secure the glass plate onto the paper with adhesive tape.

5. Using a rigid blade (e.g. a microtome blade, or single-edged razor blade) cut down the entire length of the sample lane on both sides, using the edge of the adhesive tape as a guide. Cut the stacking gel from the separating gel and discard, as this is too frail to slice effectively. Starting at the top of the separating gel, slice the line into 2 mm slices using the underlying graph paper to direct the cutting. The slicing can be done either through the adhesive tape, or alternatively, the tape can be repeatedly drawn back to expose 1−2 cm of gel which is then sliced.

6. Transfer each slice to a counting vial and depolymerize by adding 400 μl of 60% (200 volumes) hydrogen peroxide. Ensure the slices are at the bottom of the vials and immersed in hydrogen peroxide. This is facilitated by inclining the vials. Cap the vials and incubate at 50°C.

7. When the slices have fully depolymerized, allow the vials to cool, add 20 ml of a water-miscible scintillant, and count.

3.7.6 Visualizing proteins after electroblotting onto nitrocellulose

In addition to locating proteins directly in the gel, they can be visualized by the use of antibodies or lectins after being transferred from the gel onto nitrocellulose paper under the influence of an electric field, a process termed electroblotting. Electroblotting apparatus is commercially available (e.g. Trans-Blot Cell; Bio-Rad). Plastic gloves should be worn during all manipulations.

Protocol 28. Electroblotting proteins from gels

1. After electrophoresis, equilibrate the gel in a transfer buffer of 10 mM glycine, 25 mM Tris-HCl, pH 8.3, for 30 min with gentle agitation.

2. Assemble the blotting sandwich in a clamp immersed in a vessel filled with pre-cooled (4°C) transfer buffer. Into the clamp place Scotchbrite sponge and Whatman 3 mm paper cut to an appropriate size for the gel and pre-soaked in transfer buffer. Place the equilibrated gel on top of the paper and overlay

with the pre-wetted nitrocellulose sheet. Air bubbles trapped between the gel and nitrocellulose will cause poor electrical contact, thus transfer of proteins will be distorted or prevented. To avoid this, place the nitrocellulose onto the gel by holding the edges (wear gloves) and allow the middle of the sheet to sag. Place the middle section in contact with the gel then slowly lower the edges to expel the air. Overlay this with 3 mm paper, and roll the sandwich with an acid-washed test-tube to ensure that all the air is expelled. Complete the sandwich with another piece of pre-soaked Scotchbrite sponge, and tightly close the clamp.

3. Locate the clamp into the electroblotting tank with the nitrocellulose nearest the anode. As electroblotting generates heat, the blotting buffer should be cooled (4°C) by an immersed cooling coil connected to a circulating cooling bath and stirred continually. Electroblot the protein using 150−200 mA for 3−16 h, depending on the efficiency of transfer. Proteins will pass through the nitrocellulose sheet if the duration of transfer is too long. A second nitrocellulose sheet can be incorporated into the sandwich if this is suspected.

Note that a number of variants on the original nitrocellulose transfer media are now widely available, and are claimed to be superior in their qualities (e.g. zeta-probe, Bio-Rad). For discussion of the utility of different membranes in peptide sequencing after blotting, see Section 6.3.2.

Transfer of proteins is not uniform. Low molecular weight peptides transfer more rapidly than high molecular weight peptides. In addition, proteins transfer more slowly from high percentage polyacrylamide gels and thick gels. Occasionally enough protein remains in the gel to allow detection (see Section 3.7.1) after electroblotting. When a blot is obtained the protein can be probed with antibodies or lectins after non-specific protein binding sites on the nitrocellulose are blocked.

Protocol 29. Probing electroblots of gels

1. Incubate the nitrocellulose in 500 mM NaCl, 20 mM Tris-HCl, pH 7.5 (TBS) containing 3% (w/v) gelatin for 90 min at 37°C.

2. Discard the blocking buffer and add the first antibody in TBS containing 1% (w/v) gelatin. Incubate overnight shaking gently.

3. Remove the antibody, wash the nitrocellulose with water, then wash with TBS supplemented with 0.5 ml/l Tween 20 (TTBS) for 10 min. Repeat the TTBS wash. Wash with TBS for 10 min.

4. Add the radiolabelled second antibody (or Protein A) and process as for the first antibody.

5. Dry the nitrocellulose sheet and autoradiograph (see Section 3.7.3).

It should be noted that Protein A does not bind to all classes of immunoglobulin. Lectins can also be used as probes and detected by a radiolabelled anti-lectin antibody.

Conversely, the lectin itself can be radioiodinated. When lectins are utilized, control gels should be processed which have been pre-incubated with an excess of the appropriate sugar for the lectin being used. Note that radiolabelled lectins can also be used to probe the original gel (see Chapter 5, Section 7).

It is noteworthy that semi-dry blotting systems have recently been marketed (e.g. Biotrans, Gelman Sciences, BioRad, etc.) which do not require cooling, and use small (200 ml) volumes of buffer.

4. Two-dimensional peptide maps

The approach of 1D peptide maps (see Section 3) can be expanded. With 2D peptide mapping a receptor is digested by one proteolytic procedure and the products resolved by SDS-PAGE. Individual products can then be excised from the gel and proteolytically cleaved employing the same, or a different method and reanalysed on SDS-PAGE. This protocol is particularly applicable to determining precursor/product relationships. If a peptide is to be excised from a gel, its location in the gel has to be known. This is conveniently done by referring to fluoresceinated or coloured molecular weight standards run on either side of the sample lane.

4.1 Fluoresceination of standard proteins

Protocol 30. Fluoresceinating proteins

1. Dissolve 2 mg of fluorescein isothiocyanate (FITC) in 1 ml of 25 mM di-sodium tetraborate (0.953 g/100 ml).
2. Dissolve 1 mg of standard protein(s) in 1 ml of 25 mM di-sodium tetraborate containing 1% (w/v) SDS.
3. Add 200 μl of FITC solution to 1 ml of the standard protein solution, and incubate at 55°C for 60 min.
4. Add 12 μl of β-mercaptoethanol and incubate at 55°C for 15 min.

If required, free label can be separated for the labelled protein by gel filtration; however, some free fluorescein label is convenient, as it locates the dye front of sample lanes. Obviously it is helpful to let the choice of molecular weight standard be dictated by the molecular weight of the peptide being excised. For example, when excising intact mAChR we found that reduced bovine serum albumin (BSA) migrates at the leading edge of the receptor, making it an ideal marker for mAChR (see *Figure 6*). Fluoresceinated standards also ensure rapid detection of any anomalous peptide migration or defective protein focusing. Note that a variety of prestained marker proteins are now commercially available, e.g. from Bio-Rad (blue standards) or Amersham (Rainbow standards). These are extremely convenient to use, and mean that molecular weights can be determined without fixing and staining the gel, blotting can be followed easily, etc. The coloured standards can be added conveniently to the unknown lanes on gels for autoradiography, and can thus be used directly to estimate the molecular weights of labelled bands. Note that the β-mercaptoethanol concentration recommended by Amersham for use with the Rainbow standards is too high. The following procedure works better:

Protocol 31. Preparing Rainbow standards for SDS-PAGE

  1. Take a 15 μl aliquot of the standard mixture.

  2. Add 0.5 μl of β-mercaptoethanol.

  3. Heat at 100°C for 2 min.

  4. Add an equal volume of buffer and use immediately.

4.2 Re-digesting peptides generated by 1D peptide maps

Protocol 32. Excising a specific protein from a gel

  1. After proteolytic cleavage of the radiolabelled receptor protein, prepare the sample and standards for SDS-PAGE. Using only alternate sample wells of a gradient gel, load the sample flanked by fluoresceinated or coloured standards, and electrophorese.

  2. When electrophoresis is complete, remove one glass plate and place the gel, supported on the other glass plate, on a UV (366 nm) transilluminator, causing the standards to fluoresce (not necessary with coloured markers). Take safety precautions against the UV light, such as wearing safety goggles and gloves.

  3. Excise the peptide band with a scalpel blade using the fluorescing standards as guides. This excised peptide can then be cleaved further either in the gel slice, or in the stacking gel of the second SDS-PAGE. Note that very small quantities (*c*. 10 ng) of fluoresceinated standards can be detected, cutting down the risks of cross-contamination.

4.2.1 Cleavage in excised gel slices

Protocol 33. Cleaving proteins in an excised gel slice

  1. The buffer in the gel slice can be changed to suit the requirements of the second endoproteinase. Place the slice in a microcentrifuge tube, add 1 ml of the required buffer and gently mix for 40 min at room temperature.

  2. Remove the buffer and freeze-dry the slices. Add the endoproteinase in the minimal volume of suitable buffer necessary to rehydrate the slice. This buffer should not contain SDS, as enough SDS will be present in the slice to ensure that the peptide is fully denatured. As the slice rehydrates, the enzyme is drawn into contact with the peptide. Incubate as in Section 3.1 until the required degree of cleavage is attained.

3. Transfer the gel slices into the sample wells of a gradient gel using clean fine forceps. Then using a syringe and a warm hypodermic needle, seal the slice into the well with 1% (w/v) agarose in 0.1% (w/v) SDS, 125 mM Tris-HCl, pH 6.8, to ensure a good electrical contact. Take care not to entrap air bubbles under the gel slice. Keep the agarose solution at 60°C to prevent solidification. Run the gel as described in Section 3.6.1.

4.2.2 Cleavage in the stacking gel

Protocol 34. Cleaving proteins in the stacking gel during SDS-PAGE

1. Perform *step 3* of Section 4.2.1.
2. Overlay the slice in the sample well with endoproteinase in 0.1% (w/v) SDS, 10% (v/v) glycerol, 50 mM Tris-HCl, pH 6.8. Then start the electrophoresis with a low current of 4 mA. The endoproteinase and peptide stack together allowing cleavage to occur. The degree of cleavage produced by a fixed amount of endoproteinase can be increased by (a) electrophoresing the sample through the stacking gel more slowly, (b) using larger stacking gels (up to 6 cm), or (c) allowing the enzyme and peptide to stack together then stopping the electrophoresis for up to 30 min.
3. Run the peptides through the separating gel using standard conditions (see Section 3.6.1).

An alternative to cutting protein bands out is to excise the entire sample lane from the first gel without fixing, and locate all of this along the top of the stacking gel of the second gel using the protocol outlined above. Obviously sample wells are not required for this procedure. This strategy allows all the proteolytic products from the first digest to be analysed by the second digest.

4.2.3 Gel-permeation chromatography

A variant on the use of SDS-PAGE to perform the first-dimension separation is to use gel-permeation chromatography. We have used FPLC gel-filtration on a Superose 12 column (Pharmacia) in 50 mM Tris-Cl/0.1% SDS to good effect for this purpose. Although the resolution of this method is not as high as that of SDS-PAGE, it has several compensating virtues:

(a) It is rapid.
(b) It is non-destructive. The polypeptides are recovered in solution, in a state suitable for further cleavage, either chemical or enzymatic, or modification.
(c) The recoveries are excellent, *c.* 70% even for small quantities of hydrophobic peptides. The UV absorbance of the peptides at 214 nm is monitored immediately before their collection, providing a sensitive and direct measure of recovery.
(d) In contrast to SDS-PAGE, impurities are not introduced into the samples.

5. Glycosylated peptides

5.1 Using immobilized lectin to construct peptide maps

Receptors that mediate their effects by coupling to G-proteins appear to exhibit common structural elements (24), one of which is *N*-glycosylation sites near the N terminus of the receptor protein. This carbohydrate moiety therefore forms a natural label for the N terminus and may be exploited by the use of immobilized lectin for constructing simplified peptide maps. Immobilized lectin can be used to extract from a partial proteolytic cleavage of denatured receptor the glycosylated N terminus plus increasing lengths of receptor protein, which when analysed by SDS-PAGE produces a ladder of discrete products. The difference in molecular weight of each product will be dictated by the position of the cleavage sites in the amino-acid sequence. For example, the mAChR protein possesses Lys residues only in the loops between transmembrane (TM) segments. Consequently, wheat-germ, agglutinin (WGA) − Sepharose can extract from a Lys-C digest (see Section 3.1.3) of mAChR, a ladder of products corresponding to the N terminus plus increasing numbers of TM segments (*Figure 9*). When the mAChR has been irreversibly labelled with the ligand [^{3}H]propylbenzilylcholine mustard ([^{3}H]PrBCM) and this protocol followed, ^{3}H was only observed in the bands which included TM 3, thereby helping to locate the attachment site of [^{3}H]PrBCM. As a peptide-mapping technique, this strategy is applicable to glycosylated receptor proteins in general.

Protocol 35. Peptide mapping with immobilized lectin

1. Determine which lectin may be used to specifically adsorb the receptor under study. This is elucidated empirically by testing the ability of the receptor to bind to a range of different lectin − Sepharoses (marketed by Sigma) and be specifically eluted by the appropriate sugar. WGA − Sepharose has often proved useful for receptors in this regard.

2. Specifically cleave the receptor protein in a microcentrifuge tube (see Section 3.1) in a total volume of 100 μl.

3. Dilute the digest tenfold using a lectin buffer of 0.1% (w/v) digitonin, 200 mM NaCl, 50 mM Tris-acetate, pH 7.8. This dilutes out the denaturing detergent (SDS) with a non-denaturing detergent (digitonin), thereby enabling the proteolytically cleaved sample to be exposed to lectin − Sepharose without the specificity of the lectin/carbohydrate interaction being lost through lectin denaturation. The NaCl is included in the buffer to suppress non-specific binding of peptides to the lectin − Sepharose via ionic interactions.

4. Wash the lectin − Sepharose with lectin buffer, then add 100 μl of this gel to the diluted digest sample. Incubate for 2 h at 4°C, with gentle stirring.

5. Spin down the gel, remove the supernatant and wash the gel with cold lectin buffer by several resuspension/centrifugation cycles. This washing should be

done thoroughly but rapidly to reduce losses caused by glycosylated peptides dissociating from the lectin−Sepharose.

6. Elute the glycosylated peptides from the gel with 200 μl of lectin buffer containing 200 mM of the appropriate sugar for the lectin being employed.

7. Gel-filter this sugar eluate into a suitable buffer for SDS-PAGE (running buffer, Solution H in Section 3.4).

The specificity of uptake can be controlled for by blocking peptide uptake with an excess of the eluting sugar. In addition, it is advisable to perform a control elution using lectin−Sepharose that has not been exposed to protein. This is to determine if lectin leaching from the Sepharose is contaminating the sugar eluate. A low level of leaching is to be expected.

5.2 Deglycosylation of peptides

The electrophoretic properties of *N*-glycosylated peptides are anomalous, and accurate estimation of their molecular weights by SDS-PAGE requires an exhaustive and sophisticated analysis. It is useful to have a method for the deglycosylation of such peptides, in order that better estimates may be obtained. We have found that hydrophobic, N-terminal *N*-glycosylated peptides isolated from digests of muscarinic receptors can be effectively deglycosylated using an enzyme, *N*-glycanase (Genzyme, supplied by Koch-Light) which hydrolyses the substituted asparagine linkage, generating a free carboxylate group, and removing the glycan residues. The protocol that we have followed is given below.

Protocol 36. Deglycosylating peptides prior to SDS-PAGE

1. Make up 400 mM sodium phosphate buffer containing 0.5% digitonin and 0.4 mM EDTA. Adjust to pH 8.6.

2. To 10 μl of sample, e.g. in 50 mM Tris-Cl, 0.1% SDS, pH 7.8, add 40 μl of phosphate/digitonin/EDTA and 30 μl of deionized water.

3. Add 4 μl of 100 mM 1,10-phenanthroline in methanol.

4. Add 5 μl of *N*-glycanase, giving a final concentration of *c*. 15 units/ml. Incubate for 24 h at 37°C.

5. If the peptides are to be electrophoresed, add 9.5 μl of 10% (w/v) SDS, and sonicate. Gel-filter into diluted running buffer, and reconcentrate as described previously (see *Protocol 21*, step 10).

The enzyme tolerates 0.01% SDS, in the presence of a non-ionic detergent. The chelating agents are included to inactivate residual metalloproteases, which are known to be present in the initial part of the enzyme purification procedure. A variant on this procedure is given in Chapter 8. Endo−F (Boehringer) can also be used.

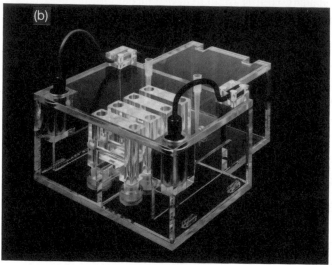

Figure 10. Apparatus for the electroelution of proteins from polyacrylamide gels, modified from Hunkapiller *et al.* (25). (a) The elution cell; (b) the elution tank with cells *in situ*.

6. Sequencing peptides after SDS-PAGE

6.1 Electroelution of peptides from gel slices

The high resolution of SDS-PAGE can be utilized as a means of obtaining purified peptides for protein sequencing after proteolytic cleavage of the receptor. A prerequisite to sequencing is recovery of these peptides from the gel. This is readily achieved by allowing the protein to migrate out of an excised gel slice under the influence of an electric field, viz. electroelution. Although commercial apparatus is available (e.g. the Biotrap marketed by Anderman, UK), we have found that the apparatus and methodology described by Hunkapiller and colleagues (25) is satisfactory. The electroelution tank and cell were made according to the specifications given in reference 25 with the following exceptions. With reference to *Figure 10*:

(a) The cell does not have the original 'H' shape but is left as a block, thereby endowing greater mechanical strength.

(b) The cross-bridge is drilled from one side through to the other vertical chamber, and the resulting drill hole sealed with a nylon screw. This construction ensures that the cell has strength without the necessity for adhesives to be used, thereby removing a source of possible contamination or leakage.

As with the original cell design, dialysis membranes are held in place at the foot of each leg, by the compression supplied by screw-caps lined with rubber washers.

6.1.1 Preparing the dialysis membrane

Protocol 37. Dialysis membrane preparation

1. Soak 20/32 dialysis tubing in 1% (w/v) NH_4HCO_3 solution at 60°C for 60 min.

2. Wash the tubing in deionized water, then soak it in 0.1% (w/v) SDS at 60°C for 60 min. The tubing may then be stored in 0.1% (w/v) SDS, 0.1% (w/v) NaN_3 solution for several days.

3. Cut a small length of tubing and wash it with water inside and outside. Place the tubing onto filter paper overlaying several layers of paper towel, and punch out discs using a sharp No. 8 cork borer. Take only the top disc, i.e. not the one in contact with the filter paper, rinse in water then use.

6.1.2 The electroelution procedure

Protocol 38. Electroelution of proteins from a gel slice

1. Clean the elution cell by sonicating it in 0.1% (w/v) SDS. Then, wearing plastic gloves, locate the dialysis membrane discs into the caps of the elution cell by careful manipulation with a clean microspatula. Secure the caps onto the cell and fill with water. By standing the cell on filter paper for 1 − 2 h prior to use, any leakage from the cell will be readily apparent.

2. Dice (not mince) the excised gel slice and rinse the pieces briefly with an elution buffer of 0.1% (w/v) SDS, 50 mM Tris-acetate, pH 7.8. Some workers argue against any mechanical interference with the slice. Empty the water from the

cell and load the washed gel slices into the cathode chamber of the cell. The entrapment of air bubbles in the cathode chamber can be prevented by placing a small volume of elution buffer into the chamber before adding the slices. Fill the cell with elution buffer, ensuring air bubbles are not trapped, especially in the cross-bridge.

3. Locate the cell in the elution tank filled with elution buffer such that the slices are on the cathode side. Remove any air bubbles trapped under the cell using a syringe with a bent needle. By placing a small piece of plastic tubing over the end of the needle as a sheath, inadvertent rupture of the membrane can be avoided.

4. While pumping buffer from the mixing chamber to the two electrode chambers, electroelute the protein at 100 V for 20 h, or until all the protein has migrated to the anode chamber. To follow this, it is convenient to have the protein 'spiked' with [^{125}I]sample (see Section 2.1). With this buffer system, electroelution can be performed at cold-room temperature (4°C).

5. The SDS concentration can be reduced by electrodialysis. Replace the elution buffer with a dialysis buffer of 0.02% (w/v) SDS, 10 mM NH_4HCO_3, and continue to electrodialyse for 36 h at 150 V, changing the buffer every 12 h.

6. The concentrated peptide sample can be removed from the anode collecting chamber using a plastic pipette tip sheathed in soft plastic tubing to prevent snagging of the membrane.

Further information on electroelution is given in reference 40, Chapter 6.

6.2 Preparing peptides for solid-phase sequencing

Protocol 39. Generating peptides for solid-phase sequencing

1. Specifically cleave the receptor protein (see Section 3.1) after full reduction and S-carboxymethylation (see Section 2.5).

2. Resolve the cleavage products by SDS-PAGE using a gradient gel, taking all the necessary precautions against possible amino-acid modification (see Section 3.3).

3. Excise the peptide bands and electroelute the protein (see Section 6.1).

4. Couple the electroeluted peptide to a solid support. This can be done using *p*-phenylene di-isothiocyanate (DITC)-glass (26) which couples to amino groups in the peptide. Although this method can be used for peptides generally, it is particularly applicable for immobilizing peptides generated by cleavage with endoproteinase Lys-C (see Section 3.1.3) which generates peptides possessing a Lys at the C terminus. The ϵ-amino group of this Lys can then be utilized in the coupling reaction. Likewise, peptides generated by CNBr cleavage (see Section 3.2.1) have a homoserine or homoserine lactone at the C terminus,

and hence can be immobilized onto aminoethylaminopropyl (AEAP)-glass, utilizing the homoserine−lactone coupling reaction described by Horn and Laursen (27).

Further details of this methodology are given in reference 42. Suitably derivatized glass beads are commercially available (Sigma, Pierce).

A great advantage of the immobilization procedure and solid-phase sequencing is that contaminating salts and detergent can be removed efficiently by washing prior to sequencing.

6.3 Preparing peptides for gas-phase sequencing

6.3.1 Sequencing peptides after electroelution

Protocol 40. Sequencing peptides after electroelution

1. Perform *steps 1−3* of Section 6.2.
2. Instead of coupling to a solid support as in Section 6.2, apply the sample directly to the sequence support matrix. If this is the quaternary amine Polybrene, it should be noted that SDS in the sample can compete with the peptide for binding sites, thereby reducing the peptide-binding capacity of the support. The SDS concentration can be reduced by electrodialysis (see Protocol 38, *Step 5*). Conversely, provided the sample peptide's molecular weight is not too low, the sample can be diluted with SDS-free buffer then concentrated in a Centricon 10 concentrator (Amicon Ltd).

6.3.2 Sequencing peptides after electroblotting

The rationale of this technique is to separate peptides generated by proteolytic cleavage of the receptor by SDS-PAGE, then directly transfer these onto derivatized paper using the electroblotting technique. Peptide bands can then be cut out and sequenced directly. As all the protein bands from the digest are transferred, all the digestion products can be conveniently sequenced. If the peptides are being electroeluted for sequencing (see Section 6.1) each peptide would have to be individually processed, which is time-consuming.

Protocol 41. Sequencing peptides after electroblotting

1. Perform *steps 1* and *2* of Section 6.2.
2. Construct a sandwich for electroblotting (see Section 3.7.6), but instead of using nitrocellulose sheets, use derivatized glass-fibre sheets. Polybrene-coated glass-fibre sheets have been advocated for protein sequencing by Vandekerckhove and colleagues (28) who found that glycine contamination from buffers was not a problem after the first few sequencing cycles. Alternative

forms of activated glass-fibre sheet have been developed for peptide sequencing, including aminopropyl-, quaternary ammonium-, and DITC-derivatized glass-fibre sheets. The manufacture and application of these to sequencing has been described in detail (29).

3. After electroblotting, excise the peptide bands transferred onto the derivatized sheets, and apply directly to the sequencer.

Hydrophobic quaternized ammonium polybase-coated glass-fibre paper is available through Janssen Life Science Products. Applied Biosystems recommend the use of Immobilon (PVDF) membranes (Millipore) for blotting followed by sequencing. Further details are given by Matsudaira (37). It is advisable to put a polybrene-coated disc under the PVDF membrane in the sequencer, to retain peptides which may be eluted from the PVDF. This applies particularly to pulsed liquid-phase sequencers.

This method is as applicable to sequencing from 2D SDS-PAGE as it is to 1D SDS-PAGE.

7. Purification of sequenceable peptides without using SDS-PAGE

7.1 Proteolytic cleavage of receptors in the absence of SDS

Frequently in receptor research the peptides generated by proteolytic cleavage of the receptor protein are resolved by methods other than SDS-PAGE. When employing these methods the presence of SDS can be detrimental to peptide resolution, e.g. reverse-phase HPLC. Consequently, we developed the following technique for generating SDS-free peptides from a receptor sample containing SDS (30).

Protocol 42. *in situ* cleavage of adsorbed protein

1. Precipitate the receptor protein in a microcentrifuge tube by adding ice-cold absolute alcohol to a final concentration of 90−95% (v/v) and incubate at −20°C overnight.

2. Spin down the protein precipitate and wash it twice with ice-cold absolute alcohol. The SDS is soluble in alcohol so is effectively removed by this precipitation procedure.

3. Rinse the pellet and tube with the appropriate buffer for the endoproteinase being employed, omitting any SDS, and digest the protein *in situ* on the tube with endoproteinase as described in Section 3.1.

When the receptor is precipitated onto the tube and SDS-free buffer is added, the hydrophobic domains of the protein adhere to the plastic, whereas the hydrophilic domains extend into the aqueous medium. Enzymatic cleavage occurs in both domains, but because of the hydrophobic interactions, only the hydrophilic peptides are released into solution. If required, both fractions can be analysed by SDS-PAGE after the hydrophobic peptides have been extracted by 1% (w/v) SDS, 60°C for 15 min. This cleavage technique has two main benefits. First, the most easily manipulated peptides are readily obtained, the strongly hydrophobic, difficult to manipulate, peptides are removed. Secondly, because only a proportion of the receptor molecule is recovered, resolving the proteolytic products is simplified. The water-soluble peptides generated can then be separated by HPLC or FPLC without further manipulation.

SDS has been removed from receptor polypeptides before enzymatic digestion by gel-filtration in 70% v/v formic acid on Superose 12, followed by evaporation of the formic acid by lyophilization. It may also be possible to reduce SDS concentrations to acceptably low values by treatment of peptides with Extractigel-D detergent-removing gel (Pierce), or with Bio-beads SM-2 or SM-4 (Bio-Rad). A key factor here is to make sure that the porosity of the gel is such that the detergent can penetrate the beads, but the peptide cannot. In practice, this may mean that the peptide molecular weight has to be in excess of $c.$ 10 kDa. In fact, SDS concentration in solutions of polypeptides can often be reduced by simple gel-filtration, provided that the SDS concentration is first reduced to below the CMC, which is about 0.2% w/v.

7.2 Isolating peptides by HPLC and FPLC

The most popular method for purifying peptides that do not contain SDS is reverse-phase HPLC. The systems of choice are microbore systems with high-precision pumps (e.g. Gilson) as these are more sensitive, and cartridge columns, which are small and relatively cheap. The quality of the pumps is critical for the performance of a microbore system. The pumps chosen should be capable of delivering a smooth, reproducible, pulsation-free gradient at a flow rate of 50 μl per min. Dual-wavelength detection is a good investment, and a fraction collector capable of collecting small fractions is also necessary. The wide pore (300 Å) beads seem more suitable for peptides encountered in receptor sequencing. The hydrophilic peptides from an *in situ* digest (see Protocol 42) can be directly injected onto a reverse-phase column equilibrated with 0.1% (v/v) HPLC-grade trifluoroacetic acid (TFA; Pierce), and the peptides eluted by a linear (0−60%) gradient of HPLC-grade acetonitrile (Romil Chemicals) containing 0.1% (v/v) TFA. The eluate is then passed through a spectrophotometer to monitor absorbance at 215 nm. As the solvents are UV-transparent, eluted peptides appear as absorbance peaks, allowing fractions of purified peptides to be identified. Large numbers of reverse-phase columns are marketed, including Anachem, Waters, LKB, and Whatman. Packing materials are bonded with an alkyl phase, the chain-length of which is variable: C_{18}, C_8, C_4, C_2, etc. Although the difference in selectivity between the various alkyl groups is not yet fully understood, in practice the longer the alkyl chain length the greater the retention. Thus peptides can be eluted with lower acetonitrile concentrations from a C_8 column than from a C_{18} column. The use of end-capped matrices prevents the peptide sample from interacting with residual exposed silanols, ensuring that retention is by virtue of the bonded alkyl phase. We have found that Zorbax C8 columns give good results. Aquapore (Brownlee) columns are also popular, and are available with a range of alkyl chain lengths, or with cationic and anionic substitutions, in a 300 Å pore size, and a cartridge format which is suitable for small amounts of peptides. The Vydac C18 column, with 300 Å pore size, also has a good reputation as an all-purpose column with the ability to cope with a very wide size-range of peptides. TSK produce the TMS 250 Cl column, with a 250 Å pore size, and 10 μM beads, which is said to give very good resolution, and can give good separation even of quite large proteins. There are also non-porous resin-particle-based materials available, which give good resolution and recovery, but take lower loads. Finally, the Pharmacia FPLC Pro reverse-phase column also seems to give good resolution. Many of the small, cartridge columns can, in principle, be run on the Pharmacia FPLC system. All solutions for HPLC should be degassed and passed through a 0.5 μ filter prior to use, then have helium continuously bubbled through them to retard regassing.

The hydrophobic peptides may also be amenable to purification by reverse-phase HPLC after extraction by 50% (v/v) TFA, followed by addition of acetonitrile and water to give the gradient starting conditions. Note that peptides that have been obtained from receptor digests and fractionated by reverse-phase HPLC have almost always turned out to be from the hydrophilic loops, and not from the hydrophobic portions of the polypeptide chain. The retention time of peptides can be calculated from the amino-acid composition, as shown by Meek (38), and gives some indication of whether or not a given peptide is likely to run. Solvent systems employing isopropanol may be more suitable for reverse-phase fractionation of more hydrophobic peptides. Some further details are given in reference 40, Chapter 6.

The HPLC system can also separate peptides by molecular weight using gel-filtration columns such as TSK G3000 SW (available from many suppliers, e.g. Anachem). Various columns are available, differing in the molecular weight range they resolve. These columns are compatible with aqueous buffers and also with water-miscible organic solvents, e.g. methanol, acetonitrile. They are also stable to aqueous solutions of denaturants such as SDS, guanidinium hydrochloride, and urea. However, it should be noted that SDS interacts with the column and is difficult to remove, thereby rendering the column permanently 'denaturing', so once exposed to SDS the column should be reserved for SDS buffers only. It should also be noted that denaturation of peptides can alter their apparent molecular weight, which may have ramifications on the choice of column used. Finally, size exclusion columns can be employed in series to improve resolution, provided that they are connected in order of decreasing pore size. Cross-linked versions of the TSK column materials are now available, and give enhanced resolution.

Ion-exchange and hydrophobic interaction chromatography is also compatible with HPLC systems, with both anionic- and cationic-exchangers available, e.g. TSK DEAE-3SW and TSK CM-3SW, respectively (available from Anachem). Anachem produce a very useful guide to HPLC separations.

The same purification procedures of reverse-phase, molecular-sieve, and ion-exchange chromatography that are performed using HPLC can be employed using the FPLC system of Pharmacia/LKB. This is an integrated system specifically designed to give rapid, high resolution of biological molecules using a lower operating pressure than HPLC. The general utility of FPLC for analysing samples is noteworthy (see Section 4.2.3 for use of gel-filtration on Superose columns). As the FPLC system is sensitive and yields are high, it can be used to monitor the species present in a sample and to quantitate the protein content.

7.3 Isolating peptides by Sephadex LH

This is available as Sephadex LH-20 and Sephadex LH-60 from Pharmacia, which differ in the molecular weight range over which they fractionate. Separation of peptides on these hydroxypropylated Sephadex matrices is predominantly based on molecular weight, although there is a hydrophobic interaction component as well. This gel has proved useful for resolving hydrophobic peptides from the sarcoplasmic Ca^{2+} ATPase (31) without the use of detergents, by using a solvent system of ethanol:formic acid:water in the ratio 70:26:4, which seems to prevent peptide aggregation. With mAChR, 40% of the hydrophobic peptides from an *in situ* digestion (see Protocol 42) could be extracted with formic acid, and after the addition of ethanol be analysed on Sephadex LH-60 (30). The same solvent system has been employed for reverse-phase HPLC (32), in which a μBondapak C_{18} column (Waters) was equilibrated with 5% (v/v) formic acid, 40% (v/v) ethanol in water. The sample was injected onto the column in 88% (v/v) formic acid, and peptides eluted with an ethanol gradient in 5% (v/v) formic acid.

Hydrophobic peptides formed by digestion of rhodopsin have been successfully fractionated by gel-filtration on Sephadex LH-20 and LH-60 in formic acid:acetic acid:ethanol:water 1:1:2:1. Further details are given in reference 40, Chapter 6. In general, as far as hydrophobic peptides are concerned, high-resolution gel-filtration fractionations, either in detergent-containing buffers (see Section 4.2.3), or in organic solvents such as formic acid/ethanol/water or formic acid/acetic acid/ethanol/water appear, at present, to be amongst the most useful separation procedures. We have made extensive use of Superose 12 FPLC columns run in Tris-Cl/0.1% SDS, pH 7.8, for fractionating mAChR-derived peptides. The range of the fractionation technique is good (*c*. 1−100 kDa) but the resolution is only moderate. Much more work needs to be done on the high-resolution fractionation of hydrophobic peptides.

Acknowledgements

I would like to express my gratitude to Dr E. C. Hulme for his constructive comments on this manuscript, to Mrs C. A. M. Curtis and Dr D. Poyner for useful discussions during its preparation, and to Mrs G. A. Chamberlain for her typing skills. Where reference is made to mAChRs, these studies were performed in the laboratories of Dr E. C. Hulme and Dr N. J. M. Birdsall, National Institute for Medical Research, Mill Hill, London, UK, where the author was supported by the Medical Research Council.

References

1. Regeoczi, E. (1984). *Iodine-labelled Plasma Proteins.* Vol. I, CRC Press, Florida.
2. Shorr, R. G. L., Strohsacker, M. W., Lavin, T. N., Lefkowitz, R. J., and Caron M. G. (1982). *J. Biol. Chem.,* **257**, 12341.
3. Greenwood, F. C., Hunter, M. W., and Glover, J. S. (1963). *Biochem. J.,* **89**, 114.
4. Markwell, M. A. K. (1982). *Anal. Biochem.,* **125**, 427.
5. Markwell, M. A. K. and Fox, C. F. (1978). *Biochemistry,* **17**, 4807.
6. Bolton, A. E. and Hunter, W. M. (1973). *Biochem. J.,* **133**, 529.
7. Laemmli, U. K. (1970). *Nature,* **227**, 680.
8. Cleveland, D. W., Fischer, S. G., Kirschner, M. W., and Laemmli, U. K. (1977). *J. Biol. Chem.,* **252**, 1102.
9. Wilcox, P. E. (1970). In *Methods in Enzymology.* (ed. G. E. Perlmann and L. Lorand), Vol. 19, p. 64. Academic Press, New York.
10. Venter, J. C. (1983). *J. Biol. Chem.,* **258**, 4842.
11. Allen, G. (1981). In *Laboratory Techniques in Biochemistry and Molecular Biology.* (ed. T. S. Work and R. H. Burdon), Vol. 9. Elsevier/North Holland, Amsterdam.
12. Cole, R. D. (1967). In *Methods in Enzymology.* (ed. C. H. W. Hirs), Vol. 11, p. 315. Academic Press, New York.
13. Jekel, P. A., Weijer, W. J., and Beintema, J. J. (1983). *Anal. Biochem.,* **134**, 347.
14. Houmard, J. and Drapeau, G. R. (1972). *Proc. Natl Acad. Sci. USA,* **69**, 3506.
15. Drapeau, G. R. (1977). In *Methods in Enzymology.* (ed. C. H. W. Hirs and S. N. Timasheff), Vol. 47, p. 189. Academic Press, New York.
16. Kobilka, B. K., Matsui, H., Kobilka, T. S., Yang-Feng, T. L., Francke, U., Caron, M. G., Lefkowitz, R. J., and Regan, J. W. (1987). *Science,* **238**, 650.
17. Gross, E. and Witkop, B. (1962). *J. Biol. Chem.,* **237**, 1856.
18. Schroeder, W. A., Shelton, J. B., and Shelton, J. R. (1969). *Arch. Biochem. Biophys.,* **130**, 551.
19. Savige, W. E. and Fontana, A. (1977). In *Methods in Enzymology.* (ed. C. H. W. Hirs and S. N. Timasheff), Vol. 47, p. 459. Academic Press, New York.
20. Koziarz, J. J., Kohler, H., and Steck, T. L. (1978). *Anal. Biochem.,* **86**, 78.
21. Burr, F. A. and Burr, B. (1983). In *Methods in Enzymology.* (ed. S. Fleischer and B. Fleischer), Vol. 96, p. 239. Academic Press, New York.
22. Morrisey, J. H. (1981). *Anal. Biochem.,* **117**, 307.
23. Sigel, E. and Barnard, E. A. (1984). *J. Biol. Chem.,* **259**, 7219.
24. Dohlman, H. G., Caron, M. G., and Lefkowitz, R. J. (1987). *Biochemistry,* **26**, 2657.
25. Hunkapiller, M. W., Lujan, E., Ostrander, F., and Hood, L. E. (1983). In *Methods in Enzymology.* (ed. C. H. W. Hirs and S. N. Timasheff), Vol. 91, p. 227. Academic Press, New York.
26. Wachter, E., Machleidt, W., Hofner, H., and Otto, J. (1973). *FEBS Lett.,* **35**, 97.
27. Horn, M. J. and Laursen, R. A. (1973). *FEBS Lett.,* **36**, 285.
28. Vandekerckhove, J., Bauw, G., Puype, M., Van Damme, J., and Van Montagu, M. (1985). *Eur. J. Biochem.,* **152**, 9.
29. Aebersold, R. H., Teplow, D. B., Hood, L. E., and Kent, S. B. H. (1986). *J. Biol. Chem.,* **261**, 4229.
30. Wheatley, M., Birdsall, N. J. M., Curtis, C. A. M., Eveleigh, P., Pedder, E. K., Poyner, D., Stockton, J. M., and Hulme, E. C. (1987). *Biochem. Soc. Trans.,* **15**, 113.
31. Green, N. M. and Toms, E. J. (1985). *Biochem. J.,* **231**, 425.

32. Gerber, G. E., Anderegg, R. J., Herlichy, W. C., Gray, C. P., Beimann, K., and Khorana, H. G. (1979). *Proc. Natl Acad. Sci. USA,* **76**, 227.
33. Mahoney, C., Smith, P. K., and Hermodson, M. A. (1981). *Biochemistry,* **20**, 443.
34. Fontana, A. (1972). In *Methods in Enzymology.* (ed. C. H. W. Hirs and S. N. Timasheff), Vol. 25, p. 419. Academic Press, New York.
35. Fernandez-Luna, J. L., Lopez-Otin, C., Soriano, F., and Mendez, E. (1985). *Biochemistry,* **24**, 861.
36. Wachter, E. and Werhahn, R. (1979). *Anal. Biochem.,* **97**, 56.
37. Matsudaira, P. (1987). *J. Biol. Chem.,* **261**, 10035.
38. Meek, J. L. (1980). *Proc. Natl Acad. Sci. USA,* **77**, 1632.
39. Laskey, R. A. and Mills, A. D. (1977). *FEBS Lett.,* **82**, 314.
40. Findlay, J. B. C. and Evans, W. H. (1987). *Biological Membranes: A Practical Approach.* IRL Press, Oxford.
41. Hames, B. D. and Rickwood, D. (ed.) (1990). *Gel Electrophoresis of Proteins, A Practical Approach.* IRL Press, Oxford.
42. Findlay, J. B. C. and Geisow, M. J. (ed.) (1989). *Protein Sequencing, A Practical Approach.* IRL Press, Oxford.

11

Expression of receptor genes in cultured cells

CLAIRE M. FRASER

1. Introduction

The cloning and sequence analysis of a number of genes encoding hormone and neurotransmitter receptors has provided a wealth of new information on the primary structure of these proteins and has revealed the existence of at least three multigene families of receptors including:

(a) those containing an integral ion channel, such as the nicotinic acetylcholine (1−4), GABA (5), and glycine (6) receptors;
(b) those that interact with guanine nucleotide regulatory proteins, such as the adrenergic (7−10) and muscarinic cholinergic (11−13) receptors; and
(c) those which stimulate mitogenesis and possess protein-tyrosine kinase activity, such as the insulin (14), and EGF (15) receptors.

Characterization of the genes encoding these receptors has necessitated their expression and assay in an appropriate system. The development of a number of techniques for gene transfer into eukaryotic cells (transfection) provides the investigator with several options for the expression of a particular gene. This chapter will consider the use of plasmid vectors for gene expression in mammalian cells in culture. While this approach requires proficiency in cell culture, it is straightforward and is not labour intensive. Depending on the choice of plasmid vector and host cell line, transfection of cultured cells can result in either transient or stable expression of genes.

2. Choosing a recipient cell line for use in transfection

Regardless of the cell line chosen for use as a recipient in transfection experiments, it is of utmost importance to make certain that one has a thorough understanding of the optimal conditions for cell growth. Excellent references on tissue culture procedures are available and the reader is referred to these for additional information (16,40). Cells should be maintained in log-phase growth at all times prior to transfection and cultures should be screened for mycoplasma contamination before use.

In selecting a recipient cell for use in transfection experiments one should ideally look for established (rather than primary) cultures that grow in monolayers. There is a high correlation between the growth rate of cells in culture and the ability to take up DNA (17); therefore, cell lines with rapid doubling times (<20 h) should be used whenever possible. Some suggested cell lines include Chinese hamster ovary (CHO) cells, HeLa cells, B-82 cells, and COS cells. Most of these lines are available from the American Type Culture Collection.

Perhaps the most important aspect in selecting a cell line for use in transfection studies is to identify a line that lacks the receptor to be expressed. While this point may seem rather obvious it can have profound implications for characterization of ligand binding to expressed receptors and an inappropriate choice of cell line can invalidate results obtained in functional assays such as stimulation of adenylate cyclase or phosphoinositide hydrolysis.

3. Vectors used in gene expression studies

3.1 General features

All plasmid vectors used for transfection of eukaryotic cells must possess four main components (17, *Figure 1*). The first elements that must be contained within the plasmid are a bacterial origin of replication (Ori) and an antibiotic-resistance marker. These features allow for amplification and selection of the plasmid in the appropriate bacterial host. The plasmid must also contain eukaryotic elements that control initiation of transcription, and these are generally derived from viruses such as SV40 or the mouse mammary tumour virus (MMTV). Similarly, the plasmid must also include sequences involved in the processing of DNA transcripts such as SV40 polyadenylation signals. The last required element in a plasmid vector is the gene for expression. This gene may be derived from either a cDNA or genomic clone. All vectors contain a multiple cloning site downstream from the eukaryotic promoter region that allow for gene insertion at one or several unique restriction sites in the vector.

3.2 Choice of promoter sequences

There are a number of eukaryotic promoter sequences available for use in plasmid vectors, and a judicious choice of promoter can greatly increase the level of expression of the receptor gene under study. Many viral promoters display a marked species preference, and therefore, the promoter sequence to be used in a plasmid expression vector should be most compatible with the recipient cell line (18). For example, it is thought that the SV40 early and late promoters function well in primate cells, such as the COS line, but are less efficient in murine cells. Similarly, a promoter of choice for high-level gene expression in murine cells has been the Moloney murine sarcoma virus (MSV) long terminal repeat (LTR) (19). Two promoter sequences

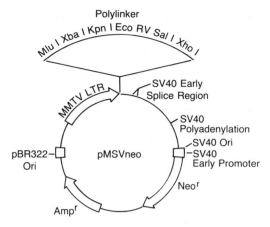

Figure 1. Schematic representation of the essential elements of a plasmid expression vector, pMSVneo. The bacterial origin of replication (pBR322 Ori) and the gene for antibiotic resistance (*Amp*r) allow for replication of the plasmid in *E. coli* strains HB101 and JM109, for example. This vector contains the dominant selectable marker, *neo*r, under transcriptional control of the SV40 early promoter that enables selection of stable transformants in culture medium containing the neomycin analogue, G418. Genes for expression are inserted into one of the six unique restriction sites in the polylinker region of the vector and are under the control of the MMTV LTR promoter. Their expression can be induced by addition of steroids to the culture medium. This plasmid vector has been utilized for expression of β-adrenergic and muscarinic cholinergic receptors in B-82 and CHO cells (21 – 23). pMSVneo was synthesized by Dr Fu-Zon Chung and is available from the author.

of wide applicability include the LTRs from the Rous sarcoma virus (RSV) and the mouse mammary tumour virus (MMTV) (20). A number of plasmid vectors are available commercially that contain either the RSV promoter and/or the MMTV promoter sequences (some examples are given in Section 9.3).

One caveat of the choice of promoter sequences has emerged from our work on the expression of the β-adrenergic receptor in B-82 cells, a murine cell line. We have found that the promoter sequence located in the first 1000 bases upstream from the coding region of the human β-adrenergic receptor functions as a strong promoter, being more efficient than the MMTV LTR but less efficient than the SV40 late promoter (21; Chung, Fraser, and Venter, in preparation). Therefore, if high-density expression of a receptor gene is desired it is suggested that more than one strategy be utilized as recommendations on the efficiency of promoter sequences should be taken as guidelines only.

3.3 Inducible promoter sequences

While it is likely that generally one would desire constitutive expression of a receptor gene expressed in a mammalian cell, in some instances it may be preferable to control the level of receptor transcription and expression via an inducible promoter.

The inducible promoters suitable for use in mammalian cells include the MMTV LTR under the control of steroid hormones and the metallothionein promoter under the control of heavy metal ions. It must be remembered that the stimulus for induction may affect other cellular systems in addition to transcription of the transfected gene. We have had considerable experience with the MMTV LTR and steroid-inducible gene expression in B-82 cells. In cells grown in the absence of steroid, receptor expression is generally very low, being in the order of $5-30$ fmol receptor/mg membrane protein (22,23). Addition of dexamethasone to cell cultures results in a three- to tenfold induction in the level of receptor expression, with the average density of receptors in stimulated cells being in the range of $100-150$ fmol/mg protein. Thus, use of the MMTV LTR promoter offers the advantage of manipulation of receptor density; however, the level of receptor expression following steroid induction is still low, especially when compared to the levels that we have achieved utilizing the β-receptor-associated promoter sequence (>1 pmol/mg protein) (21). Obviously, the choice of promoter will depend on the goals of the experiment.

4. Transient expression of receptor genes in cultured cells

Within a few hours of transfection, plasmid vectors are translocated to the cell nucleus and transcription of plasmid DNA begins. Cells can be harvested and assayed for specific gene products, either RNA or protein, within $24-72$ h following transfection. The advantages of performing transient expression of receptor genes is that the suitability of a particular plasmid vector or transfection protocol can be assessed within a few days. Modifications can be made, if necessary, before undertaking stable expression of genes. The obvious disadvantage to performing transient expression of genes is that one has only a limited supply of cells for assay. Furthermore, only a small fraction of the transfected cells take up the DNA and these cannot be grown as a continuous source of material.

5. Stable expression of receptor genes in cultured cells

In order to achieve stable expression of genes in cultured cells one requires a selection system that allows for survival of only those cells that have taken up the plasmid DNA and integrated it into the cell chromatin. The efficiency of cell transfection is usually in the order of $1/10^4 - 1/10^5$ cells. Surviving cells grow into discrete colonies that can be isolated, expanded, and assayed for the presence of a particular gene product. This entire procedure from transfection to cell assay generally requires $4-6$ weeks. When performed successfully one obtains clonal cell lines that continuously express a gene of interest, and this allows for an essentially unlimited supply of the gene products.

Two basic approaches for stable expression of genes in cultured cells have been successfully utilized. The first approach involves the expression of dominant selectable markers from *Escherichia coli* such as *gpt* or *neo*[r]. The second approach requires the use of mutant recipient cells lacking the genes for either thymidine kinase (tk) or dihydrofolate reductase (dhfr). The second scheme is less versatile than the first as it imposes considerable restrictions on the choice of recipient cells for gene transfer. Each of these systems is discussed in more detail below.

It is preferable if the plasmid vector containing the gene for expression also contains the selectable marker to be used for cell selection. This virtually ensures that any cells surviving in selective medium will be expressing the gene of interest. The supply of plasmid vectors available commercially is still somewhat limited, and one may find it necessary to construct a custom plasmid vector for use in expression studies. This provides for design of a vector most suitable to a particular laboratory's needs but does require a certain level of competence in molecular biology techniques. Alternatively, if a plasmid vector containing the desired marker is not available, one can perform a co-transfection in which one vector contains the gene to be expressed and a second vector contains the marker. The obvious disadvantage to this approach is that some cells may integrate only the plasmid containing the marker and survive in selective medium without expressing the gene of interest. This situation makes screening of clonal cell lines more time-consuming. In our hands, the efficiency of co-transfection of plasmid vectors has ranged from $20-95\%$.

5.1 *gpt* selection

This selection scheme involves manipulation of the pathways involved in purine biosynthesis and requires the expression of the bacterial enzyme, hypoxanthine phosphoribosyl transferase (HPRT), encoded in the *gpt* gene (24). Transfected cells are grown in selective medium containing hypoxanthine, aminopterin, thymidine, xanthine, and mycophenolic acid. Aminopterin blocks the pathway converting precursors to inosine monophosphate (IMP), the precursor of guanine monophosphate. Mycophenolic acid (Sigma), an inhibitor of IMP dehydrogenase, prevents the formation of guanosine monophosphate. With both of these inhibitors present, cells are unable to grow in the presence of hypoxanthine alone. However, inhibition can be overcome if both guanine and hypoxanthine are present in the medium. Rather than adding guanine directly, cells are selected for their ability to convert xanthine to guanine via the bacterial enzyme, HPRT. Only those cells expressing the bacterial enzyme will survive in the selective medium since the mammalian enzyme cannot convert xanthine to guanine. Many plasmid vectors containing the *E. coli gpt* gene are available (examples are given in Section 9.3). This selection scheme has been used with 3T3 cells and CHO cells; however, it is not recommended for use with B-82 cells as transfected cells grow very slowly in this selective medium.

5.2 Neomycin resistance selection

This dominant selection scheme requires transfer of a gene conferring drug resistance (25). Neomycin is a bacterial antibiotic that interferes with prokaryotic ribosomes.

A neomycin analogue, G418 (Geneticin, GIBCO), affects the function of 80S ribosomes and protein synthesis in mammalian cells. G418 can be inactivated by a bacterial phosphotransferase encoded by the *neo*[r] gene. Cells which express this gene under eukaryotic control will survive in medium supplemented with G418. This selection scheme can be utilized with virtually any recipient cell. Recipient cells should be assayed for sensitivity to G418 before transfection as cells display a wide range in the lethal dose of the antibiotic. It is suggested that a dose range of 100 μg/ml to 1 mg/ml be tested. With the appropriate conditions, untransfected cells should die in 5−8 days. Recently, several vectors containing the *neo*[r] gene have become available commercially (examples are given in Section 9.3).

5.3 Thymidine kinase selection

This selection scheme requires the use of recipient cells deficient in the expression of thymidine kinase (tk) such as tk⁻ts13 cells (26). These cells are unable to convert thymidine to IMP because of the lack of the enzyme. When placed in medium containing hypoxanthine, aminopterin, and thymidine (HAT), the tk⁻ cells die as the aminopterin inhibits formation of IMP from other precursors. Cell survival in HAT medium depends on the reconstitution of the thymidine pathway by thymidine kinase contained in the plasmid vector.

5.4 Dihydrofolate reductase selection

In order to employ this selection scheme, one must have recipient cells deficient in dihydrofolate reductase (dhfr). The most commonly used cells are mutant Chinese hamster ovary (CHO) cells (27). These cells require glycine, a purine, and thymidine for growth, therefore, dhfr⁻ cells placed in medium lacking ribonucleosides will die. MEM alpha medium (GIBCO) supplemented with dialysed fetal bovine serum is used for selection. Plasmid vectors containing the dhfr gene restore the ability of transfected cells to grow in the selective medium. One advantage of this selection scheme is that it allows gene amplification in transfected cells by methotrexate selection (28). Methotrexate, a tetrahydrofolate analogue, binds to dhfr in a ratio of 1:1. By growing transfected cells in the presence of increasing concentrations of methotrexate, cellular resistance to the growth-inhibitory effects of this agent can be achieved by overproduction of the transfected gene sequence encoding dhfr and the receptor of interest. Using methotrexate selection following transfection, the m2 muscarinic acetylcholine receptor expressed in CHO cells was amplified to a level of 2.5×10^6 receptors/cell, or approximately 10−25 pmol receptor/mg protein (29).

6. Baculovirus expression systems

An increasingly popular approach for the expression of foreign genes at high levels in insect cells is the use of baculovirus expression vectors (30−32). The baculovirus vector utilizes the highly expressed and regulated *Autographa californica* nuclear polyhedrosis virus (AcMNPV) polyhedrin promoter which has been modified for the insertion of foreign genes. One of the main advantages of this expression system over mammalian cell systems is the high level of foreign protein expression, ranging from 1 to 500 mg/l for soluble proteins (32) and from approximately 5 to 20 pmol/mg membrane protein for integral membrane proteins. For additional information on the biological properties and molecular biology of baculoviruses, the reader is referred to two comprehensive reviews (33,34). The baculovirus expression system is available upon written request from Dr Max Summers, Department of Entomology, Texas A&M University and Texas Agricultural Experimental Station, College Station, Texas 77843-2475.

7. General methodology for expression of receptor genes in cultured cells

7.1 Preparation of plasmid DNA

The first step in the preparation of plasmid DNA for transfection involves the ligation of a receptor gene of interest into an appropriate expression vector, transformation of a suitable bacterial host, and isolation of an individual bacterial colony containing the plasmid vector. The details of each of these steps will vary tremendously depending upon the restriction endonuclease sites present in a particular DNA clone and the choice of plasmid vector for expression studies.

Once a bacterial colony containing plasmid vector is in hand, isolation of plasmid DNA involves three basic steps: growth of bacteria containing plasmid, harvesting and lysis of the bacteria, and purification of the plasmid DNA. The purity of the plasmid DNA is an important factor for successful transfection of cultured cells. DNA preparations harvested from bacterial cultures should be free of contaminating RNA and bacterial chromosomal DNA. Closed circular plasmid DNA behaves differently than chromosomal DNA and these differences are exploited in the purification of plasmids harvested from bacteria. The preferred method for purification of plasmid DNA for use in transfections involves a Triton−lysozyme lysis of bacteria followed by banding of the plasmid DNA in a CsCl/ethidium bromide equilibrium gradient. A detailed protocol for preparation of plasmid DNA using this method can be found in references 25 and 41, and the reader is referred to these references if necessary.

7.2 Transfection of monolayer cells by the calcium phosphate method

The calcium phosphate method is an efficient means for introduction of cloned genes in plasmid vectors into mammalian cells (25,35). It can be used to obtain both transient and stable expression of receptor genes. Cells to be used in transfection experiments should be subcultured for no more than 2−3 weeks and should be maintained in log-phase growth at all times prior to transfection. The ability of the recipient cells to survive a glycerol shock should be established before transfection using the procedure outlined in *step 7* below. If cells can withstand this treatment it may be included in the transfection protocol. If cells die following glycerol shock then this step should be omitted.

Protocol 1. Transfection by the calcium phosphate method

1. On the day before transfection (day 1), trypsinize and replate cells at a density of 10^4 cells/cm^2.

2. On day 2 replace the culture medium with fresh medium and return the cells to the incubator for 3 h.

3. Prepare the calcium phosphate−DNA precipitates using the protocol described in Section 7.4.

4. Add the precipitate to the cells and gently rock the dish to distribute the DNA precipitate. Return the cells to the incubator immediately to avoid any change in pH.

5. Incubate the cells for 3−4 h.

6. At this point examine the cells under the microscope and note the appearance of the precipitate. Ideally, the precipitate should appear as small dark grains covering the cells. If the pH of the medium is too acidic, no precipitate will form and, if it is too alkaline, large pieces of DNA will be seen floating in the medium.

7. Wash the cells in serum-free medium or buffered saline solution. A glycerol shock may be used at this time to increase the transfection efficiency. The shock is performed by adding 1 ml of 15% glycerol in HBSP per 30 cm^2 and incubating the cells at 37°C for 30 sec to 3 min, depending on the tolerance of the cells for this treatment. The glycerol solution is aspirated from the cells and they are washed once with serum-free medium or buffered saline and fed with complete medium.

8. For transient expression experiments, harvest the cells 2−3 days following transfection and assay for receptor expression. For stable expression, incubate cells for 48 h in complete medium and then trypsinize and reseed the cells at a 1:3 dilution in selective medium.

9. Monitor cultures for cell death daily and replace selective medium as necessary to remove cell debris. Cell death should be complete within 7 days at which time small colonies of transfected cells should become visible.

10. When cell colonies have reached a diameter of 2−4 mm, isolate individual colonies using cloning rings and expand cultures for assay. Frozen stocks of positive clonal cell lines should be prepared as soon as possible to insure against loss of a desirable culture.

7.3 Solutions for calcium phosphate precipitation of DNA

Protocol 2. Solutions for calcium phosphate precipitation of DNA

1. 2 M $CaCl_2$ solution:
   This solution should be prepared in distilled, deionized water and sterilized with a 0.22 μm filter. The solution should be stored frozen at $-20°C$.

2. 2X HBSP buffer:
   1.5 mM Na_2HPO_4;
   10 mM KCl;
   280 mM NaCl;
   12 mM glucose;
   50 mM Hepes, pH 7.12.
   This solution should be prepared in distilled, deionized water and sterilized with a 0.22 μm filter. The pH of this solution is critical as it affects the formation of the DNA precipitate. Aliquots of 2X HBSP buffer should be stored at $-20°C$.

3. 15% glycerol/HBSP:
   This solution is made by mixing 30 ml of 50% glycerol (w/v), 50 ml of 2X HBSP (pH 7.12), and 20 ml of distilled, deionized water. The solution is sterilized by filtration and stored at 4°C.

7.4 Preparation of DNA – CaPO$_4$ precipitate

All stock solutions should be at room temperature. The final volume of the solution is 1 ml. Place the appropriate amount of sterile H_2O in the bottom of a sterile conical centrifuge tube. Gently add plasmid DNA (10−50 μg) to the H_2O. Do not mix. Add 248 μl 2 M $CaCl_2$ (final concentration of 0.5 M). Do not mix. Add the DNA solution dropwise to a second sterile tube containing 1 ml 2X HBSP. The order of addition is important. A translucent precipitate should form immediately. If the precipitate appears dense and opaque then the pH of the HBSP buffer is incorrect and the DNA should not be used for transfection. Add 500 μl of the DNA−CaPO$_4$ precipitate to each 5 ml of culture medium as described in Section 7.2, *step 4.*

7.5 Other methods for transfection of cultured cells with plasmid DNA

A number of other methods have been described for the transfection of cultured cells with plasmid vectors. These include the DEAE dextran method (36), lipofection (37), and electroporation (38,39). As compared to the calcium phosphate method each of these techniques presents certain advantages or disadvantages. The reader is referred to the appropriate references for more information on each of these techniques.

8. Summary and conclusions

Establishment of a system for the continuous expression of genes in cultured mammalian cells allows for the study of the pharmacology, biochemistry, and processing of a single homogeneous population of receptors in a defined membrane environment. This approach lends itself to an examination of the behaviour of a particular receptor in a number of different cell types and also to a comparison of different receptor subtypes in the same cell. The expression of human receptors in cultured cells may have a marked impact on the identification and screening of new therapeutic agents as well a providing a source of receptor for structural analysis. We have extended our initial work on the expression of neurotransmitter receptors in cultured cells (21) to produce new clonal cell lines expressing oligonucleotide-directed mutant receptors (22,23). These types of studies will allow for identification of important structural and functional domains of receptor proteins and elucidation of the mechanisms of receptor action.

9. Appendix

9.1 Medium for thymidine kinase selection

To complete culture medium add to a final concentration:
0.1 mM hypoxanthine;
0.02 mM aminopterin;
0.1 mM thymidine.

These reagents can be made as $500\times$ stock solutions and should be filter sterilized and stored at $-20°C$.

9.2 Medium for *gpt* selection

To complete medium add to a final concentration:
0.25 mg/ml xanthine;
0.01 mg/ml glycine;
0.0136 mg/ml hypoxanthine;
1.18 μg/ml thymidine;
1.98 μg/ml aminopterin;
25 μg/ml mycophenolic acid.

The xanthine, hypoxanthine, and aminopterin are dissolved in 0.1 M NaOH and the pH is then adjusted to 9.5 with HCl. Mycophenolic acid is dissolved in absolute alcohol and stored in a glass tube. Solutions should be filter sterilized and stored at $-20°C$. When making selective medium, dialysed serum should be used. This can be purchased or one can dialyse serum against 20 volumes of 0.9% NaCl with four changes of saline solution. Serum must be sterilized by filtration under positive pressure before use.

9.3 Partial list of plasmid vectors suitable for transfection of cultured mammalian cells

Plasmid vector	Promoter	Antibiotic resistance	Selectable marker	Comments
pEUK-Cl[a]	SV40 late promoter	ampicillin	none	Suitable for transient expression in COS cells
pMAM[a]	MMTV promoter	ampicillin	E. coli gpt	Suitable for stable expression in mouse fibroblasts; inducible promoter
pMAMneo[a]	MMTV promoter	ampicillin	E. coli neo	Same as pMAM with alternate selectable marker
pdBPV-MMTneo[a]	Mouse metallothionein I promoter	ampicillin	E. coli neo	Suitable for stable expression in mouse cells; transcription induced by heavy metal ions
pMSG[b]	MMTV promoter	ampicillin	E. coli gpt	Similar to pMAM
pSVL[b]	SV40 late promoter	ampicillin	none	Suitable for transient expression in COS cells

[a] Plasmid vectors are available from:
Clontech Laboratories Ltd,
4055 Fabian Way,
Palo Alto, CA 94303, USA.
Tel: 415 – 424 – 8188.

[b] Plasmid vectors are available from:
Pharmacia,
Molecular Biology Division,
Centennial Avenue,
Piscataway, New Jersey, USA.
Tel: 800 – 558 – 7110.

9.4 Cell lines available from the American Type Culture Collection[a] for use in receptor expression studies

COS-1 (SV40 transformed African Green monkey kidney)
COS-7 (SV40 transformed African Green monkey kidney)
CHO-K1 (Chinese hamster ovary)
tk⁻ts13 (thymidine kinase mutant of BHK-21, Syrian hamster)
3T3-Swiss albino (mouse embryo)

[a] Address for American Type Culture Collection:
12301 Parklawn Drive,
Rockville, MD 20852, USA.

References

1. Noda, M., Takahashi, H., Tanabe, T., Toyosato, M., Furutani, Y., Hirose, T., Asai, M., Inayama, S., Miyata, T., and Numa, S. (1982). *Nature*, **299**, 793.
2. Claudio, T., Ballivet, M., Patrick, J., and Heinemann, S. (1983). *Proc. Natl Acad. Sci. USA*, **80**, 1111.
3. Noda, M., Takahashi, H., Tanabe, T., Toyosato, M., Kikyotani, S., Furutani, Y., Hirose, T., Takashima, H., Inayama, S., Miyata, T., and Numa, S. (1983). *Nature*, **302**, 528.
4. Noda, M., Furutani, Y., Takahashi, H., Toyosato, M., Tanabe, T., Shimizu, S., Kikyotani, S., Kayano, T., Hirose, T., Inayama, S., and Numa, S. (1983). *Nature*, **305**, 818.
5. Schofield, P. R., Darlison, M. G., Fujita, N., Burt, D. R., Stephenson, F. A., Rodriguez, H., Rhee, L. M., Ramachandran, J., Reale, V., Glencorse, T. A., Seeburg, P. H., and Barnard, E. A. (1987). *Nature*, **328**, 221.
6. Grenningloh, G., Rienitz, A., Schmitt, B., Methfessel, C., Zensen, M., Beyreuther, K., Gundelfinger, E. D., and Betz, H. (1987). *Nature*, **328**, 215.
7. Chung, F.-Z., Lentes, K.-U., Gocayne, J. D., FitzGerald, M. G., Robinson, D. A., Kerlavage, A. R., Fraser, C. M., and Venter, C. M. (1987). *FEBS Lett.*, **211**, 200.
8. Dixon, R. A. F., Kobilka, B. K., Strader, D. J., Benovic, J. L., Dohlman, H. G., Frielle, T., Bolanowski, M. A., Bennett, C. D., Rands, E., Diehl, R. E., Mumford, R. A., Slater, E. E., Sigal, I. S., Caron, M. G., Lefkowitz, R. J., and Strader, C. D. (1986). *Nature*, **321**, 75.
9. Kobilka, B. K., Matsui, H., Kobilka, T. S., Yang-Feng, T. L., Francke, U., Caron, M. G., Lefkowitz, R. J., and Regan, J. W. (1987). *Science*, **238**, 650.
10. Frielle, T. *et al.* (1987). *Proc. Natl Acad. Sci. USA*, **84**, 7920.
11. Gocayne, J. D., Robinson, D. A., FitzGerald, M. G., Chung, F.-Z., Kerlavage, A. R., Lentes, K.-U., Lai, J.-Y., Wang, C.-D., Fraser, C. M., and Venter, J. C. (1987). *Proc. Natl Acad. Sci. USA*, **84**, 8296.
12. Kubo, T., Fukuda, K., Mikami, A., Maeda, A., Takahashi, H., Mishina, M., Haga, T., Haga, K., Ichiyama, A., Kangawa, K., Kojima, M., Matsuo, H., Hirose, T., and Numa, S. (1986). *Nature*, **323**, 411.
13. Peralta, E. G., Winslow, J. W., Peterson, G. L., Smith, D. H., Ashkenazi, A., Ramachandran, J., Schimerlik, M. I., and Capon, D. A. (1987). *Science*, **236**, 600.
14. Ullrich, A., Bell, J. R., Chen, E. Y., Herrera, R., Petruzzelli, L. M., Dull, T. J., Gray, A., Coussens, L., Liao, Y.-C., Tsubokawa, M., Mason, A., Seeburg, P. H., Grunfeld, C., Rosen, O. M., and Ramachandran, J. (1985). *Nature*, **313**, 756.
15. Ullrich, A., Coussens, L., Hayflick, J. S., Dull, T. J., Gray, A., Tam, A. W., Lee, J., Yarden, L. Y., Libemann, T. A., Schlessinger, J., Downward, J., Mayes, E. L. V., Whittle, N., Waterfield, M. D., and Seeburg, P. H. (1984). *Nature*, **309**, 418.
16. Paul, J. (1973). *Cell and Tissue Culture*. (4th edn), Churchill Livingston, London.
17. Gorman, C. (1985). In *DNA Cloning, Volume II. A Practical Approach*. (ed. D. M. Glover), p. 143. IRL Press, Oxford.
18. Khoury, G. and Gruss, P. (1983). *Cell*, **33**, 313.
19. Laimons, L. A., Khoury, G., Gorman, C., Howard, B., and Gruss, P. (1982). *Proc. Natl Acad. Sci. USA*, **79**, 6453.
20. Gorman, C. M., Merlino, G. T., Willingham, M. C., Pastan, I., and Howard, B. H. (1982). *Proc. Natl Acad. Sci. USA*, **79**, 6777.
21. Fraser, C. M., Chung, F.-Z., and Venter, J. C. (1987). *J. Biol. Chem.*, **262**, 14843.

22. Chung, F.-Z., Potter, P. C., Wang, C.-D., Venter, J. C., and Fraser, C. M. (1988). *J. Biol. Chem.*, **263**, 4052.
23. Fraser, C. M., Chung, F.-Z., Wang, C.-D., and Venter, J. C. (1988). *Proc. Natl Acad. Sci. USA*, **85**, 5478.
24. Mulligan, R. and Berg, P. (1981). *Proc. Natl Acad. Sci. USA*, **78**, 2072.
25. Davis, L. G., Dibner, M. D., and Battey, J. F. (1986). *Basic Methods in Molecular Biology.* Elsevier, New York.
26. Wigler, M., Silverstein, S., Lee, L.-S., Pellicer, A., Cheng, Y.-C., and Axel, R. (1977). *Cell*, **11**, 223.
27. Urlaub, G. and Chasin, L. A. (1980). *Proc. Natl Acad. Sci. USA*, **77**, 4216.
28. Kaufman, R. J., Wasley, L. C., Spiliotes, A. J., Gossels, S. D., Latt, S. A., Larsen,G . R., and Kay, R. M. (1985). *Mol. Cell. Biol.*, **?**, 1750.
29. Ashkenazi, A., Winslow, J. W., Peralta, E. G., Peterson, G. L., Schimerlik, M. I., Capon, D. J., and Ramachandran, J. (1987). *Science*, **238**, 672.
30. Smith, G. E., Fraser, M. J., and Summers, M. D. (1983). *J. Virol.*, **46**, 584.
31. Smith, G. E., Summers, M. D., and Fraser, M. J. (1983). *Mol. Cell. Biol.*, **3**, 2156.
32. Luckow, V. A. and Summers, M. D. (1988). *Biotechnology*, **6**, 47.
33. Doerfler, W. (1986). In *The Molecular Biology of Baculoviruses, Current Topics in Microbiology and Immunology,* (ed. W. Doerfler and P. Bohm), Vol. 131, p. 51. Springer-Verlag, Berlin.
34. Granados, R. R. and Federici, B. A. (1986). *The Biology of Baculoviruses,* Vol. I. *Biological Properties and Molecular Biology,* CRC Press, Boca Raton, Florida.
35. Graham, F. and van der Eb, A. (1973). *Virology*, **52**, 456.
36. Sompayrac, L. and Danna, L. (1981). *Proc. Natl Acad. Sci. USA*, **78**, 7575.
37. Felgner, P. L., Gadek T. R., Holm, M., Roman, R., Chan, H. N., Wenz, M., Northrop, J. P., Ringold, G. M., and Danielson, M. (1987). *Proc. Natl Acad. Sci. USA*, **84**, 7413.
38. Falkner, F. and Zachau, H. (1984). *Nature*, **310**, 71.
39. Potter, H., Weir, L., and Leder, P. (1984). *Proc. Natl Acad. Sci. USA*, **81**, 7161.
40. Freshney, R. I. (ed.) (1986). *Animal Cell Culture, A Practical Approach,* IRL Press, Oxford.
41. Sambrook, J., Fritsch, E. F. and Maniatis, T. (1989) *Molecular Cloning. A Laboratory Manual.* Cold Spring Harbor Laboratory Press, Cold Spring Harbor.

12

Structural deductions from receptor sequences

N. M. GREEN

1. Introduction

Receptors are a class of proteins, expressed mainly on the extracellular surface of the plasma membrane, which transmit effects of chemical stimuli to the cytoplasm. This may involve gating of ions by the receptor, modulation of systems coupled to GTP-binding proteins, direct activation of a protein kinase domain of the receptor, or internalization of the stimulating ligand (virus receptors). A few receptors for lipid-soluble molecules (steroids, retinoic acid) exist as soluble cytoplasmic proteins, but neither these nor the virus receptors will receive explicit consideration here. The methods discussed are nevertheless generally applicable to all types of protein.

Receptors are, in general, difficult to obtain in sufficient amounts for detailed characterization of their properties as proteins, since there are rarely more than a few hundred copies per cell of any one receptor. For the most part they have been studied by analysis of their biological function and, more recently, by cloning and sequencing the corresponding DNA. Most of the available information about their molecular properties has come from analysis of the deduced amino-acid sequence, except for a few naturally abundant examples.

In principle the folding pattern and hence the structure is determined by the sequence, but there is no immediate prospect of a general solution to the complex relationship between them. A stepwise empirical approach has to be adopted, in which the new protein is related to other proteins either at the level of sequence or of secondary structure, as indicated in *Figure 1*. If no relationship can be established, one is limited to the right-hand pathway on the diagram and then it is very much a matter of luck—some protein sequences have clear structural implications but many do not.

In this chapter I will consider what can be learnt by the various approaches underlined in *Figure 1*. There is insufficient space for detailed evaluation of each of the methods involved, for which reference will be given to more specialized reviews. Two general reviews (1,2) are highly recommended for a broader introduction to the subject. Many of the procedures require access to a computer with appropriate software for searching data bases and for structural analysis of

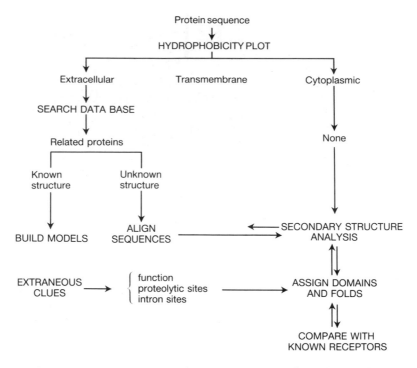

Figure 1. General strategy for analysis of receptor sequences. Operations are in upper-case letters. Note the reverse arrows indicating feedback. For example, secondary structure assignments can often be improved if domain boundaries and folding patterns can be discerned.

sequences (133). A list of these is given in Section 6. Another view of structure prediction from sequences is given by Argos (144).

2. Subdivision of the sequence

2.1 Hydrophobicity

Structurally speaking, two main classes of membrane-bound receptor can be distinguished. Those with a single transmembrane segment and those with multiple segments. The latter are at present confined to two types of neurotransmitter receptors, resembling either the nicotinic acetylcholine receptor, a ligand-gated ion channel, or the muscarinic acetylcholine receptor, which undergoes a ligand-controlled interaction with GTP-binding G-proteins. The β-subunit of the IgE receptor (3) is a new member of this class, which undoubtedly will expand. Receptors with a single transmembrane segment have diverse structures and functions.

It is not difficult to detect long hydrophobic runs by visual inspection but it is quicker and less subjective to use one of the many programs that plot a running

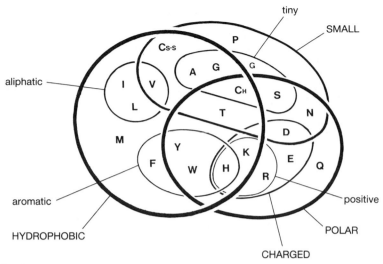

Figure 2. Grouping in of amino acids in overlapping sets. These sets, proposed by W. R. Taylor (25), are useful when the pattern of substitution of residues at specific structural sites of homologous sequences is being evaluated. The single letter code is much the most efficient way of conveying information about sequences and enables significant patterns to be recognized quickly, especially if appropriate sets are colour coded.

average of the hydrophobicity over a window of optional length. A variety of hydrophobicity scales have been proposed. They are based either on physical characteristics of the amino-acid side-chains, such as solubility in ethanol and vapour pressure, or on a statistical study of the distribution of each residue between the interior and the surface of globular proteins of known structure. A synthesis of several of these scales with a number of *ad hoc* adjustments to optimize prediction of transmembrane segments was proposed by Kyte and Doolittle (4) and has been the scale most widely used for this purpose. The relative merits of the different scales have been discussed (1,5,6,8) and two similar scales, which have a sounder theoretical basis, have been proposed. They are based on calculation of the area of hydrophobic surface that is removed from water when residues on an α-helix are transferred into the lipid bilayer, using a value of 25 cal/mole/Å^2 of surface transferred. A further scale (9), based on the frequencies of residues found in the putative transmembrane segments of a set of transmembrane proteins, gives similar results.

Although there are significant differences for individual residues between these scales, the practical difference between them when the hydrophobicity is averaged over the usual span of 20 residues is small. To locate the boundaries of the transmembrane segments, a smaller span should be used. A polarity plot proposed for this purpose (7) used a three-residue span, since a run of three of the polar residues RKDENQ(ST) (see *Figure 2*) is likely to prevent insertion of a helix into the lipid phase.

There is no objective way of deciding their relative merits because few membrane proteins have been crystallized and little hard structural information is available to provide a data-base for evaluating other sequences. The only detailed structures known are those of two multisubunit photosynthetic reaction centres (10,11) but, since the proteins were crystallized from detergents, no direct information was obtained about the nature of the lipid−protein interactions. These should become clearer from the structure of two-dimensional membranous arrays of bacterial rhodopsin, at present limited to a resolution of 7 Å (12). In both of these structural types the transmembrane segments consist of hydrophobic α-helices (eleven in the reaction centres, seven in bacterial rhodopsin) which extend for 20−24 residues and match in length the hydrophobic segments of the lipid bilayer (13).

Many of these hydrophobic helices, particularly in bacterial rhodpsin, contain a few polar groups, aligned on one face of the helix. They are involved in interactions with the photosynthetic pigments or with other polar side-chains rather than with the lipid. There are no intramembranous charged residues in the reaction centres, but there are several in bacterial rhodopsin which are thought to form ion pairs (12). It appears that the nature of the transmembrane residues varies with the function of the protein so that it is not possible to define clear rules about what can or cannot be in the membrane.

Almost all intrinsic membrane proteins have been found to contain long runs of 20−25 hydrophobic residues (e.g. 9) similar to those just described and, by analogy, these are assumed to be transmembrane α-helices. Such long runs are very rare in soluble globular proteins, although marginal cases (18−20 residues) occur in lactic dehydrogenase and tobacco mosaic virus. The only authentic membrane proteins that lack long hydrophobic segments are the bacterial porins, found in the outer envelopes of bacteria and mitochondria. They form large channels from amphipathic β-strands.

2.2 Unusual features of transmembrane helices

Examination of hydrophobic segments reveals a number of differences from α-helices of globular proteins.

(a) At least 75% of the residues are hydrophobic (LIVMFWYCAPG). Serine and threonine are also compatible with lipids since their hydroxyl groups can hydrogen bond to the neighbouring peptide carbonyls, as they do in the interior of globular proteins (14). Each is present at the 4−8% level in most transmembrane segments.

(b) Valine and isoleucine, which in globular proteins favour β-strands, are as common as the helix-promoting leucine, methionine, and alanine residues.

(c) Proline and glycine, which are rare in helices of globular proteins (except for proline near the N terminus), are not uncommon in membrane helices. Proline is sterically compatible with a straight helix, but since it lacks the hydrogen bonding NH group it weakens the helix and usually induces a 20°−30° bend (15). A special role for prolines in membrane transport has been proposed, but direct evidence for it is lacking (16). Glycine residues in globular proteins often play specific roles in bends and are then highly conserved, whereas transmembrane glycine residues are not particularly well conserved, (N. M. Green, unpublished observations on 20 ion-transport ATPases).

(d) Tyrosine, tryptophan, and phenylalanine residues are often concentrated near the hydrophobic−polar interface, either because of their bulk or because they have hydrogen-bonding requirements.

(e) Isolated transmembrane segments are almost always more hydrophobic than those which form part of a helical cluster, and they rarely contain charged residues or amides (6,16). The only examples with a charged residue are found in the subunits of the T-cell receptor complex, where the charges are likely to be buried in the multisubunit complex (17). The hydrophobic segments of such receptors therefore stand out very clearly.

(f) Multihelical clusters may give rise to problems of assignment. It has been suggested that quite highly charged amphipathic helices may be buried in such a cluster in the acetylcholine receptor and that they provide a charged ion channel (18,19), but although this is possible in principle, direct evidence for it is lacking.

(g) Most secondary structure predictions assign a β-strand conformation to a strongly hydrophobic segment, unless helix-promoting leucine and alanine predominate. This follows from the use of a data-base of globular proteins in which the most hydrophobic segments are buried β-strands.

3. Homology searches and pattern matching

It is much easier to interpret a new sequence if it can be related to other sequences of known function or structure. Several data-bases are now available, containing over 10 000 sequences and doubling in size every two or three years. One or more of these are included in the sequence-handling packages listed at the end of the chapter. They are usually searched by one of the rapid methods introduced by Pearson and Lipman (20). Discussion of the mode of operation and merits of different search methods will be found in more extended reviews and in the documentation of comprehensive packages (UWGCG and PIR). A few useful hints are given here.

Always search the protein banks first. Searches of DNA banks are much slower and more difficult to interpret, partly because of the degeneracy of the genetic code and partly because patterns in a four-letter language are difficult to recognize. The only reason for searching DNA banks for protein-coding sequences is that they may contain sequences that have not yet reached the protein banks. Some banks are updated every three months but, even so, most are a year or more behind the published data.

The searches will reveal matches provided they persist over most of the sequence. A problem arises with long sequences (>500 residues) which contain limited, local matches. They may not appear in the high-scoring list because even a close match will not contribute much to the total score of a long sequence. This can be avoided by dividing the sequence into segments of $50-100$ residues before searching and scoring these independently. Any good matches will give a lead to significantly related proteins which can then be compared individually over their full lengths.

In the not infrequent absence of a close match, there will be a large number of moderate to low scores and their significance has to be evaluated. The simplest method is to use the percentage identity (unitary matrix) and although it leads to lower scores than those based on the Dayhoff relatedness matrix, it gives essentially the same order of merit for alignments (21). In view of the ease with which it can be both calculated and comprehended, independently of any extraneous assumptions, it is probably the best measure for general use. The question of what percentage identity guarantees an evolutionary or structural relationship is also important. Provided that the aligned regions cover a domain-sized segment (50−150 residues), 25% identity is usually significant, whereas the 15−25% region has been called the 'twilight zone' (2,21) where a cautious approach is required. It has been shown that in families of related proteins an almost linear relation between the sequence identity and the structural relationship persists, even in the region of 15% identity (23). Since structure is generally better conserved than sequence (24), secondary structure analysis provides a useful check on significance. If the predicted structures of aligned segments are incompatible, allowing for the uncertainties of the method, then the alignment should be regarded with suspicion. Similarly, good matches of short segments are unlikely to be significant unless they are found in a similar structural context.

4. Functional sites in receptors

A rapidly increasing number of short sequences have been recognized as characteristic of specific binding sites, domains, or sites of covalent modification. If it is known from functional studies that a particular receptor includes such regions, then their sequences should be searched either by eye or by using an appropriate pattern-matching algorithm. Ideally, patterns and scoring systems are devised by adjusting the consensus pattern and the scoring until all authentic examples in a data-base score significantly higher than the background. An approach of this type, initially used for identifying and subclassifying immunoglobulin domains, is described by Taylor (25,26). A simpler, though less specific, approach acessible in the UWGCG package, is that of profile analysis (27). This aligns sequences or segments thereof and produces a consensus profile which is used to search a data bank or a new sequence, with scoring based on the Dayhoff relatedness matrix. This is a good general method which is easy to use but has the disadvantage that the weightings cannot be adjusted for important matches at specific positions. In consequence, insignificant alignments can score quite highly. For example, a sequence consisting of a run of hydrophobic residues followed by about 12 acidic residues followed by more hydrophobics would score quite highly against a calmodulin profile (see *Table 2*). It is therefore necessary to examine critically each match produced by the program and not to rely blindly on the score. The alternative is to use methods that allow more flexible scoring (26,28). Examples of the better-characterized 'fingerprints'

Table 1. Sites for specific covalent modification.

Covalent modification	Enzyme	Sequence	Reference
N-glycosylation		N.(ST)	31
Phosphorylation	cAMP kinase	+ +.(ST)	32
		+ +..S	
	Protein kinase C	+.(TS).(RKP)	33
		+..S	
	Glycogen synthetase kinase 3	(ST)P..S	34
	Casein kinase II	(ST) (DE)$_{3-5}$	35
N-terminal acylation		NH$_2$ G.......	36

Residues in parentheses are alternatives at a single site; + = R or K; . = no specific requirement. Some residues in the latter positions may have a modulatory influence; a second N in the glycosylation site is inhibitory.

(*Tables 1* and *2*) are discussed below. A more extensive survey for all types, including those associated with particular secondary structure classes, will be found elsewhere (29,30).

4.1 Potential sites for covalent modification

N-glycosylation sites are characterized by the sequence N.(ST), which is not very definitive (31). On a random basis about 10% of asparagines will be followed by S or T and this is indeed observed in intracellular proteins, which are not glycosylated. Extracellular proteins show a lower incidence of such sites and of these about 30% are actually glycosylated. Serine and threonine residues provide sites that can be phosphorylated by many protein kinases if they have two or more lysines or arginines in the immediate neighbourhood. Sequences such as RKS provide the usual substrate site for cAMP-dependent kinases (32), though occasionally a second residue is interposed between the basic residues and serine. More rarely, a single R suffices to promote phosphorylation. Other kinases have similar requirements at the sequence level, so the fact that they are, nevertheless, specific for their protein substrates implies that the structural environment of the sequence is critical for determining specificity. Protein kinase C often requires an extra lysine following the S or T (33). In contrast, glycogen synthetase kinase 3 (34) and casein kinase II (35) require acidic sequences. The same appears to be true for the recently characterized β-adrenergic receptor kinase (132).

4.2 Proteolytic cleavage sites

Sites of proteolytic cleavage are defined primarily by a single amino acid, usually on the carbonyl side of the cleaved bond. A critical survey of the subject is available (37). It is possible to obtain maps of potential cleavage sites from most sequence-handling packages, but these do not indicate the actual specificity, which for native proteins is determined mainly by the structure of the protein substrate. Sites in β-strands or α-helices are not sufficiently exposed to interact with proteases,

Table 2. Sites with characteristic sequence and secondary structure patterns (29)

Ligand	Structural unit		References
Dinucleotide	1st β α	$\overset{\beta}{+\text{h.h.}}$ G. $\overset{\alpha}{\boxed{\text{G . . G . . . h . . h}}}$ $(.)_{1-5}$ $\overset{\beta}{\boxed{\text{H . h . -}}}$	46,47
Mononucleotide	1st β α	$\overset{\beta}{++\text{hhhh}}$ Gss $\overset{\alpha}{\boxed{\text{GsGk\$ . . hh . hh . .}}}$ $(.)_{1-7}$ $\overset{\alpha}{\boxed{}}$	49
	3rd or 4th β α	$\overset{\alpha}{\boxed{\text{hh . hh + p}}}$ G $\overset{\beta}{\boxed{\text{phhhh}}}$ (Dss)	46
Calcium	EF hand	$\overset{\alpha}{\boxed{\text{hpphph}}}$ h-pBGBG . l-pE $\overset{\alpha}{\boxed{\text{hpphh . hh}}}$	40,41
Calmodulin	Amphipathic helix	No clear consensus; known sequences are characteristic of amphipathic helices with two clusters of basic residues separated by a short hydrophobic segment	42

Sets of amino acids that can occupy specified sites are represented below: Upper-case letters refer to unique amino-acid residues. The sequence of residues within parentheses does not matter. In the following list residues in parentheses are less common members of the set:

h, hydrophobic   LIVFM (APWYGTKH);
p, polar   SNDEQRK (HTCYW);
s, small   GASCV (DNT);
+,   KR;
−,   DE (ST);
B,   DN (ST);
\$,   ST;
.,   no specific requirement.

284

and cleavage is limited to the potential sites, often only one or two in the whole molecule, which contain accessible bonds. There is no sure way to identify these without experiment. When cleavage is limited to two or three sites, the molecular weights of fragments may suffice to identify susceptible loops.

If complete proteolysis is required, with cleavage at all potential sites, the protein must be unfolded. This is best done by urea or extremes of pH, provided that this is compatible with activity of the protease. Detergents may be necessary to maintain a membrane protein in solution, but since many bind to the protein they may interfere with cleavage. This is less likely to be a problem if non-ionic detergents or bile salts are used, but SDS binds extensively, particularly to cationic groups, where it may interfere with tryptic digestion. Replacement of SDS by a cationic detergent (e.g. DTAB) can give more extensive tryptic digestion (N. M. Green, unpublished experiments). SDS (and possibly DTAB) has the further disadvantage that it stabilizes α-helical rather than unfolded configurations of a protein (38) and this may also limit the proteolysis. Another way to maintain a protein in the unfolded state is to succinylate all the lysine residues and increase the total negative charge. This has a similar effect to that of high pH without the problem of inhibiting the protease. Experimental details for these and other strategies for handling membrane proteins are given by Allen (39).

4.3 Calcium-binding sites

Many different types of calcium-binding site have been identified but only one of these has been associated with a clear consensus sequence. This is the so-called EF hand identified in the crystal structure of parvalbumin and then extended by sequence homology to troponin, calmodulin, and other intracellular proteins (40,41). The consensus sequence forms a 12-residue loop between α-helical segments and is characterized by a conserved glycine, a hydrophobic residue, usually isoleucine, and an alternation of oxygen-containing residues, usually aspartic acid or asparagine. Such sequences are easily recognized but it is possible to obtain misleading matches. For example, the plasma membrane Ca^{2+}-ATPase contains two quite high-scoring regions, identified by the 'profile' method mentioned above, which, although they may be Ca^{2+}-binding sites, would certainly not form EF hands (64). The presence of strong α-helices on either side of the loop can be used as an additional criterion for a genuine EF hand. If these are absent then the match can be regarded with scepticism, although there is one structure in which a typical loop is supported by two β-strands rather than by helices (43). Other Ca^{2+}-binding patterns of unknown structure have been found in a class of lipid-binding proteins, the annexins (44), and in osteonectin (45). Tentative structural predictions were made.

4.4 Nucleotide-binding sites

Many nucleotide-binding sites include short glycine-rich sequences, characterized by the pattern G . (G.) . GKT (46,47). Examination of known structures shows that the sequence forms a loop at the C terminus of the first β-strand of a parallel β-sheet, the last two glycines being incorporated into the following α-helix. Two subtypes characteristic of dinucleotide and mononucleotide sites can be defined (49). The simplest form of the template should be used with caution since it picks out many regions that do not bind nucleotides (48), but a more stringent, extended form has given reliable results (50). Any match should be checked for preceding β-strand and following α-helix as well as for further β-α-β- patterns which together could form a parallel sheet. All these characteristics are required if a nucleotide-binding site is to be assigned with confidence. Frequently the third or fourth β-strand and bend has an additional characteristic sequence (46) which can provide further confirmation. This again has a defined structural context, forming a second loop which contributes to the site. A recent analysis of the sequences of ion-transporting ATPases provides an example of the use of these clues to define a nucleotide-binding domain that differs from the standard type (51).

4.5 Cysteine-rich domains

Fingerprints for EGF-like domains, cringles, fibronectin, and immunoglobulin domains, which are abundant in receptors, are more difficult to define. Although they have characteristic patterns of disulphide bonds, the loop sizes vary and there are few well-conserved sites (29). Specific examples are provided in *Table 4* below.

4.6 Unusual repetitive elements

Repetitions of short segments are not uncommon in protein sequences. Some, such as the x-Pro-Gly repeats of collagen, have a defined structural context, but most do not. Until they are found in proteins of known structure one can only speculate about their function. Many regions rich in repeated glutamic acids will always be predicted as α-helical. They will not, however, adopt a helical configuration unless the charge is neutralized either by protonation, by metal-ion binding, or by clusters of positively charged amino-acid residues. Long segments rich in bend-promoting residues (P,G,S,T,D,N) are not uncommon. Some, such as those in the immunoglobulin hinge region, may be genuinely structureless loops linking domains together. They form invisible, disordered segments in a crystal structure (143). A segment of this type is found in the long $M_5 - M_6$ loop of the muscarinic receptors (126).

5. Higher-order structure

5.1 Secondary structure prediction

Although in principle it should be possible to determine structure from sequence, computation from first principles using energy minimization procedures has not yet proved possible. Most practical methods are based on a statistical analysis of 30−60 sequences out of 300 or more proteins of known structure. Here we will consider only the two most commonly used approaches, due to Chou and Fasman (52) and Garnier, Osguthorpe and Robson [GOR method (53)], which are available in most packages. More detailed reviews provide comparisons with other approaches (1,54).

Analysis of the frequency of each amino acid in different regions of protein secondary structure has shown that many amino acids have preferred locations, but that the preferences are not strong. Most methods of prediction calculate the probability of forming α-helix, β-strand, or β-turn, based on the distribution of α, β, or turn-promoting residues. Any region not predicted as one of these is regarded as unstructured (coil region). In practice, the distinction between β-turn and coil prediction is rarely very clear and a more realistic three-state prediction is commonly used. A few methods are based on pattern recognition rather than statistics. The earliest, due to Lim (55), has proved as accurate as the best statistical methods but it implements a large number of complex rules which make it difficult to program. Related methods (e.g. 56) match all short peptide sequences (3 – 7 residues, depending on the method) with similar sequences of known structure found in structural data-bases, such as that of Kabsch and Sander (67). A comparison of the GOR method with those of Chou and Fasman and of Lim showed that their accuracy was no better than 50 – 55%, with the GOR method marginally the best (58). It is also one of the easiest to run on a computer and, being based on information theory, it has the advantage that the probabilities for the different states are all on the same numerical scale. Since the raw numerical data are difficult to evaluate by eye over long stretches of sequence, a graphical output, in which the three plots are superimposed, is essential (*Figure 3*). This enables the significance of very short helices or strands to be better evaluated by taking into account maxima in the curves. Any critical evaluation requires the sequence to be written immediately beneath the graphical output, to allow a check on location of actual sequence pattern (e.g. of hydrophobicity or bend-promoting residues) in relation to the predicted structure.

Many attempts have been made to improve these methods, but the increase in accuracy has been limited to 5 – 10%. The use of combined predictions has given very little improvement (54,59). In conclusion, the GOR method is the most widely accessible in software packages and it is recommended for an initial survey. A recently updated version (59), using a larger data-base, has improved the proportion of correctly predicted residues to 63%. Incorporation of methods based on structures of homologous seven-residue peptide segments and on hydrophobicity patterns improved the results by only 2.5%, hardly justifying the further complication. The disappointing low accuracy of secondary structure prediction blocks any progress towards a simple hierarchical progress from secondary to supersecondary to tertiary structure. If no clues are available from sequence relationship to other proteins, it is unlikely that one will be able to progress beyond subdivision of a protein into domains, as described in the next section, some of which may have an approximately assignable fold according to one of the recognized patterns.

How then should one proceed after the initial prediction of secondary structure? Since successive elements of secondary structure tend to fold in an antiparallel fashion, the presence or absence or a particular helix or strand is more important than an accurate estimate of its length, if one is trying to visualize the overall folding pattern. It is therefore encouraging that the chance of predicting a run of α or β residues correctly has been shown to be about 70%, significantly higher than the accuracy for an individual residue (59). The problem of what to do with very short runs (<5 α or <3 β) remains unsolved. Either they are the nucleus of an actual element which was underpredicted or they are incorrect. Inspection of the profile may help. A clearly defined maximum in the prediction of one conformation which barely rises above the background of the others could indicate a viable structural unit. Systematic evaluation procedures have been described (61).

The availability of several related sequences can be of great assistance, particularly if the percentage identity lies between about 30 and 75, since it can be assumed that they will adopt a common fold (23). Strong predictions will almost certainly be reinforced and the significance of vestigial ones may be clarified. This approach is likely to provide more information than the use of several different prediction methods on the same sequence (51,72,77,78).

Eventually it should be possible to improve the prediction further by searching for plausible units of supersecondary structure by template fitting or other pattern matching techniques (28,61), but the programs for doing this are not yet included in standard packages. Much depends upon the judgement of the analyst, and this applies equally to the next problem of subdividing the sequence into domains and trying to assign folding patterns.

5.2 Hydrophobicity patterns

Since the main driving force behind the folding process is the formation of a stable hydrophobic core (15), the patterns of hydrophobicity on the potential secondary structure elements are of prime importance. It is therefore legitimate to take hydrophobicity into account twice, implicitly in the initial secondary structure prediction and explicitly by considering the patterns which emphasize either overall hydrophobicity or amphipathicity of strands or helices. The most systematic approach to defining the hydrophobicity patterns of elements of secondary structure is the use of the hydrophobic moment concept of Eisenberg (5). Unfortunately, although theoretically precise, it is not easy to perform a mental integration of a hydrophobic moment plot with a secondary structure prediction in any productive fashion.

Figure 3. Sequence analysis of laminin segment. The plot shows the boundary between one of the EGF-like disulphide cross-linked domains and the C-terminal α-helix, which continues for a further 400 residues. The helical predictions are strong and markedly amphipathic. Three of the β-strand predictions are very short and they include four disulphide-linked cysteines. No confident prediction can be made for this region solely on the basis of secondary structure, but disulphide assignments and NMR evidence for EGF itself provide a basis for model building (85,86). The display, provided by the MGS package (*Table 5*), gives most of the useful structural information in a readily comprehensible form, enabling the following questions to be answered quickly: Are there long hydrophobic segments? Are there dominant patterns of supersecondary structure (α t α, α t β t α, β t β) which extend over domain-sized segments? If not, is it possible to produce them by re-evaluating segments which are not clearly predicted? Are the local hydrophobicity patterns consistent with the predicted folds or do they suggest alternative interpretations? Are the predictions consistent with the CD measurements or other extraneous evidence?

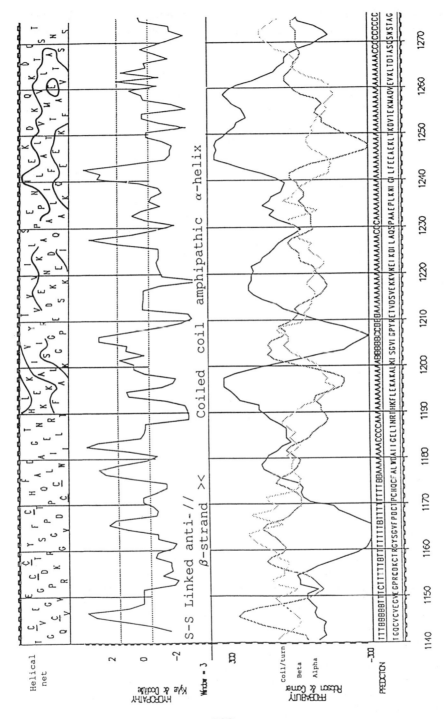

The same information can be more easily assimilated if the sequence is displayed on an α-helical net (62) with individual residues colour-coded to emphasize their hydrophobic, charged-polar, or α-helical character. This net can be analysed wherever a peak in the helical prediction occurs (e.g. *Figure 3*). The net is more informative than the commonly used α-helical wheel, especially for long sequences. It is possible to employ a similar device for β-strands, writing the sequence in an alternating fashion A^BC^DE but, since strands are short, a normal linear representation combined with colour-coding is adequate to provide a mental image of the hydrophobicity pattern associated with the putative strand.

Given this input of strands and helices of defined amphipathicity, attempts should be made to integrate them into supersecondary and domain structures (61). Unfortunately, there are many approximate solutions to this problem and only one biologically unique solution (25,54).

5.3 Domain types and boundaries

If a new sequence has no relation to a known sequence, it becomes difficult to go much beyond the initial secondary structure prediction. The best hope of further progress is to use both extrinsic and intrinsic clues to subdivide the sequence into domain-sized units and then to try and assign folding patterns, which may eventually be correlated with function.

A domain can be defined conceptually as a unit that folds independently of neighbouring sequences, but this is not a useful operational criterion. Originally they were recognized as major subdivisions in known three-dimensional structures. Automatic procedures have been devised for assigning subdivisions (63), but inspection of the structure is usually sufficient.

Recognition of domain boundaries and types from sequence data alone is much more difficult. Extrinsic clues are provided by:

(a) intron sites;

(b) proteolytic cleavage sites;

(c) affinity labelling sites; and

(d) amino-acid composition (57).

In ancient proteins of prokaryotic origin, intron sites are not a reliable guide to domain boundaries, but receptors involved in cellular recognition processes are of much more recent origin and coincidence between intron site and domain boundary has been frequently noted (22,60). Proteolytic cleavages are mostly confined to exposed loops, but although many domain boundaries fall in this class most loops are within domains. Affinity-labelling sites identify functional regions and may be useful in domain demarcation, particularly if several are known. Intrinsic clues include local changes in amino-acid composition, secondary structure composition, and supersecondary structure patterns (57,61).

The approach is successful only because the number of observed domain types is remarkably limited (61,65). About 80% of domains in some 300 known structures can be assigned to one of four structural classes:

(a) 'alpha', characterized by α, turn, α-motifs, and a helix content usually greater than 35%.

(b) 'alternating alpha beta' (parallel β-sheet) characterized by β, β-turn, α, β-turn, β motifs with >35% helix and >15% β.

(c) 'beta' (antiparallel β-sheet) characterized by β, β-turn, β motifs with >15% β and <15% α; and

(d) a relatively poorly defined (alpha + beta) class in which α and β regions are segregated and the helix content is lower than the 'alpha beta' class.

The unclassified 20% of domains includes a further class which is poorly defined in secondary structure but which is particularly common in the extracellular domains of receptors. It is characterized by a high content of cysteine residues, which form networks of disulphide bonds. The different structural types are abundantly illustrated in two reviews of protein taxonomy (65,66), which should be consulted before making any attempt at structure prediction.

When attempting to assign domains and folding patterns it is important to consider domain-sized units of sequence, since their folds can be expected to make sense without reference to other segments. Any protein segment of more than 200–250 residues is likely to include a domain boundary. The smallest domains, excluding transmembrane segments, are disulphide-rich (50–80 residues) or small β-barrels containing up to eight β-strands (e.g. γ-crystallin). Bundles of four α-helices (e.g. TMV protein, uteroglobin) have about 100 residues, whereas parallel β-sheet proteins, and particularly eight-stranded α-β-barrels, are much larger (150–200 residues). More complex α-helical domains (globin) and large β-sandwich domains (Concanavalin A) are almost as large. Sometimes a large domain may be made from two non-contiguous segments of sequence with an intervening region folding into a more or less independent domain or subdomain. Adenylate kinase forms a parallel β-sheet domain with an inserted α-helical diversion and the α-β-barrel of pyruvate kinase is interrupted by an antiparallel β-sandwich (100 residues) which forms a separate domain. Such split domains will be difficult or impossible to recognize in the uncertain pattern of helices and strands present in the usual predicted secondary structure. Although they are less common than single segment domains their occasional appearance renders many assignments uncertain.

Provisionally ignoring such complications, the best approach is to take the most strongly predicted elements and attempt to fit them to one of the simpler domain types. The more weakly predicted elements can be included at the edges of domains or, in the case of an ambiguous prediction, β and α may be changed into each other if doing so makes more sense of the domain as a whole.

The central segments of strands that belong to a parallel sheet are buried and will contain runs of three or more hydrophobic residues, unless they happen to be at

Table 3. Structures predicted by sequence analysis[a]

Protein	Predicted structure	Confirmatory evidence	References
Interferon	α-helical bundle	–	70
Interleukin-2	α-helical bundle	–	71
Spectrin	repeated helical bundles	CD	72
Viral protease (HIV)	anti-// β-sandwich	X-ray	73,74
Protein kinase	β α β		75,68
SV40 T-antigen	// β		76
Cation transport ATPase	// β		51
Tryptophan synthetase	eight-stranded α-β-barrel	X-ray	77,78

[a] Cysteine-rich domains appear in *Table 4*.

the edge of the sheet. Strands of antiparallel sheets lie on the surface so that polar and hydrophobic residues tend to alternate. Patterns required to nucleate particular folds were analysed by Ptitsyn and Finkelstein (69). However, it is unlikely that this information will be sufficient to determine a folding pattern without clues from sequence-related domains of known structure.

A further check on the plausibility of the domains and folds assigned can be made if several homologous sequences are available. If the domains are aligned then most of the hydrophobic sites of predicted strands and helices should be conserved. Surface features are more variable and include polar surfaces of strands or helices as well as most bends. Highly conserved bends are candidates for ligand binding or for domains or subunit interaction sites. Examples of this sort of analysis will be found in the more successful examples of structure prediction listed in *Table 3*.

5.4 Domains of receptors

Most receptors are mosaic structures containing several different types of domain in their extracellular structures, often separated by intron sites. Many of the domains are rich in cysteine and few can be related to proteins of known structure. Most disulphide-bonded proteins are β-sheet structures, in which the strands are often short (e.g. wheat-germ agglutinin). Many of the strands would not be regarded as viable in a secondary structure prediction, which may reveal little more than β-bends and unstructured loops (*Figure 3*).

A guide to this relatively uncharted field of cysteine-rich domains is provided in *Table 4*. Although the oxidation state of the cysteines has rarely been determined directly (79), it is almost certain that they form disulphide bonds, first because they are extracellular (80), and secondly by analogy with similar domains of soluble proteins in which the disulphide pattern is established. Apart from the domains of the immunoglobulin type (81,82,83), and the cringle domain of prothrombin (84), none of these disulphide-containing regions have had their structure determined by crystallography. Some small disulphide-rich proteins have been studied by NMR

(85,86). The explosion of information in the field of complement proteins (87), blood-clotting factor (60), and basement membrane proteins (88), is beginning to provide insight into the structure of related domains found in receptors (*Table 4*).

The EGF domain provides an example of the insight that can emerge. Tandem repeats of EGF-like domains have been found in a number of receptors, in products of certain developmental genes, and in the basement membrane protein, laminin. Electron microscopic studies of laminin (90) and of integrin (91) have shown a correlation between rod-like segments and the tandem repeats. Taken together with the structures of the EGF unit provided by NMR (85,86) this is leading to an understanding of this type of receptor structure. Tandem arrays of the short repeats (SCRs) found in many complement components (87), and the multiple immunoglobulin domains of cell adhesion proteins (89), provide further examples of extended structures. It should not be long before many of the gaps in structural knowledge represented by the entry 'Seq' (sequence analysis) in *Table 4* are filled. This will provide a much firmer basis for future sequence analysis of new receptors.

6. Software for sequence analysis

Many commercial and academic packages are available, most of them designed primarily for handling DNA sequences, but no one has attempted a comparative evaluation of the protein modules. The following comments on the list given in *Table 5* provide some idea of what is available. Descriptions of individual programs and packages including advertisements for commercial packages are collected in several volumes of *Nucleic Acids Research* (134) (also available as separate publications from IRL Press) and in reference 135. Limited comparative evaluations will be found in references 1 and 144, and in two books cited in other contexts (54,133).

Packages are of two main types, those which run on microcomputers (DNA Star, PC-Gene, Microgenie, Staden−Amersham) and those designed for mini or mainframe computers with multi-user access. The latter are obtainable from Intelligenetics, University of Wisconsin Genetics Computer Group (UWGCG), R. Staden, and the National Biomedical Research Foundation (NBRF). The latter has particularly fast and convenient search programs for peptide and protein identification and is inexpensive. The UWGCG package has the advantage of a comprehensive manual, and regular updates of data-bases and improvements in programs. Packages specifically for protein structure prediction are obtainable from Leeds University Biophysics group, which provides a consensus prediction using eight different methods, and from the University of Illinois, Biotechnology Department. The latter has a good colour graphics facility, runs on an IBM PC and can also display known three-dimensional structures of up to 7000 atoms using coordinates from the Brookhaven data-base (as can DNA Star). Sequences can be formatted as helical wheels or nets.

Most of the other packages are primarily for the handling of DNA sequences and, although they include most of the standard options for proteins, their displays are not well designed for further analysis. In judging the merits of a particular system the following considerations are pertinent.

Table 4. Domain types recognized in receptors and some non-membranous relatives.

I Single TM segments
Extracellular domain

Domain type	Structural type[a]	Structural information	Ref.	Receptor for:	Copy no.[b]	Ref.	Occurrence in Soluble protein	Copy no.[b]	Ref.
EGF	6 Cys/44	NMR, EM	85,90	LDL	x 3	91	complement C9	x 1	95
	Seq.			EGF precursor	x10	92	laminin B₁	x13	96
		EM	93	fibronectin	x 4	94	factor XII	x 2	97
Complement C9 SCR	6 Cys/40	Seq. IR	98	LDL	x 7	91	complement C9	x 1	95
	4 Cys/60	EM	131	IL-2	x 2	99	complement	x 2	87
							C4bp,CR,H,I	x30	
Fibronectin type II	2 Cys/60	X-ray (cringle)	84	mannose-P	x 1	101	fibronectin	x 2	102
			100	IGF-II	x 1	130	factor XII	x 1	97
EGF receptor segment	8 Cys/44	Seq.	103	EGF	x 2	105			
			104	Insulin	x 1	106			
				IGF-I	x 1	129			
Immunoglobulin	anti-/β	X-ray	81	Polymeric Ig	x 5	108			
			82	MHC I	x 2	109			
	2 Cys/95	Seq.	25,83	MHC II	x 2	83			
			107	T-cell	x 2	110			
			89	N-CAM	x 7	111			
			89	Fc	x 1	112			
			89	CSF-1	x 5	114			
				PDGF	x 5	113			

294

					Ligand/receptor		Ref.	Oncogene
	Ca²⁺		Seq.		Fibronectin	x 5	94	
	Unclassified	23 Cys/210			NGF		115	
		7 Cys/240			growth hormone		116	
		6 Cys/160			asialoglycoprotein		117	
		7 Cys/640			transferrin		118	
		4 Cys/80			T-cell		17	
		4 Cys/105			T-cell		17	
Cytoplasmic domain	Protein kinase	Iβ	Seq.	68,75	EGF		105	erb B
					Insulin		106	v-ros
					PDGF		113	v-kit
		300 residues			CSF-1		114	v-fms
					mannose-P		101	
II Multiple T segments								
Extracellular domain	Nicotinic acetylcholine	anti-Iβ	EM	119	acetylcholine		122	
			Seq.	120	γ-aminobutyrate		123	
				121	glycine		124	
	Opsin	Undefined short loops	Seq.	125	acetylcholine (muscarinic)		126	
					catecholamine		127,128	
					light (rhodopsin)		125	
	IgE	short loops			IgE		3	

The fifth column lists receptors, grouped according to the domains (column 1) that they contain.
The eighth column shows non-membranous proteins which contain similar domains.
[a] Where the domains are not defined the total number of cysteines in the extracellular region is given.
[b] Shows the number of times the domain type is repeated in the receptor or soluble protein.
[c] No known receptor activity.

Table 5. Software for sequence analysis.

Package	Hardware	Address	Cost (approx.)	References
DNA Star	IBM PC	1810 University Avenue, Madison, Wisconsin 53711, USA; and	£4600 (protein only)	1,54
		565 Chiswick High Road, London W4 3AY, UK	£8200 (complete)	
Staden[b]	VAX or	Dr R. Staden, LMB, Hills Road, Cambridge.		137,133
	IBM PC	Amersham International	£2000	
SEQANAL	IBM PC	A. R. Crofts & H. Robinson University of Illinois, Biotechnology Center, 105 Observatory Buildings, 901 South Madison Avenue, Urbana, IL 61801, USA	$300	140
(Protein Data Viewer)			$150	
Leeds[b] Protein Structure Prediction Suite	VAX IBM PC VM/CMS	E. Eliopoulos & A. I. Geddes, Department of Biophysics, Leeds University, Leeds LS2 9JT, UK	£50	72
PIR[a,b]	VAX/VMS	National Biomedical Research Foundation (NBRF), 3900 Reservoir Road, Washington DC 20007, USA		133,138
MGS	SUN/UNIX	P. Gillett, National Institute for Medical Research, Mill Hill, London NW7 1AA, UK		*Figure 3*
UWGCG[b]	VAX/VMS SUN/UNIX	University of Wisconsin, Genetics Computer Group, Biotechnology Center, 1710 University Avenue, Madison, Wisconsin 53705, USA	$3000 p.a.	27
Intelligenetics[a]		1975 El Camino Real, Mountain View, California 94040, USA	$50 000	141

[a] These programs can be accessed via BIONET (133,141,142).
[b] Programs accessible in U.K. by SEQNET (UIG, Room C12, Daresbury Laboratory, Warrington WA4 4AD).
The content and availability of data banks is considered in references 1, 54, 133, and 134.

There is little point in giving large amounts of information if it cannot be integrated into a meaningful picture. Redundant information should be eliminated and the essentials displayed in a clearly differentiated fashion alongside the sequence, which is essential to any critical evaluation. The output shown in *Figure 3* (MGS from National Institute for Medical Research) fulfils most of these criteria, although it lacks the colour-coding of amino-acid residues present in the original. The three secondary structure profiles are superimposed in different colours or line types, a hydrophobicity plot is given, and an α-helical net allows an assessment of amphipathicity of the helical segments. Hopp and Wood antigenic site parameters and Chou and Fasman prediction are additional options. The UWGCG programs display more profiles, including several for identifying antigenic sites, but it is less easy to integrate the information into a structural prediction. Some of the programs, as well as the main data banks, are accessible on line via the SEQNET or BIONET facilities (141,142). More information about networks and data banks is provided in the books already cited (1,54,133,135,139).

References

1. von Heijne, G. (1987). *Sequence Analysis in Molecular Biology,* Academic Press, London.
2. Doolittle, R. F. (1986). *Of Urfs and Orfs,* Mill Valley, California University Science Books and Oxford University Press.
3. Kinet, J.-P., Blank, V., Ra, C., White, K., and Metzger, H. (1988). *Proc. Natl Acad. Sci. USA,* **85**, 6483.
4. Kyte, J. and Doolittle, R. F. (1982). *J. Mol. Biol.,* **157**, 105.
5. Eisenberg, D. (1984). *Ann. Rev. Biochem.,* **53**, 595.
6. von Heijne, G. (1985). In *Current Topics in Membranes and Transport.* (ed. P. A. Knauf and J. S. Cook), Vol. 24, p. 151. Academic Press, London.
7. MacLennan, D. H., Brandl, C. J., Korczak, B., and Green, N. M. (1985). *Nature,* **316**, 696.
8. Engelman, D. M., Steitz, T. A., and Goldman, A. (1986). *Ann. Rev. Biophys. Biophys. Chem.,* **15**, 321.
9. Argos, P., Rao, J. K. M., and Hargrave, P. A. (1982). *Eur. J. Biochem.,* **128**, 565.
10. Deisenhofer, J., Epp, O., Miki, K., Huber, R., and Michel, H. (1985). *Nature,* **318**, 618.
11. Allen, J. P., Feher, G., Yeates, T. O., Komiya, H., and Rees, D. C. (1987). *Proc. Natl Acad. Sci. USA,* **84**, 6162.
12. Engelman, D. M., Henderson, R., McLachlan, A. D., and Wallace, B. A. (1980). *Proc. Natl Acad. Sci. USA,* **77**, 2023.
13. Yeates, T. O., Komiya, H., Rees, D. C., Allen, J. P., and Feher, G. (1987). *Proc. Natl Acad. Sci. USA,* **84**, 6438.
14. Gray, T. M. and Matthews, B. W. (1984). *J. Mol. Biol.,* **175**, 75.
15. Barlow, D. J. and Thornton, J. M. (1988). *J. Mol. Biol.,* **201**, 601.
16. Brandl, C. J. and Deber, C. (1986). *Proc. Natl Acad. Sci. USA,* **83**, 917.
17. Gold, D. P., Puck, J. M., Pettey, C. L., Cho, M., Coligan, J., Woody, J. N., and Terhorst, C. (1986). *Nature,* **321**, 431.
18. Finer-Moore, J. and Stroud, R. (1984). *Proc. Natl Acad. Sci. USA,* **81**, 155.
19. Lodish, H. (1988). *TIBS,* **13**, 332.
20. Pearson, W. R. and Lipman, D. J. (1988). *Proc. Natl Acad. Sci. USA,* **85**, 2444.
21. Feng, D. F., Johnson, M. S., and Doolittle, R. F. (1985). *J. Mol. Evol.,* **21**, 112.
22. Doolittle, R. F., Feng, D. F., Johnson, M. S., and McClure, M. A. (1986). *Cold Spring Harbor Symp. Quant. Biol.,* **51**, 447.

23. Chothia, C. and Lesk, A. M. (1986). *EMBO J.*, **5**, 823.
24. Bajaj, M. and Blundell, T. (1984). *Ann. Rev. Biophys. Bioeng.*, **13**, 453.
25. Taylor, W. R. (1986). *J. Mol. Biol.*, **188**, 233.
26. Taylor, W. R. (1988). *Protein Engineering*, **2**, 77.
27. Gribskov, M., McLachlan, A. D., and Eisenberg, D. (1987). *Proc. Natl Acad. Sci. USA*, **84**, 4355.
28. Patthy, L. (1987). *J. Mol. Biol.*, **198**, 567.
29. Thornton, J. and Taylor, W. R. (1989). In *Protein Sequencing*, (ed. J. B. C. Findlay and M. J. Geisow), p. 147−190. IRL Press, Oxford.
30. Hodgman, T. C. (1989). *Comput. Appl. Biosci.*, **5**, 1.
31. Hubbard, S. C. and Ivatt, R. J. (1981). *Ann. Rev. Biochem.*, **50**, 555.
32. Cohen, P. (1983). In *Post Translational Modification of Proteins*, (ed. B. C. Johnson), p.19. Academic Press, New York.
33. Aitken, A. (1987). *Botan. J. Linn. Soc.*, **94**, 247.
34. Hemmings, B. A., Aitken, A., Cohen, P., Taymond, M., and Hofmann, F. (1982). *Eur. J. Biochem.*, **127**, 473.
35. Cohen, P., Yellowlees, D., Aitken, A., Donella-Deana, A., Hemmings, B. A., and Parker, P. J. (1982). *Eur. J. Biochem.*, **124**, 21.
36. Schultz, A. M., Henderson, L. E., and Oroszlan, S. (1981). *Ann. Rev. Cell Biol.*, **4**, 612.
37. Mihalyi, E. (1978). In *Application of Proteolytic Enzymes to Protein Structural Studies*, CRC Press, Boca Raton, FL. Vols 1 and 2.
38. Green, N. M. (1963). *Arch. Biochem. Biophys.*, **101**, 186.
39. Allen, G. (1981). In *Laboratory Techniques in Biochemistry*, (ed. T. S. Work and R. H. Burdon), Vol. 9, p. 30. Elsevier, London and Amsterdam.
40. Kretsinger, R. H. (1976). *Ann. Rev. Biochem.*, **45**, 239.
41. Gariepy, J. and Hodges, R. S. (1983). *FEBS Lett.*, **160**, 1.
42. Harris, A. S., Croall, D. E., and Morrow, J. S. (1988). *J. Biol. Chem.*, **263**, 15754.
43. Quiocho, F., Vyas, N. K., Sack, J. S., and Vyas, M. N. (1987). *Cold Spring Harbor Symp. Quant. Biol.*, **52**, 453.
44. Taylor, W. R. and Geisow, M. G. (1987). *Protein Engineering*, **1**, 183.
45. Engel, J., Taylor, W. R., Paulsson, M., Sage, H., and Hogan, B. L. M. (1987). *Biochemistry*, **26**, 6958.
46. Walker, J. E., Saraste, M., Runswick, M. J., and Gay, N. J. (1982). *EMBO J.*, **1**, 945.
47. Wierenga, R. K. and Hol., W. G. J. (1983). *Nature*, **302**, 842.
48. Argos, P. and Leberman, R. (1985). *Eur. J. Biochem.*, **152**, 651.
49. Moller, W. and Amons, R. (1985). *FEBS Lett.*, **186**, 1.
50. Wierenga, R. K., Terpstra, P., and Hol, W. G. J. (1986). *J. Mol. Biol.*, **187**, 101.
51. Taylor, W. R. and Green, N. M. (1989). *Eur. J. Biochem.*, **179**, 241.
52. Chou, P. Y. and Fasman, G. D. (1974). *Adv. Enzymol.*, **47**, 45.
53. Garnier, J., Osguthorpe, D. J., and Robson, B. (1978). *J. Mol. Biol.*, **120**, 97.
54. Taylor, W. R. (1987). In *Nucleic Acid and Protein Sequence Analysis*, (ed. M. J. Bishop and C. J. Rawlings), pp. 285−301. IRL Press, Oxford.
55. Lim, V. I. (1974). *J. Mol. Biol.*, **88**, 873.
56. Levin, J. M., Robson, B., and Garnier, J. (1986). *FEBS Lett.*, **205**, 303.
57. Sheridan, R. P., Dixon, J. S., Venkataraghavan, R., Kuntz, I. D., and Scott, K. P. (1985). *Biopol.*, **24**, 1995.
58. Kabsch,W. and Sander, C. (1983). *FEBS Lett.*, **155**, 179.

59. Biou, V., Gibrat, J. F., Levin, J. M., Robson, B., and Garnier, J. (1988). *Protein Engineering*, **2**, 185.
60. Patthy, L. (1987). *FEBS Lett.*, **214**, 1.
61. Taylor, W. R. and Thornton, J. M. (1984). *J. Mol. Biol.*, **173**, 487.
62. McLachlan, A. D. and Stewart, M. (1975). *J. Mol. Biol.*, **98**, 293.
63. Rose, G. (1985). *Meth. Enzymol.*, **115**, 430.
64. Verma, A. K., Filoteo, A. G., Stanford, D. R., Wieben, E. D., and Penniston, J. T. (1988). *J. Biol. Chem.*, **263**, 14152.
65. Richardson, J. S. (1981). *Adv. Protein Chem.*, **34**, 167.
66. Richardson, J. S. (1985). *Meth. Enzymol.*, **115**, 341.
67. Kabsch, W. and Sander, C. (1983). *Biopol.*, **22**, 2577.
68. Taylor, S. S., Bubis, J., Toner-Webb, S., Saraswat, L. D., First, E. A., Buechler, J. A., Knighton, D. R., and Sowadski, J. (1988). *FASEB J.*, **2**, 2677.
69. Ptitsyn, O. B. and Finkelstein, A. V. (1980). *Quart. Rev. Biophys.*, **13**, 339.
70. Sternberg, M. J. and Cohen, F. (1982). *Int. J. Biol. Macromol.*, **4**, 137.
71. Cohen, F. E., Kosen, P. A., Kuntz, I. D., Epstein, L. B., Ciardelli, T. L., and Smith, K. A. (1986). *Science*, **234**, 349.
72. Davison, M. D., Baron, M. D., Critchley, D. R., and Wootton, J. C. (1989). *Int. J. Biol. Macromol.*, **11**, 81.
73. Pearl, L. H. and Taylor, W. R. (1987). *Nature*, **329**, 351.
74. Navia, M. A., Fitzgerald, P. M. D., McKeever, B. M., Leu, C.-T., Heimbach, J. C., Herber, W. K., Sigal, I. S., Darke, P. L., and Springer, J. P. (1989). *Nature*, **337**, 615.
75. Sternberg, M. J. E. and Taylor, W. R. (1984). *FEBS Lett.*, **175**, 387.
76. Bradley, M. K., Smith, T. F., Lathrop, R. H., Livingston, D. M., and Webster, T.A. (1987). *Proc. Natl Acad. Sci. USA*, **84**, 4026.
77. Crawford, I. P., Niermann, T., and Kirschner, K. (1987). *Proteins*, **2**, 118.
78. Hyde, C. C., Ahmed, S. A., Padlan, E. A., Miles, E. W., and Davies, D. R. (1988). *J. Biol. Chem.*, **263**, 17857.
79. Janatova, J., Reid, K. B. M., and Willis, A. C. (1989). *Biochemistry*, **28**, 4754.
80. Thornton, J. M. (1981). *J. Mol. Biol.*, **151**, 261.
81. Amzel, L. M. and Poljak, R. J. (1979). *Ann. Rev. Biochem.*, **48**, 961.
82. Bjorkman, P. J., Saper, M. A., Samraoui, B., Bennett, W. S., Strominger, J. L., and Wiley, D. C. (1988). *Nature*, **329**, 506.
83. Brown, J. H., Jardetzky, T., Saper, M. A., Samraoui, B., Bjorkman, P. J., and Wiley, D. C. (1988). *Nature*, **332**, 845.
84. Tulinsky, A., Park, C. H., and Skrzypcak-Jankun, E. (1988). *J. Mol. Biol.*, **202**, 885.
85. Montelione, G. C., Wuthrich, K., Nice, E. C., Burgess, A. W., and Scheraga, H. A. (1987). *Proc. Natl Acad. Sci. USA*, **84**, 5226.
86. Cooke, R. M., Wilkinson, A. J., Baron, M., Pastore, A., Tappin, M. J., Campbell, I. D., Gregory, H., and Sheard, B. (1987). *Nature*, **327**, 339.
87. Campbell, R. D., Law, S. K. A., Reid, K. B. M., and Sim, R. B. (1988). *Ann. Rev. Immunol.*, **6**, 161.
88. Martin, G. R., Timpl, R., and Kuhn, K. (1988). *Adv. Protein Chem.*, **39**, 1.
89. Williams, A. F. and Barclay, A. N. (1988). *Ann. Rev. Immunol.*, **6**, 381.
90. Engel, J. and Furthmayr, H. (1988). *Meth. Enzymol.*, **145**, 3.
91. Yamamoto, T., Davis, C. G., Brown, M. S., Schneider, W. J., Casey, M. L., Goldstein, J. L., and Russell, D. W. (1984). *Cell*, **39**, 27.

92. Scott, J. M., Urdea, M., Quiroga, M., Sanchez-Pescador, R., Fong, N., Selby, M., Rutter, W. J., and Bell, G. I. (1983). *Science,* **221**, 236.
93. Nermut, M. V., Green, N. M., Eason, P., Yamada, S. S., and Yamada, K. M. (1989). *EMBO J.,* **7**, 4093.
94. Argraves, W. S., Suzuki, S., Arai, H., Thompson, K., Pierschbacher, M. D., and Ruoslahti, E. (1987). *J. Cell Biol.,* **105**, 1183.
95. Stanley, K. K., Kocher, H. P., Luzio, J. P., Jackson, P., and Tschopp, J. F. (1985). *EMBO J.,* **4**, 375.
96. Sasaki, M., Kato, S., Kohno, K., Martin, G. R., and Yamada, Y. (1987). *Proc. Natl Acad. Sci. USA,* **84**, 935.
97. McMullen, B. A. and Fujikawa, K. (1985). *J. Biol. Chem.,* **260**, 5328.
98. Perkins, S. J., Haris, P. I., Sim, R. B., and Chapman, D. (1988). *Biochemistry,* **27**, 4004.
99. Leonard, W. J., Depper, J. M., Kanchisa, M., Kronke, M., Peffer, N. J., Svetlik, P. B., Sullivan, M., and Greene, W. C. (1985). *Science,* **230**, 633.
100. Holland, S. K., Harlos, K., and Blake, C. C. F. (1987). *EMBO J.,* **6**, 1875.
101. Lobel, P., Dahms, N. M., Breitmeyer, J., Chirgwin, J. M., and Kornfeld, S. (1987). *Proc. Natl Acad. Sci. USA,* **84**, 2233.
102. Skorstengaard, K., Jensen, M. S., Sahl, P., Petersen, T. E., and Magnusson, S. (1986). *Eur. J. Biochem.,* **161**, 441.
103. Fishleigh, R. V., Robson, B., Garnier, J., and Finn, P. W. (1987). *FEBS Lett.,* **214**, 219.
104. Bajaj, M., Waterfield, M. D., Schlessinger, J., Taylor, W. R., and Blundell, T. (1987). *Biochim. Biophys. Acta,* **916**, 220.
105. Ullrich, A., Coussens, L., Hayflick, J. S., Dull, T. J., Gray, A., Tam, A. W., Lee, J., Yarden, Y., Liberman, T. A., Schlessinger, J., Downward, J., Mayes, E. L. V., Whittle, N., Waterfield, M. D., and Seeburg, P. H. (1984). *Nature,* **309**, 418.
106. Ullrich, A., Bell, J. R., Chen, E. Y., Herrera, R., Petruzzelli, L. M., Dull, T. J., Gray, A., Coussens, L., Liao, Y.-C., Tsubokawa, M., Mason, A., Seeburg, P. H., Grunfeld, C., Rosen, O. M., and Ramachandran, J. (1985). *Nature,* **313**, 756.
107. Chothia, C., Boswell, D. R., and Lesk, A. M. (1988). *EMBO J.,* **7**, 3745.
108. Mostov, K. E., Friedlander, M., and Blobel, G. (1984). *Nature,* **308**, 37.
109. Ploegh, H. L., Orr, H. T., and Strominger, J. L. (1981). *Cell,* **24**, 287.
110. Saito, H., Kranz, D. M., Takagi, Y., Hayday, A. C., Eisen, H. N., and Tonegawa, S. (1984). *Nature,* **309**, 757.
111. Cunningham, B. A., Hemperley, J. J., Murray, B. A., Prediger, E., Brackenbury, R., and Edelman, G. M. (1987). *Science,* **236**, 799.
112. Stuart, S. G., Trounstine, M. L., Vaux, D. J. T., Koch, T., Martens, C. L., Mellman, I., and Moore, K. W. (1987). *J. Exp. Med.,* **166**, 1668.
113. Yarden, Y., Escobedo, J. A., Kuang, W.-J., Yang-Feng, T. L., Daniel, T. O., Tremble, P. M., Chen, E. Y., Ando, M. E., Harkins, R. N., Francke, U., Fried, V. A., Ullrich, A., and Williams, L. T. (1986). *Nature,* **323**, 226.
114. Coussens, L., Van Beveren, C., Smith, D., Chen, E., Mitchell, R. L., Isacke, C. M., Verma, I. M., and Ullrich, A. (1986). *Nature,* **320**, 277.
115. Radeke, M. J., Misko, T. P., Hsu, C., Herzenberg, L. S., and Shooter, E. M. (1987). *Nature,* **325**, 593.
116. Leung, D. W., Spencer, S. A., Cachianes, G., Hammonds, R. G., Collins, C., Henzel, W. J., Barnard, R., Waters, M. J., and Wood, W. I. (1987). *Nature,* **330**, 537.
117. Drickamer K. (1984). *J. Biol. Chem.,* **256**, 5827.
118. Schneider, L., Owen, M. J., Banville, D., and Williams, J. G. (1984). *Nature,* **311**, 675.

119. Kubalek, E., Ralston, S., Lindstrom, J., and Unwin, N. (1987). *J. Cell Biol.*, **105**, 9.

120. Popot, J. L. and Changeux, J. P. (1984). *Physiol. Rev.*, **64**, 1162.

121. McCrea, P. D., Popot, J. L., and Engelman, D. M. (1987). *EMBO J.*, **6**, 3619.

122. Noda, M., Takahashi, H., Tanabe, T., Toyosato, M., Kikyotani, S., Furutani, Y., Hirose, T., Takashima, H., Inayama, S., Miyata, T., and Numa, S. (1983). *Nature*, **302**, 528.

123. Schofield, P. R., Darlison, M. G., Fujita, N., Burt, D. R., Stephenson, F. A., Rodriguez, H., Rhee, L. M., Ramachandran, J., Reale, V., Glencorse, T. A., Seeburg, P. H., and Barnard, E. A. (1987). *Nature*, **328**, 221.

124. Grenningloh, G., Rienitz, A., Schmitt, B., Methfessel, C., Zensen, M., Beyreuther, K., Gundelfinger, E. D., and Betz, H. (1987). *Nature*, **328**, 215.

125. Findlay, J. B. C. and Pappin, D. J. C. (1986). *Biochem. J.*, **238**, 625.

126. Kubo, T., Fukuda, K., Mikami, A., Maeda, A., Takahashi, H., Mishina, M., Haga, T., Haga, K., Ichiyama, A., Kangawa, K., Kojima, M., Matsuo, H., Hirose, T., and Numa, S. (1986). *Nature*, **323**, 411.

127. Dixon, R. A. F., Kobilka, B. K., Strader, D. J., Benovic, J. L., Dohlman, H. G., Frielle, T., Bolanowski, M. A., Bennett, C. D., Rands, E., Diehl, R. E., Mumford, R. A., Slater, E. E., Sigal, I. S., Caron, M. G., Lefkowitz, R.J., and Strader, C. D. (1986). *Nature*, **321**, 75.

128. Dixon, R. A. F., Sigal, I. S., Rands, E., Register, R.B., Candelore, M. R., Blake, A. D., and Strader, C. D. (1987). *Nature*, **326**, 73.

129. Ullrich, A., Gray, A., Tam, A. W., Yang Feng, T., Tsubokawa, M., Collins, C., Henzel, W., Le Bon, T., Kathuria, S., Chen, E., Jacobs, S., Franke, U., Ramachandran, J., and Fujita-Yamaguchi, Y. (1986). *EMBO J.*, **5**, 2503.

130. MacDonald, R. G., Pfeffer, S., Coussens, L., Tepper, M. A., Brocklebank, C. M., Mole, J. E., Anderson, J. K., Chen, E., Czech, M. P., and Ullrich, A. (1988). *Science*, **239**, 130.

131. Dahlback, B. and Muller-Eberhard, H. J. (1984). *J. Biol. Chem.*, **259**, 11631.

132. Benovic, J. L., Stone, W. C., Caron, M. G., and Lefkowitz, R.J. (1989). *J. Biol. Chem.*, **264**, 6707.

133. George, D. G., Hunt, L. T., and Barker, W. C. (1988). *Computational Molecular Biology*, (ed. A. M. Lesk), pp. 100−115. Oxford University Press.

134. Soll, D. and Roberts, R. J. (ed) (1982, 1984, 1986, 1988). *Applications of Computers to Research on Nucleic Acids*, IRL Press, Oxford, Vols. I−IV.

135. (1988). *Comput. Appl. Biosci.*, **4**, 1.

136. Gribskov, M., Burgess, R. R., and Devereux, J. (1986). *Nucl. Acids Res.*, **14**, 327.

137. Staden, R. (1986). *Nucl. Acids Res.*, **14**, 217.

138. Sidman, K. E., George, D. G., Barker, W. C., and Hunt, L. T. (1988). *Nucl. Acids Res.*, **16**, 1869.

139. Wolf, H., Modrow, S., Motz, M., Jameson, B. A., Hermann, G., and Fortsch, B. (1988). *Comput. Appl. Biosci.*, **4**, 187.

140. Robinson, H. and Crofts, A. R. (1988). *Biophys. J.*, **53**, 404a.

141. Maulik, S. (1989). *Prot. Seq. Data Anal.*, **2**, 111.

142. Roode, D., Liebschutz, R., Maulik, S., Fridemann, T., Benton, D., and Kristofferson, D. (1988). *Nucl. Acids Res.*, **16**, 1857.

143. Bennett, W. S. and Huber, R. (1984). *CRC Crit. Rev. Biochem.*, **15**, 291.

144. Argos, P. (1989). In *Protein Structure, a Practical Approach*, (ed. T. E. Creighton), p. 169. IRL Press, Oxford.

A

Receptor-binding studies, a brief outline

E. C. HULME

Ligand binding provides a direct approach to the *in vitro* assay of receptors. A full overview is provided in reference 1.

1. Basic principles

The principle of receptor-binding assays is deceptively straightforward. The basic binding assay protocol is as follows:

Protocol 1. Steps in the performance of a binding experiment

1. Choose and make a tissue preparation containing the receptor.
2. Select a suitable labelled ligand.
3. Incubate the receptor preparation with an appropriate concentration of a labelled ligand for a defined time at a defined temperature.
4. Separate the bound from free ligand, using an appropriate separation technique.
5. Measure the bound and free ligand concentrations.
6. Repeat *steps 3 − 5* with the addition of unlabelled ligands or modulating agents.
7. Analyse the data to extract quantitative estimates of rate constants and/or affinity constants.
8. Relate the estimates of the binding parameters to pharmacologically determined values.

In the context of the present book, the receptor preparation may be:

(a) a membrane fraction;
(b) a solubilized preparation;
(c) a purified preparation.

Table 1. Comparison of rapid filtration technique with other methods for separation of the free and the bound ligand[a].

Method	Complete separation	Simplicity of operation	Time of separation	Specific to non-specific ratio	Reproducibility of results
Rapid filtration	good	good	sec	high	good
Centrifugation	fair	fair	min	fair	fair
Dialysis	poor	poor	day	low	poor
Gel-filtration	fair	fair	min	high	fair
Precipitation	fair	fair	min	fair	fair
Absorption	fair	fair	min	fair	poor

[a] This table was kindly provided by Dr W. R. Roeske (reference 1, Chapter 6).

The labelled ligand is almost always radiolabelled to high specific activity with ^{3}H or ^{125}I. Separation of bound from free ligand usually entails the following. For particulate preparations:

(a) Centrifugation, in which the receptor−ligand complex is pelleted and thus separated from the free ligand. This requires a bench-top microfuge giving at least 12 000 *g* and preferably more than 14 000 *g*.

(b) Filtration, in which the receptor−ligand complex is retained on a filter (most often, a glass-fibre filter, e.g. Whatman GF/F) (0.6 *μ*m pore size), GF/B (1.0 *μ*m pore size), or GF/C (1.2 *μ*m pore size) while the free ligand passes through.

For solubilized preparations:

(a) Gel-filtration, e.g. on a small column of Sephadex G50 (fine or medium) in which the high molecular weight receptor−ligand complex is separated from the low molecular weight free ligand by gel exclusion.

(b) Precipitation of the receptor−ligand complex, e.g. with polyethylene glycol (mol. wt 6000−8000) followed by centrifugation, or filtration.

(c) Adsorption of the free ligand, e.g. by adsorption onto activated charcoal, followed by centrifugation, leaving the receptor−ligand complex in solution.

(d) Adsorption of the receptor−ligand complex onto an ion-exchange filter (e.g. a DEAE 81 filter disc, or a polyethylenimine-pretreated glass-fibre filter), which allows the free ligand to pass through.

Equilibrium dialysis has also been applied to receptor-binding assays, but is usually not practical for multiple assays and is not widely used. The same information can be obtained by equilibrium gel-filtration (1). Some basic comments on these various methods are given in *Table 1*.

The binding of unlabelled ligands is studied by examining their effects on the receptor-specific binding of labelled ligands. Unlabelled ligands may either compete directly with the tracer or may modulate the tracer binding by non-competitive interaction, i.e. by interaction with a site distinct from but linked to the tracer binding site. A non-competitive binding site may be a separate, allosteric site located on the receptor itself, or may be on an effector molecule, or on another interacting macromolecule.

While the protocol for binding experiments may seem straightforward, complexities arise

(a) from the open-ended nature of binding experiments with their large number of variable experimental parameters;

(b) from the complex nature of the trade-offs that need to be made between ligand affinities, ligand concentrations, concentrations of binding sites, receptor stability, ligand stability, and incubation times if artefacts are to be avoided;

(c) from the fact that the extraction of numerical values for rate constants, binding constants, and number of sites requires the application of curve-fitting, and other statistical techniques.

These complexities are fully covered in reference 1.

2. The ligand off-rate may determine the assay method

The compatibility of a given receptor—ligand interaction with a particular assay method is determined by the off-rate of the ligand. Filtration, gel-filtration, or ligand adsorption assays all reduce the concentration of free ligand in the vicinity of the receptor-binding site, and will therefore initiate ligand dissociation. Whether the amount of dissociation that ensues is acceptable or not depends on

(a) the separation time;

(b) the dissociation rate constant.

If no more than 10% of the specifically bound ligand is to be permitted to dissociate during the separation, then the dissociation rate constant must be less than $0.105/t$ where t is the separation time. The separation time depends on the assay method. It is $c.$ 10 s for filtration assays (if the filters are washed; Chapter 6), $c.$ 30 s for charcoal adsorption assays (Chapter 3), and $c.$ 100 s for gel-filtration assays (Chapter 1).

From *Table 2*, it is evident that performing the filtration, adsorption or gel-filtration step at low temperature improves the performance of the assay method (see reference 1). This accounts for the utility of quenching the reaction by dilution with cold buffer before filtration.

Even so, there may be complete failure to detect a rapidly dissociating component of binding by these methods. If this is suspected, it is advisable to perform centrifugation assays, in which the binding equilibrium is not disturbed, and compare the results with those from the filtration or other assays.

Table 2. Maximum values of dissociation rate constants (k_{21}) and minimum values of equilibrium affinity constants (K) compatible with non-equilibrium assay methods.

| | k_{21} (s^{-1}) | K^a (M^{-1}) | |
		30°C	0°C
Filtration assay	0.01	10^9	3×10^7
Adsorption assay	0.003	3×10^9	10^8
Gel-filtration assay	0.001	10^{10}	2.5×10^8

^{a} K_d (dissociation constant) $= 1/K$.

Unfortunately, centrifugation assays suffer from limitations imposed by the relatively unfavourable ratio of specific : non-specific binding. If the minimum requirement is that there should be equal amounts of specific and non-specific binding at a receptor occupancy of 10%, we find that the minimum ligand affinity that is usable in a centrifugation assay is approximately 10^8/M. Under favourable circumstances, i.e. high ligand specific activity, high receptor concentration, and low non-specific binding, an affinity of 10^7/M may just about be compatible with a centrifugation binding assay. However, there are few credible examples of this.

3. Criteria for receptor-specific binding

The normal criteria for the definition of receptor-specific binding must be rigorously applied. Thus:

(a) The binding of the tracer ligand should be rapid enough to allow equilibration within the period of the assay.

(b) The binding of the tracer ligand should be fully reversible, except in the specific case of site-directed affinity labels.

(c) A component of the tracer ligand should be inhibited by pharmacologically appropriate concentrations of unlabelled ligands, allowing the definition of a specific and a non-specific component of binding. A set of different ligands should yield the same estimate of non-specific binding.

(d) Specific binding should be saturable, so defining a finite concentration of binding sites

(e) There should be a quantitative correlation between the affinities determined by binding assays, and by pharmacological or functional assays on the same receptor type.

These criteria are developed in detail in reference 1.

4. Tracer ligand saturation curve

In receptor purification studies, the measurements of greatest immediate interest are usually the total concentration of binding sites (R_t) which defines the success of the purification protocol, and the affinity constant (K) or dissociation constant $(K_d = 1/K)$ of the receptor-specific radioligand, which gives some indication of the integrity of the binding site. These basic parameters are obtained by determination of a tracer ligand saturation curve, and can be supplemented with measurements of inhibition of receptor-specific binding of the primary receptor-specific ligand by a range of unlabelled ligands, further defining the pharmacological specificity of the binding sites.

The aims of a tracer ligand saturation experiment are the following:

(a) to define the concentration dependence of total binding;

(b) to define the concentration dependence of non-specific binding;

(c) to demonstrate the saturability of specific binding;

(d) to ensure the accumulation of sufficient data to allow extraction of the affinity(s) of the tracer ligand for the sites so defined, and to determine the absolute concentration(s) of these sites;

(e) to ensure that the binding measurements are free of artefacts;

(f) to analyse the binding curve quantitatively, extracting valid estimates of the parameters and statistically correct estimates of the associated errors.

Tracer ligand binding curves ('direct' binding curves) may be determined either in the presence, or in the absence, of competing ligands or modulators. The effect of added unlabelled ligands on the tracer binding curve provides information about the interactions of the former as well as the latter with the receptor. However, the experimental protocol described below is not fundamentally affected by this extension of the scope of the basic saturation experiment.

4.1 Receptor preparation

Dilute the receptor preparation to give a concentration of binding sites which, ideally, will be $c.\ 0.1 \times K_d$ of the tracer ligand. Concentrations up to $0.5 \times K_d$ may just be acceptable, provided that the tracer ligand is chemically and radiochemically pure. These restrictions are relaxed if the aim is simply to determine R_t by making a single measurement of binding at a near receptor-saturating concentration of radioligand ($>10 \times K_d$). In such measurements, receptor concentrations of $0.1-1$ pmol/ml are usually appropriate. Methods of testing the radiochemical and chemical purity of the tracer ligand are given in (1). The minimum concentration of binding sites for use with [^{3}H]ligands ($c.\ 80$ Ci/mmole) is approximately 5×10^{-11} M, and for use with [^{125}I]ligands ($c.\ 2000$ Ci/mmole) the minimum is 2×10^{-12} M if the saturation curve is to be measured accurately in the $10-90\%$ saturation range using 1 ml aliquots of receptor preparation. Using this concentration of sites it should be possible to obtain 1000 bound dpm at 10% saturation, which will be measurable with 1% counting error over 10 min. If the receptor preparation is to be diluted further, it is advisable to harvest larger volumes. Otherwise counting and, in practice, other handling errors will be increased. Store the receptor preparation on ice until it is needed.

4.2 Setting up the binding curve

The usual protocol for measuring direct ligand saturation curves involves making a series of dilutions of the radioligand which are then pipetted out into replicate tubes. The advantage of this method is that SEMs are calculable for each point on the saturation curve. The disadvantage is that errors due to the pipetting of the tracer ligand are not minimized. The second method involves making a series of serial dilutions of the tracer ligand in the receptor preparation itself. This method minimizes the errors due to pipetting of the tracer, and allows the determination of a larger number of points per decade. It does, however, demand full reversibility of specific and non-specific binding within the incubation time employed. The latter protocol is described in reference 1.

4.3 Replicate incubations

Protocol 2. Direct binding curve using replicate incubations

1. Dilute the stock tracer ligand in assay buffer to give a concentration $100\times$ the value desired for the highest concentration to be assayed. This should be at least $10\times$ and preferably $30-100\times$ K_d. For example, if the tracer ligand has a K_d of 10^{-9} M, the saturation curve should extend at least to 10^{-8} M and the stock tracer ligand solution should be 10^{-6} M. The stock solution should be sampled for assay. In the case of a radioligand, the concentration present in the stock solution is dpm $\times$ $10^{-9}/(V \times 2220 \times SA)$ M, i.e. dpm/$(V \times 2220 \times SA)$ nM, where dpm is the dpm measured in V ml of radioligand of specific activity SA (Ci/mmole).

2. Set up a series of $M \times N$ microfuge tubes, or glass or plastic culture tubes, Beckman Biovials, etc., depending on the volume to be incubated, and whether or not additions need to be made to the tubes during processing, e.g. of excess cold buffer, before filtration. M is the number of concentrations of radioligand to be used per decade of concentration and N is the number of decades of concentration to be covered. N is minimally 2 and preferably up to 4. M should be at least 2 for accurate definition of the saturation curve and can be up to 6.

3. Pipette volumes of buffer into these sets of tubes as shown in *Table 3*. Make up one such set for each decade of concentration.

4. Pipette 0.1 ml of stock tracer ligand into each of the first set of tubes, e.g. for 2 points per decade, obtaining 0.1 ml + 0.216 ml (1 : 3.16-fold dilution) and 0.1 ml + 0.9 ml (1 : 10-fold dilution). Discard the pipette tip and take a fresh one. Vortex-mix the tubes.

5. Take 0.1 ml from the 1 : 10 dilution and repeat the above process with the next set of tubes. Do not forget to discard the pipette tip. Repeat the process, using the previous 1:10 dilution as the starting point until the dilutions are completed. If the tubes are laid out in a row, it is a good idea to displace them one space to the side as the dilutions are made, to keep count. If your preference

is to work from left to right, you will end up with the highest concentration of tracer ligand on the LHS and the lowest on the RHS. Reverse this order, so that the lowest concentration is on the LHS and the highest on the RHS.

6. Take small replicate samples (10 μl) from each tube to count to ascertain that the dilutions have been made correctly. At this stage, any problems with handling the tracer, owing to adsorption onto plasticware or glassware will become apparent. For a discussion of how to avoid handling losses see reference 1.

7. Set up two racks of tubes each containing $M \times N$ rows of R tubes so that the number of rows of tubes in each rack corresponds to the number of tracer ligand dilutions made. R is the number of replicates. Into each of the R tubes in a given row, pipette an appropriate volume (e.g. 10 μl) of the corresponding concentration of the radioligand. Having completed this process, pipette in competitor or modulator ligands, if appropriate. Finally, pipette the unlabelled ligand used for definition of non-specific binding into the corresponding set of tubes. 10 μl is a convenient volume. Aim for a concentration of 1000 $\times$ K_d of the unlabelled, competing ligand in the final incubation to ensure 99% blockade up to 90% tracer ligand occupancy.

8. Divide the receptor preparation into two separate aliquots, one for measurement of total binding, the other for measurement of non-specific binding. If necessary, pre-incubate until the appropriate temperature is attained. The volume of receptor preparation needed will be $2 \times R \times M \times N \times V$, where R is the number of replicates per point (a bare minimum of 2; 4 for a reasonable estimate of the SEM), and the factor of 2 comes from the need to measure the concentration dependence of non-specific as well as total binding. V is the volume of each assay (typically 0.5 – 1.0 ml).

9. Dispense aliquots of the pre-incubated preparations into the assay tubes, starting with those set up for total binding. Work from the lower tracer ligand concentrations towards the higher, to minimize problems due to carry-over of small volumes of tracer. Do this SEPARATELY for the total and non-specific binding measurements.

10. Cap and vortex-mix the tubes. Transfer to a water bath if necessary. Incubate for a time sufficient for tracer equilibration to occur under off-rate-limited conditions. To be safe, incubate for 5 times the half-time for tracer dissociation. This time can be cut down to c. 0.5 times the half-time for tracer dissociation at a 90% receptor-saturating ligand concentration (see reference 1 for a full discussion). Note that the presence of competitors or modulators may slow the approach to equilibrium. Shake if necessary to maintain homogeneity.

11. Process the tubes to obtain separation of bound from free ligand. If performing a microcentrifugation assay, sample the supernatants for determination of the free tracer ligand concentration. If performing a filtration assay on membranes, or precipitated soluble receptors, check the free ligand concen-

tration by supplementary centrifugation assays. If necessary, use equilibrium dialysis or equilibrium gel-filtration to establish the free ligand concentration.

12. Measure the counter background (4 replicates).

Table 3. Volumes for making ligand dilutions.

Points per decade	Volume/tube (ml)
1	0.9
2	0.216, 0.9
3	0.115, 0.364, 0.9
4	0.0778, 0.216, 0.462, 0.9

---

5. Preliminary analysis and interpretation

Protocol 3. Analysis of direct binding curve

1. Convert measurements of bound and free ligand to molar concentrations. In the case of radiolabelled ligands

$$RL^* = 10^{-9} \times B/(V \times SA \times 2220) \text{ M}$$

where B is the bound dpm, V is the assay volume, and SA is the radioligand specific activity. In simple cases, L^* (free) $= L^*$ (total)$-RL^*$.

2. Plot total and non-specific binding v. L^* (free). If replicates are available, calculate means and SEMs (SEM $= \sigma/\sqrt{R}$). If depletion of the tracer ligand was kept to a minimum ($<10\%$), calculate specific binding by subtraction of non-specific from total binding. The SEM of the specific binding is given by

$$SE_{spec} = \sqrt{(SE_t^2 + SE_{ns}^2)}.$$

If depletion was more marked, it will be necessary to use the non-specific binding curve to read off the non-specific binding appropriate to a given value of total binding from the corresponding free ligand concentration.

Plot RL^* (specific binding) against L^* (free concentration) (*Figure 1a*) and against Log L^*(1), to gain a visual impression of its saturability

3. Plot RL^*/L^* v. RL^* (Scatchard plot, *Figure 1b*). If the data allow it, fit a straight edge through the points to determine R_t (intercept on x-axis in Scatchard plot) and $K = 1/K_d = -$slope. It is better not to attempt linear regression analysis of the Scatchard plot, but simply to draw a line by eye, to obtain initial estimates of the binding parameters.

Note that the ability of the Scatchard plot to lead the eye on is notorious and may entice one to make a large extrapolation from a limited linear segment of the plot

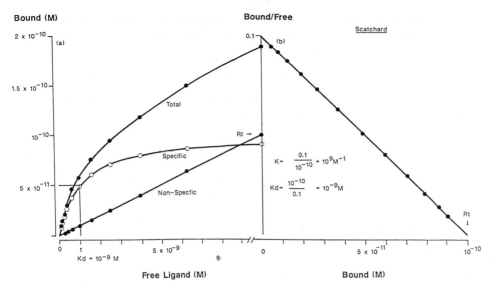

Figure 1. (a) Tracer ligand saturation curve showing total, specific, and non-specific binding. (b) Scatchard plot of specific binding, showing calculation of initial estimates of K and R_t.

to yield an apparent R_t value which the data do not really justify. Thus Scatchard analysis should be regarded as a means of visualizing, or presenting, data. It is not a substitute for a non-linear least-squares fit to a properly justified model of the binding process, which should incorporate terms for both specific and non-specific binding. The analysis of binding data by non-linear least-squares fitting is described in reference 1. A variety of commercial packages is available, including Accufit (Beckman), Enzfitter (Elsevier Biosoft), EBDA and LIGAND (Cambridge Biosoft), and LUNDON-1, LUNDON-2 (Lundon software).

A linear Scatchard plot implies a uniform, non-interacting population of binding sites. If this is not the situation, or if certain binding artefacts occur, the plot will deviate from linearity. A Scatchard plot which is convex upwards may indicate the occurrence of positive co-operativity between binding sites in an oligomeric structure (cf. the binding of acetylcholine to nicotinic acetylcholine receptors). Alternatively, a variety of binding artefacts can give rise to the appearance of positive co-operativity, for instance:

(a) incomplete equilibration with the lower but not the higher concentrations of the tracer ligand;

(b) overestimation of the free tracer ligand concentration. This is a particular hazard under depletion conditions at the low end of the saturation curve if the tracer ligand is impure, if part of the bound ligand is recovered in the free ligand pool, or if non-specific binding is overestimated.

A Scatchard plot which is concave upwards may have a variety of genuine explanations. It may arise from site − site negative co-operativity within an oligomer, from the presence of multiple independent populations of receptors with different affinities for the tracer ligand, or from the presence of receptor−effector complexes which manifest different affinities for the ligand, as in the case of receptor−G-protein interactions.

Alternatively, an upwardly concave Scatchard plot can arise from the presence of an undiagnosed artefact:

(a) from the improper definition of non-specific binding with the recovery of part of the non-specific binding in the apparently specific component;

(b) from incomplete equilibration of the ligand with an occluded component of the receptor population, which would lead to an underestimation of its contribution at low, but not at higher ligand concentrations;

(c) under certain circumstances, from ligand−ligand interactions or ligand heterogeneity; fortunately, these are relatively infrequent phenomena.

A fuller discussion of such artefacts is presented in reference 1.

Several different models are available for the analysis of ligand binding curves which deviate from the simple Langmuir isotherm. However, in practice, the most useful of them is the multiple sites model:

$$RL^* = \Sigma\, R_{it} \cdot K_i \cdot L^*/(1 + K_i \cdot L^*) + C \cdot L^*.$$

For fitting purposes, this should be recast in the form:

$$RL^* = \Sigma\, R_{it} \cdot 10^{(P_i+x)}/(1 + 10^{(P_i+x)}) + C \cdot 10^x$$

where $P_i = \log_{10}(K_i)$ and $x = \log_{10}(L^*)$.

According to this model, the bound ligand, RL^*, is made up of the sum of contributions from sites with different abundances, R_{it}, exhibiting different affinities, K_i, for the ligand, and a linear term describing non-specific binding. The latter can be replaced by a saturable function if required. The use of expressions such as these to fit saturation binding curves is described in reference 1.

6. Separation methods: particulate preparations

6.1 Filtration assays

Protocol 4. Filtration binding assay

1. Set up a filtration manifold (e.g. Millipore, Schleicher and Schüll) with 2.5 cm Whatman glass-fibre discs (GF/C, GF/B or GF/F depending on the size of the membrane particles), or use a cell-harvester (e.g. Brandel).

2. Pre-wet the filters with ice-cold wash buffer (ideally, the same as the incubation buffer), and apply vacuum using a vacuum pump connected via a reservoir. Keep filters not in use wet, or blocked off to maintain vacuum.

3. Apply the incubation mixture to the filter and follow up immediately with 2 or more washes with ice-cold wash buffer, which can be used to wash out the incubation tubes and complete the transfer. The effect of number and volume of rinses on the total level of specific binding, and the ratio of specific: non-specific binding, must be determined empirically. Not more than 10 mg of tissue (1 mg of protein) should be applied to each filter. Filtration times are reduced at lower loadings, which are thus better.

4. Remove the filters. Ensure that no hanging drops adhere to the reverse side. Place the filters in 10 ml of aqueous scintillator, and vortex-mix. Leave to extract for several hours, then vortex again before counting. Some workers prefer to dry the filters in an oven or under a lamp before counting. This runs the risk of baking the ligand onto the filter.

Examples are given in Chapters 5 and 6.

6.2 Centrifugation assays

Protocol 5. Microcentrifugation binding assay

1. Perform the assays in capped microcentrifuge tubes.

2. At the end of the incubation period, spin the tubes at *c*. 14 000 *g* or greater for 1–5 min. The length of the spin depends on the buffer composition, and is reduced for buffers containing divalent cations, or physiological salt concentrations, and greater for dilute, low-ionic strength media, particularly those containing EDTA.

3. Carefully uncap the tubes and pour off the supernatant from the pellets. Wash the pellets, the sides of the tubes, and the caps by carefully serially immersing the tubes in a large excess of deionized water, or wash buffer (e.g. 3 × 1 litre). Leave the tubes in an inverted position in a test-tube rack to drain. Blot off any remaining droplets of buffer onto Kleenex.

4. Add 100 μl of Soluene 350 (Packard) to each tube. Cap and leave at room temperature overnight to solubilize.

5. Add 1.2 ml of non-aqueous scintillator to each tube. Vortex, and count the tubes by using them as inserts in plastic scintillation vials.

7. Separation methods: soluble preparations

7.1 Gel-filtration

Protocol 6. Gel-filtration binding assay

1. Pour a series of disposable polypropylene columns containing 2 ml of Sephadex G50 (fine) or G50 (medium), depending on the nature of the ligand. Ionic

ligands (e.g. quaternary ammonium compounds) can usually be separated on G50 M, but less-polar ligands, which may partition into detergent micelles, usually require the greater resolving power of G50 F. The columns can be conveniently poured by pipetting out 4 ml of a 1:1 stirred slurry of swollen Sephadex in deionized water (containing 0.03% azide for storage at 4°C). The use of Kontes Disposaflex columns is convenient.

2. Equilibrate the columns with 2−3 bed volumes of cold detergent-buffer (which may be the incubation buffer). Work at 4°C. Stand the columns in scintillation vials.

3. Apply 0.1 ml of incubation mixture to the gel, and allow it to run in. Wash it in with 2 × 100 μl of ice-cold detergent buffer, allowing each wash to run in completely. Elute the receptor−ligand complex with 0.7 ml of detergent buffer.

4. Remove the column for re-equilibration (5 ml of deionized water, then repeat *step 2*). Add 10 ml of aqueous scintillator to the contents of the scintillation vial, and count.

Variants on this technique occur throughout the book.

7.2 Filtration on polyethyleneimine-treated filters

This method relies on the acidic isoelectric point of most receptors, which means that they are negatively charged at neutral pH and therefore retained by anion exchangers.

Protocol 7. PEI filter binding assay

1. Pretreat glass-fibre filters by soaking with 0.3% aqueous polyethyleneimine for 2−3 h.

2. Place the filters without washing in the filtration manifold or cell harvester, and filter as in Protocol 4 above.

3. An alternative is to use DEAE 81 filter discs (Whatman). These are less robust than glass-fibre filters. Note that quaternary ammonium derivatized glass-fibre paper is now also available.

7.3 Precipitation by polyethylene glycol

Protocol 8. (cf. Chapter 8) PEG precipitation assay

1. Perform the binding assay in a volume of 200 μl.

2. Add 15 μl of 3.3% (w/v) bovine γ-globulin (Sigma), and 85 μl of 36% (w/v) polyethylene glycol (mol. wt 6000−8000, Sigma) in binding buffer. Vortex vigorously. The solution should go cloudy. Stand on ice for 12 min.

3. Filter on 2.5 cm GF/C glass-fibre filters or on Millipore cellulose acetate (EH) filters as above. Wash the filters with 3×3 ml 7.5% (w/v) PEG in binding buffer.

4. Count the filters for radioactivity.

Note that PEG can also precipitate polypeptide ligands, and that this can render the assay method invalid. The assay works best between pH 7 and 8.

7.4 Charcoal adsorption

A full protocol for this method is given in Chapter 3.

Reference

1. Hulme, E. C. (ed.) *Receptor–Ligand Interactions, A Practical Approach.* IRL Press, Oxford, in press.

Index

A431 cells, culture of 205−6 *see also under*
　　EGF receptor
ABT 51−5
　agarose 54−5
　identification 53−5
　synthesis 52−3
acetylcholine binding assay 168
acetylcholine receptors, *see* muscarinic *and*
　　nicotinic receptors
Achromobacter lyticus 224
adrenergic receptors, *see* α_2- *and also*
　　β_2-adrenergic receptors
Affi-Gel, *see* EGF−Affi-Gel
affinity chromatography
　acetylcholine−agarose 172−3
　α_2-adrenergic receptors 149−52
　β_2-adrenergic receptors 136−8
　D_2 dopamine receptor 91−5
　EGFRs 201−2
　$GABA_A$ receptor 182−4
　inverse 151−2
　ligands 29−32
　mAChRs 60−1
　nAChRs 172−5
　opioid receptors 116−18
　polypeptide ligands 38−40
　　monoclonal antibodies 38−9
affinity cross-linking, and opioid
　　receptor 111−16
　photoreactive ligands 113−15
　reagents, bifunctional 115−16
　receptor irreversible opiate ligands 111−13
affinity gel development 36−7
affinity gel preparation 32−5, 51−5
　ABT 51−5
　mAChRs 51−78
affinity gel receptor elution 35−6
　immobilized polypeptide ligands 39−40
affinity gel testing 36−8
agarose−acetylcholine affinity
　　chromatography 172−3
agarose−heparin chromatography 151
agaroses
　ABT- 54−5
　epoxy-activated 54
　Ro-1986/1 180−1

agarose−WGA chromatography 152
α_2-adrenergic receptor, purification 141−61
α-bungarotoxin 165−8
　binding assay 166−8
　iodination 165−6
alprenolol, and β-adrenergic receptor
　　purification 136−8
amino acids
　mAChRs 66
　receptors 42−3
　see also under peptides *and also* proteins
2-aminobenzhydrol 51−2
3-(2′-aminobenzhydryloxy)tropane, *see* ABT
antibodies, *see* monoclonal antibodies
atrial synaptic membranes 57
autoradiography 242−3
　[^{3}H]PrBCM-labelled mAChRs 65−6
　[^{125}I]iodinated mAChRs 65

baculovirus expression systems 268−9
β-adrenergic receptors 125−40
　membrane preparation 128−9
　purification 135−9
　photoaffinity labelling 129−33
　radioligand binding 126−8
　solubilization 133−5
　tissue sources of 125−6
benzodiazepines, *see under* $GABA_A$ receptor
bifunctional cross-linking agents 115−16
binding studies, and solubilized
　　receptor-site identification 20−1
BNPS skatole 231−2
Bolton−Hunter reagent 217−18
brain synaptosomes, and opioid
　　receptors 102−3
bungarotoxin, *see* α-bungarotoxin

calcium binding sites 285
calcium phosphate transfection method 269−70
carbohydrates, and $GABA_A$ receptor 195−9
caudate nuclei, D_2 dopamine receptors 80−1
cDNA 43−6
　mAChRs 75−7

centrifugation assays 313
cerebral synaptic membranes 56−7
CHAPS 12
  and opioid receptors 102
chemical cleavage methods, and peptide
    mapping 228−34
  BNPS skatole 233−4
  cyanogen bromide 229−32
  DMSO/HBr 233
  iodosobenzoic acid 233
  at tryptophan 232
chloramine-T 215
cholate/sodium chloride solubilized D_2
    receptors, protein assay 83−4
chromatography, *see specific forms of*
Cibacron Blue Sepharose columns 171−2
cloning, *see* cDNA
computer software, for sequence analysis 293,
    296−7
concanavalin, *see* [^{125}I]concanavalin A lectin
concentration methods, for dilute receptor
    solutions 26−9
Coomassie Brilliant Blue R-250 241
covalent receptor modification 283
cross-linking, *see* affinity cross-linking, and
    opioid receptors
cyanogen bromide cleavage 229−32
cysteine/cysteic acid oxidation 227−8
cysteine-rich receptor domains 286

Datura stramonium lectin 24
D_1 dopamine receptor solubilization 79−80
D_2 dopamine receptor 80−6
  affinity chromatography 91−2
    column washing procedure 92−3
  mixed mitochondrial−microsomal
      membranes 80−2
  protein assay for 83−4
  solubilization of 82−3
  [^{3}H]spiperone 84−6
  WGA−Sepharose affinity
      chromatography 87−8
  see also lectin affinity *and* ligand affinity
      chromatography *and also* ligand binding
      assay *of*
deglycosylation, of peptides 251
detergents
  and nAChR extraction 171
  opioid receptors 102−5
    CHAPS 102
    evaluation 105−7
    non-ionic 103−5
    zwitterionic 102
  receptor solubilization 14
    choice of 8−10
    group 1 receptors and 10−11
    group 2 receptors and 11−12
    group 3 receptors and 12−14

  testing 15−19
  screening 15−19
  solubilized mAChR receptors 67−75
    gel filtration 72−5
    sucrose-gradient centrifugation 67−72
digitonin 13
  opioid receptors 104
    with Mg^{2+} 107−8
dihydrofolate reductase selection 268
dimethylsulphoxide, *see* DMSO/HBr cleavage
DMSO/HBr cleavage 232−3
DNA 43−6,75−7
  c-, *see* cDNA
  plasmid 269−71
  precipitation, calcium phosphate 271
domains 290−2
  of receptors 292−3
dopamine receptors 79−97
  fractionation of soluble 86−95
  molecular properties of 95−6
  solubilization 79−86

EGF−Affi-Gel 204−5,207−11
EGF receptor 203−11
  and EGF−Affi-Gel 204−5, 207−11
  extraction 206−7
  purification 203−4
  solubilization 205−11
electroblotting 245−7
electroelution 251−4
  procedure for 253−4
elution, and affinity gels 35−6
  see also electroelution
endoproteinases 224−7
  Arg-C 224−5
  Asp-N 227
  Glu-C (V8) 226−7
  Lys-C 225−6
enzymatic cleavage methods 223−8
  cysteine/cysteic acid oxidation 227−8
  endoproteinases 224−7
    Arg-C 22−3
    Asp-N 227
    Lys-C 224−5
  immobilized proteases 228
  pepsin 228
  thermolysin 228
  trypsin 223−4
  V8 proteinase 226−7
enzyme inhibitors 22−3
enzymes, and hydrodynamic techniques,
    mAChR 69−70
epidermal growth factor, *see* EGF receptor
epoxy-activated agarose 54
ethylenediamine-activated agarose 145−7
exclusion chromatography, *see* molecular
    exclusion chromatography

Index

filtration assays 310−11
fluorescamine protein determination 61
fluorescein, and two-dimensional peptide
    mapping 247−8
fluorography 243−4
FPLC 258
functional identification, of receptors 44−5
    sites 282−3

GABA$_A$ receptor 177−200
    carbohydrate properties 195−9
    [^{125}I]concanavalin A lectin binding 198−9
    isoelectric focusing 185−8
    molecular characterization 185−99
    molecular weight determination 188−91
    N-glycanase treatment 196−7
    polyacrylamide gel isoelectric
        focusing 186−8
    radioiodination 185
    SDS-PAGE 191−5
    sequence-specific, polyclonal antibodies 195
    solubilization of 178−85
    subunit composition of 191−5
γ-aminobutyric acid-a receptor, see GABA$_A$
    receptor
gas-phase peptide sequencing 253−4
gel filtration 311−12
    detergent-solubilized mAChRs 72−5
gel-permeation chromatography 24, 249
gel-permeation HPLC, and β-adrenergic
    receptors 138−9
gels, and peptide mapping 236−41
    slicing 244−5
    see also affinity gels and also specific types
        of
gene promoter sequences 264−5
genes 261−73
    inducible promoter sequences 265−6
    stable expression 266−7
    transient expression 266
    vectors 264−6
genomic clones 43−6
glycosylated peptides 250−1
GOR method 286−7
G-proteins 7−8,45
    see also group 3 receptors
gpt selection 267
    medium for 272
gradient gels 240−9
group 1 receptors 3−4
    detergent choice 10−11
group 2 receptors 3
    detergent choice 11−12
group 3 receptors 12−14
    digitonin 13
GTP-binding protein coupled receptors, see
    group 3 receptors
GTP-binding proteins 7

haloperidol−Sepharose affinity matrix, D$_2$
    receptor 89−91
heparin adsorption 24−5
heparin−agarose chromatography 151
[^{3}H]ligand binding assay, opioid
    receptors 107−8
HPLC, and peptide isolation 257−8
[^{3}H]propyl-benzilylcholine mustard 63−4
    autoradiography of 65−6
[^{3}H]spiperone, and D$_2$ receptors 84−6
hydrobromic acid, see DMSO/HBr cleavage
hydrophobic chromatography 24−5
    of opioid receptors 110−11
hydrophobicity
    patterns of 286−8
    structure 276−8
hydroxyapatite chromatography 24−5, 60

[^{125}I]concanavalin A lectin 198−9
[^{125}I]iodination 62−63,65,158−60
    GABA$_A$ receptor 185−8
    oxidizing methods 214−17
    peptide mapping 213−17
immobilized lectins 250−61
immobilized polypeptide ligands 39−40
inverse affinity chromatography 151−2
iodination, see [^{125}I]iodination
Iodo-beads 215−16
Iodo-gen 216−17
iodosobenzoic acid 233
ion-exchange chromatography 24−5
irreversible opiate ligands 111−13
islet-activating protein, see pertussis toxin
isoelectric focusing 25−6
    GABA$_A$ receptor 185−8

Laemmli method 64−5, 234−241
lectin affinity chromatography
    α$_2$ adrenergic receptors 152
    D$_2$ dopamine receptors 86−8
    opioid receptors 108,110
lectin binding, [^{125}I]concanavalin A and
    GABA$_A$ receptors 198−9
lectin chromatography 23−4
lectins
    and peptide mapping 250−1
    WGA 23−4
    see also Datura stramonium
ligand affinity chromatography
    D$_2$ receptors 88−95
    haloperidol−Sepharose affinity matrix 89−91
    immobilized polypeptide 39−40
    low-molecular-weight 29−38
    polypeptide 38−9
ligand-binding activity 40
    α$_2$ adrenergic receptor 156−7
    β$_2$ adrenergic receptor 127−8

319

D$_2$ dopamine receptors 84−6
EGFRs 208
GABA$_A$ receptors 183−4
mAChRs 58−9,77
    solubilized 58−9
nAChRs 165−7
ligand off-rate 305−6
ligands
    affinity gel preparation 32−4
    irreversible opiate 111−13
    photoreactive 113−15
    radio-, and β-adrenergic receptors 126−7
    synthesis, functional groups 29−32
lithium diiodosalicylate 171
Lysobacter enzymogenes 225

mAChRs 51−78
    affinity gels and 51−5
    amino acids 66
    autoradiography of 65−6
    cDNAs and 75−7
    electrophoretic analysis 62−6
    fluorescamine protein determination 61
    hydrodynamic properties, and molecular
        size 67−75
    Laemmli method 64−5
    ligand-binding activity 77
        solubilized 58−9
    membrane yield 57−8
    molecular-size determination 66−75
    purification 59−62
    SDS-PAGE 62−6
    silver staining 65
    specific binding activity 61−2
    subtype I 75−6
    subtype II 75−6
    sucrose-gradient centrifugation 67−72
magnesium ions, and digitonin 107−8
membrane preparations 21
    AChR-rich, pure 169−70
    α$_2$-adrenergic receptors 148,156
    β-adrenergic receptors 128−9
    buffer systems, and opioid receptors 100−1
    GABA$_A$ 181
    mAChR solubilization 55−9
    trans-, helices 280−1
membrane solubilization, *see* solubilization
membrane ultrafilters 26−9
microsomal−mitochondrial membranes, D$_2$
        receptors 80−2
mitochondrial−microsomal membranes, bovine
        caudate nuclei 80−1
    protein assay 81−2
molecular characterization, of receptors 40−6
    amino acid sequence 42−3
    cDNA expression 43−4
    function 44−5
    new strategies for 45−6

oligonucleotide sequence 42−3
polypeptide composition 40−2
protein estimation 40−2
reconstitution 44−5
molecular exclusion chromatography, and
        opioid receptors 108,110
molecular-size determination, mAChRs 66−75
muscarinic Ach receptors, *see* mAChRs

N-acetylglucosamine 23−4
nAChRs 163−76
    affinity chromatography 172−5
    α-bungarotoxin iodination 165−8
    assays of 165−9
    binding assay 168
    Cibacron Blue Sepharose
        chromatography 171−2
    crude receptor extracts 169
    molecule 163−4
    proteins 171
    pure membranes 169−70
    purification 169−75
    Torpedo californica 163−4
Naja naja siamensis toxin 174−5
N-alkylglycosides, *see under* detergents
neomycin resistance selection 267−8
N-glycanase, and GABA$_A$ receptors 196−8
N-glycosylation 250−1
nicotinic Ach receptors, *see* nAChRs
nitrocellulose electroblotting 245−7
non-ionic detergents, and opioid
        receptors 103−5
nucleotide-binding sites 285−6

O-glycanase 198
oligomeric ligand-binding/ion-channel receptors,
        see group 2 receptors
oligonucleotides, receptors 42−3
one-dimensional peptide mapping 222−47
    autoradiography 242−3
    chemical cleavage 228−34
    Coomassie Brilliant Blue R-250   241
    enzyme cleavage 223−8
    fluorography 243
    gels for 236−41
    nitrocellulose electroblotting 245−7
    post-SDS-PAGE 241−7
    protein visualization 245−7
    SDS-PAGE and 234−5
    silver staining 241−2
    slicing gel and 244−5
    stock solution for 235
    two-dimensional peptide mapping and 248−9
opiate ligands, irreversible 111−13
opioid receptors 99−124
    buffer systems 100−1
    CHAPS solubilization 102−3

detergent evaluation 105−6
digitonin/Mg^{2+} solubilization 107−8
heterogeneity 19−20
[^{3}H]ligand-binding assay 107−8
ion-detergent solubilization 101−2
molecular biology of 120−1
molecular studies on 118−19
non-ionic detergent solubilization 103−5
purification 108−118
solubilization 99−108

particulate β-adrenergic receptors
    photoaffinity labelling 132
    radioligand binding 127−8
pepsin 228
peptide deglycosylation 251
peptide gel-slice electroelution 251−4
peptide mapping 213−61
    Bolton−Hunter reagent 217−18
    free radioiodine removal and 218−21
    glycosylated 250−1
    lectin 250−1
    non-SDS-PAGE sequencing 256−9
    one-dimensional 222−47
    oxidizing methods, of
        radioiodination 214−17
    post-SDS-PAGE sequencing 251−6
    radioiodination 213−17
    sample preparation 213−22
    S-carboxymethylation 221−2
    two-dimensional 247−50
peptide re-digestion 248−9
peptides
    FPLC 258
    gas-phase sequencing 255−6
    glycosylated 250−1
    HPLC 257−8
    Sephadex LH 258−9
    sequencing, non-SDS-PAGE 256−9
    solid-phase sequencing 254−5
    see also amino acids and also proteins
pertussis toxin 7−8
photoaffinity labelling, β-adrenergic
    receptors 132−3
    particulate 132
    soluble 133
photoreactive ligands 113−15, 130
plasmid DNA 271
    preparation 269
plasmid vectors, for transfection 273
polyacrylamide gel isoelectric focusing,
    GABA$_A$ 186−8
polyclonal antibodies, and GABA$_A$
    receptor 195
polyethylene glycol precipitation 312−13
polyethyleneimine filters 312
polypeptide ligands, and affinity
    chromatography 38−9

immobilized 39−40
monoclonal antibodies 38−9
porcine atrial membranes 57
porcine cerebral synaptic membranes 56−7
pouring gels 236−41
PrBCM, see [^{3}H]propylbenzilylcholine mustard
propylbenzilylcholine mustard, see
    [^{3}H]propylbenzilylcholine mustard
protease inhibitors 22−3
proteases, immobilized 228
proteins
    assays 80−4, 93−5, 157−8
        α$_2$-adrenergic receptors 157−8
        D$_2$ dopamine receptors, affinity
            columns 93−5
        mitochondrial−microsomal
            membrane 80−4
    estimation 40−2
    fluorescamine 61
    nAChRs 171
        lithium diiodosalicylate extraction 171
        pH 11 extraction 171
    post-electroblotting 245−7
proteolytic cleavage sites 283−5
proteolytic non-SDS-PAGE receptor
    cleavage 256−7
purification, of receptors
    α$_2$-adrenergic 141−61
        affinity chromatography 149−51
        assay procedure 156−60
        ethylenediamine-activated agarose 145−7
        heparin−agarose chromatography 151
        inverse affinity chromatography 151−2
        membrane preparation 148, 156
        methods 149−52
        protein determination 157−8
        radioiodination 158−60
        solubilization 148−9,156−7
        WGA−agarose chromatography 152
        yohimbinic acid−Sepharose 145−8
    β-adrenergic 135−9
        gel-permeation HPLC 138−9
        Sepharose-alprenolol affinity resin 136−8
    EGF 201−2; see also A431 cells and also
        EGF receptor of GABA$_A$Rs 177−201
    of mAChRs 59−62
    of nAChRs 169−75
        acetylcholine−agarose columns 172−6
        Cibracron Blue Sepharose columns 171−2
        crude extract 169
        peripheral proteins 171
        pure membranes 169−71
        toxin-affinity chromatography 174−5
    opioid 108−18
        affinity chromatography 116−18
        affinity cross-linking 111−16
        hydrophobic chromatography 110−11
        lectin-affinity chromatography 110
        molecular exclusion chromatography 110

sucrose-gradient centrifugation 108,110
techniques, general 21−9
  approaches 23
  heparin adsorption 24−5
  hydrophobic chromatography 24−5
  hydroxyapatite chromatography 24−5
techniques (*cont.*)
  ion-exchange chromatography 24−5
  isoelectric focusing 25−6
  lectin chromatography 23−4
  protease inhibitors 22−3
  SDS-PAGE 25−6

radioiodine, free, removal of 218−21
radioligand binding, *see also* ligand binding
  β-adrenergic receptors 126−8
    particulate 127−8
    soluble 128
  interpreting 310−12
  ligand off-rate 305−6
  principles 303−5
  purified $GABA_A$ 183−4
  tracer ligand saturation curve 306−10
radioligands, *see specific names of
receptors*
  affinity chromatography 29−40
  binding studies, and solubilized 20−1
  concentration, of dilute solutions 26−9
  domains 292−3
  functional sites 282−3
  group 1 10−11
  group 2 11−12
  group 3 12−14
  genes, *see* genes
  homology searches 281−2
  molecular characterization 40−6
  molecular size estimation 19−20
  peptide mapping 213−61
  purification 21−9
    approaches 23
  solubilization 8−10
sources, and membrane preparations
structural deductions, and sequencing 277−301
see also membranes, purification *and also*
    solubilization of receptors *and also other
    specific aspects of*
reconstitution, of receptors 44−5
Ro7-1986/1 agarose 180−1
running gels, *see under* pouring gels

S-carboxymethylation 220−22
SCH23390 80
SDS-PAGE 25−6, 232−9
  composition 234
  $GABA_A$ receptor 191−4
  Gradient 240
  and [^{3}H]PrBCM 63−4

iodination, of mAChRs 62−3
Laemmli methods 64−5, 234−41
mAChR and 62−6
  molecular size 66−7
peptide mapping, cleavage products 232−3
peptide sequencing after 251−56
stock solutions for 64
secondary structure, of receptors 286−8
Sephadex LH 258−9
Sepharose
  alprenolol affinity resin 136−8
    solubilized β-adrenergic receptor
      chromatography 138
  Cibacron Blue, and nAChRs 171−2
  haloperidol 89−91
  WGA- 87−8
  and yohimbinic acid, *see* yohimbinic
    acid−Sepharose
sequence analysis, software for 293, 296−7
silver staining 65, 241−2
  and SDS-PAGE 191−2
single transmembrane segment receptors, *see*
    group 1 receptors
skatole, *see* BNPS skatole
SKF101253, *see under* α_2-adrenergic receptor
    purification
slicing gels 244−5
solid-phase peptide sequencing 254−5
solubilization 55−9
  β-adrenergic receptors 133−4
    photoaffinity labelling 133
    radioligand binding 128
  D_1 dopamine receptors 79−80
  D_2 dopamine receptors 80−6
  EGF receptor 204−11
  culture of A431 cells 205−6
    EGF−Affi-Gel 204−5, 207−11
    extraction of EGF receptor and 206−7
  $GABA_A$ receptor 178−85
    affinity chromatography 182−4
    method for 181−2
    radioligand binding assay 183−4
    Ro7-1986/1 agarose 180−1
  mAChRs 55−60
  nAChRs 171
  opioid receptors 99−108
    buffer systems for 100−1
    CHAPS 102−3
    detergent evaluation 105−7
    digitonin 104
    digitonin with Mg^{2+} 107−8
    [^{3}H]ligand-binding assay 107−8
    ionic detergents 101−2
    non-ionic detergents 103−5
    sonication 104−5
    Triton X-100 103−4
  receptors, general techniques 8−21
    binding studies, sites 20−1
    detergent choice 8−10, 14

Index

detergent testing 15−19
  group 1 10−11
  group 2 11−12
  group 3 12−14
  molecular size 19−20
sonication, of opioid receptors 104−5
spacer arms, *see under* affinity chromatography
spiperone, *see* [^{3}H]spiperone, and D$_2$ dopamine
  receptors
Staphylococcus aureus V8 226
steroid-nucleus detergents, *see* detergents
sucrose-gradient centrifugation
  GABA$_A$ R 188−91
  mAChR 67−72
  opioid receptors 108,110
synaptic membranes 55−7
synaptosomes, and opioid receptor
  solubilization 102−3

1,3,4,6-tetrachloro-3′,6′-diphenylglycouril, *see*
  Iodogen
thermolysin 228
thymidine kinase selection 268
  medium for 272
Torpedo californica 163−4
toxin-affinity chromatography 174−5
tracer ligand saturation curve 306−7
transfection

monolayer cells 269−70
recipient cell lines 263−4
transmembrane helices 280−1
Tritons, *see under* detergents
trypsin 223−4
two-dimensional peptide mapping 247−50
  and one-dimensional re-digested
    peptides 248−9
  standard protein fluoresceination 247−8

ultrafiltration 26−9

V8 proteinase 226−7

WGA−agarose chromatography 152
  Sepharose affinity chromatography, D$_2$
    receptor 87−8
  see also under lectin chromatography

yohimbinic acid−Sepharose 145−8
  coupling technique 147−8
  ethylenediamine-activated agarose 145−7

zwitterionic detergents, *see* CHAPS

Receptor – Effector Coupling: A Practical Approach

Edited by E. C. Hulme

1 Preparation of G-proteins and their subunits

P. C. Sternweis and Iok-Hou Pang

2 Receptor – G-protein complexes in solution

D. Poyner

3 Reconstitution of hormone-sensitive adenylate cyclase and tyrosine kinases

R. A. Cerione

4 Reconstitution of the interactions of muscarinic acetylcholine receptors (mAChRs) with G-proteins

K. Haga and T. Haga

5 Polyphosphoinositide turnover

K. G. Oldham

6 Measurement and control of intracellular calcium

A. M. Gurney

7 Molecular pharmacology of ion channels using the patch clamp

A. M. Gurney

8 Phosphate-labelling studies of receptor tyrosine kinases

L. C. Mahadevan and J. C. Bell

Appendix. Receptor binding studies, a brief outline

E. C. Hulme

Receptor – Ligand Interactions
A Practical Approach
Edited by E. C. Hulme

1 Selection and Synthesis of Receptor-Specific Radioligands

K. G. McFarthing

2 Anti-Receptor Antibodies as Ligands

A. D. Strosberg

3 Neurotoxins as Receptor Ligands: Dendrotoxin and β-bungarotoxin

J. O. Dolly

4 Strategy and Tactics in Receptor Binding Studies

E. C. Hulme

5 Receptor Preparations for Binding Studies

E. C. Hulme

6 The Use of the Filtration Technique in In Vitro Radioligand Binding Assays for Membrane-Bound and Solubilised Receptors

J- X. Wang, H. I. Yamamura, W. Wang and W. R. Roeske

7 Centrifugation Binding Assays

E. C. Hulme

8 Charcoal Adsorption for Separating Bound and Free ligand in Radioligand Binding Assays

P. G. Strange

9 Gel Filtration Assays for Solubilised Receptors

E. C. Hulme

10 Receptor Binding Kinetics

H. Prinz

11 Experimental Design and Data Analysis
J. W. Wells

Appendix 1. Radioligands for Receptor Binding: Amersham International
K. G. McFarthing

Appendix 2. Radioligands for Receptor Binding: New England Nuclear
R. L. Young